£24·95
94

CW00499547

Introductory Quantum Mechanics

Other Titles in the Series

Pascal: an Interactive Text
S J Abas and J R Mondragon

An Introduction to Groups
D Asche

Introduction to Statistics
A W Bowman and D R Robinson

Introduction to Regression and Analysis of Variance
A W Bowman and D R Robinson

An Introduction to the Digital Analysis of Stationary Signals
I P Castro

Fourier Series and Transforms
R D Harding

A Simple Introduction to Numerical Analysis
R D Harding and D A Quinney

A Simple Introduction to Numerical Analysis
Volume 2: Interpolation and Approximation
R D Harding and D A Quinney

From Number Theory to Secret Codes
T Jackson

An Introduction to Unconstrained Optimisation
J J McKeown, D Meegan and D Sprevak

Electric Circuit Theory
B E Riches

Introduction to Probability
D R Robinson and A W Bowman

Of Related Interest

Scientific Programmer's Toolkit: Pascal Edition
M H Beilby, J R Harding and M R Manning

A Computer Illustrated Text

INTRODUCTORY QUANTUM MECHANICS

R C Greenhow

Department of Physics,
University of York

Institute of Physics Publishing
Bristol and Philadelphia

British Library Cataloguing in Publication Data

Greenhow, R. C.
Introductory quantum mechanics.
1. Quantum theory
I. Title II. Series
530.1′2

ISBN 0-7503-0010-8 (Text/disk)
ISBN 0-7503-0014-0 (Site licence pack)

Library of Congress Cataloging-in-Publication Data are available

First printed 1990
Reprint 1994

Series Editor: **R D Harding,** University of Cambridge

Published by Institute of Physics Publishing, wholly owned by The Institute of Physics, London

Institute of Physics Publishing, Techno House, Redcliffe Way, Bristol BS1 6NX, UK

US Editorial Office: Institute of Physics Publishing, The Public Ledger Building, Suite 1035, Independence Square, Philadelphia, PA 19106, USA

Printed in Great Britain by J W Arrowsmith Ltd, Bristol

〉Contents

Preface

This text and the accompanying computer programs have grown out of an introductory course in quantum mechanics taught to our second-year undergraduates at the University of York. It was intended from the outset that the book would include all the essentials of an introductory course in quantum mechanics and that it would be illustrated wherever appropriate with the aid of computer programs. The computer aspect has influenced the content only in that one or two topics have been included which are more usually to be found in more advanced courses. Thus the partial-wave treatment of scattering is included. This lends itself particularly well to illustration by means of computer programs and is readily accessible at the level of the rest of the book.

The essential mathematical prerequisites are mechanics, differential and integral calculus, complex numbers, vectors and matrices. Much of the book is concerned with the solution of partial differential equations but the methods used are explained as required. Undoubtedly some familiarity with Fourier series and transforms will be of considerable aid in comprehension, but very little direct use is made of them.

It is assumed that students will have already studied introductory modern physics, so only a brief chapter covering the absolute essential preliminaries is included. Then follows a chapter on the backbone of the entire book, Schrödinger's equation. Here and elsewhere I have been concerned to make clear what is postulated and what follows logically. Despite their mathematical complexity, I have felt the necessity to include here a brief description of wavepackets, because they provide a far better link with everyday experience than stationary states. Some material is included here on probability flux, which some may prefer to skip on a first reading, and return to when the chapters on scattering are tackled.

The computer programs' first serious applications come with the material of Chapter 3, which deals with one-dimensional potential wells. With a few notable exceptions such as the harmonic oscillator, finding energy eigenvalues usually involves, at best, the solution of a transcendental equation, and more generally the numerical integration of Schrödinger's equation. By hand this is pure drudgery. The computer excels here, and the programs provided allow the student to concentrate on the physics and, it is hoped, to derive some pleasure as well as insight from the graphical displays of the wavefunctions.

An important aspect of the computer programs is their ability to open up the range of problems which can be studied. An example is the tilted potential well, a square potential well subjected to an electric field. The interest in these stems from current developments in fabricated quantum-well devices. They are difficult to deal with satisfactorily at the undergraduate level because they require solution of Airy's equation, which cannot be done by elementary means. The computer brushes this limitation aside and makes the study of these important structures quite straightforward. The eigenvalue-finding routines which are provided in the programs illustrate how such calculations are accomplished in the real world of the professional theoretical physicist.

Chapter 4 introduces the three-dimensional world whilst avoiding dealing with angular momentum. A treatment of the spherically symmetric states of the hydrogen atom leading to the energy spectrum is included because I feel it essential to get to grips with a real, familiar physical system at the earliest possible moment in order to provide some validation for the theory.

The next chapter deals with the formal development of quantum mechanics; it introduces the language of operators, the expansion postulate and the elements of the quantum mechanical theory of measurement. This material is used extensively in the following chapter to develop the theory of angular momentum. Computer programs are less applicable here, although one is provided which integrates Legendre's equation so that spherical harmonics of any (reasonable) order can be plotted.

The chapter on spherically symmetric potential wells includes more than the usual complement of examples, including the isotropic oscillator, the spherical square well and the Morse potential for diatomic molecules as well as the full treatment of the hydrogen atom, all backed up by computer solving and plotting routines.

After a brief look at the properties of systems of identical particles, the final two chapters deal with scattering in one and three dimensions. For one-dimensional systems with a net particle flux, e.g. tunnelling phenomena, the computer programs are particularly valuable. Because the wavefunction is necessarily complex in such cases it is essential to show both real and imaginary parts of the function as well as their resultant probability density, and if and how all these vary with time. This is readily accomplished by the computer displays. Also included in this chapter, on the grounds of mathematical affinity, is a treatment of the Kronig–Penney one-dimensional solid. Special programs enable the plotting of the eigenfunctions of this periodic potential, and the calculation of the eigenvalues and hence of the band structure.

The final chapter deals with scattering in three dimensions. The significance of the scattering phase shift, its relation to resonance and the Ramsauer effect can be demonstrated particularly effectively by means of the computer programs.

Computer exercises are included in the text whenever appropriate, and often call for considerable interaction on the part of the student in order to obtain quantitative validation of points made in the theoretical development†. We have found that some of the programs lend themselves admirably to third-year project work for our theoretical students (although for extensive use it is a considerable advantage to use versions which run on a fast machine such as the Archimedes!). The author would be very pleased to hear of any interesting applications of the programs.

A set of problems is included at the end of each chapter, often with hints for their solution. Some of them can be solved or their results checked with the aid of the programs. The answers, if not given in the question, are to be found at the end of the book.

R C Greenhow
September 1989

† It is suggested that Appendix 4, on using the computer programs, should be read before any serious use of the programs is attempted.

〉 Chapter 1

〉 Introduction

〉 1.1 The need for quantum mechanics

Quantum mechanics originated in the early years of this century. During the latter part of the nineteenth century, physicists began to feel that classical physics, founded on Newton's theories of mechanics and gravitation and Maxwell's theory of electromagnetism, provided an essentially complete description of the natural world. This confidence began to crumble with the discovery that various branches of physics, seemingly described by elegant and complete theories, were inconsistent with each other. Thus the laws of electromagnetism and thermodynamics led to an absurdity when applied to radiation, a defect only remedied by the realization, prompted by the ideas of Planck, that radiant energy must be quantized. Similarly, the laws of mechanics and electromagnetism were found to be incompatible, and this was remedied by Einstein who showed in his special theory of relativity how mechanics needed to be modified. Further problems arose when attempts were made to apply the laws of mechanics and electromagnetism to models of the atom. Electromagnetism tells us that static equilibrium of the atomic charges is not possible, whereas in a dynamic system the motions of the charges would necessarily lead to the continuous radiation of energy. We must therefore conclude that matter is inherently unstable. The first steps in the resolution of this difficulty were taken by de Broglie and Bohr using the revolutionary idea that all matter exhibited a wave nature. The equation which governs these waves was discovered by Schrödinger and forms the principal subject of this book. While it is not appropriate to provide an exhaustive historical account of all these matters in a book of this nature, the basic results are so crucial to the development and interpretation of quantum mechanics that a brief review is essential. Only those quantitative results which are required in the later developments are given.

〉1.2 Quantization of radiation

Thermodynamic arguments lead to the belief that radiation confined in a cavity should have properties quite independent of the nature of the cavity itself apart from its temperature. Consequently, these properties must be inherent in the nature of radiation. However, attempts to calculate the energy density of such radiation led to the absurd result that it was infinite. The wave nature of the radiation requires that only certain wavelengths of radiation can be supported in a cavity of finite size, and for each wavelength only certain configurations of the wave are possible. The allowed configurations and wavelengths constitute the normal modes of the cavity. The smaller the wavelength, the more modes will fit the cavity. Classical physics requires an equal quantity of energy to be present in each mode when thermal equilibrium prevails and, since the number of modes has no limit if we allow the wavelength to approach zero, this leads to the famous 'ultraviolet catastrophe': the energy density becomes infinite.

The resolution of this paradox was provided by Planck in 1900. He showed that if energy could only be exchanged with the walls of the cavity in quanta whose energy is proportional to the frequency of the radiation, then he could devise an expression for the distribution of energy with frequency which was in agreement with experiment. This fundamental departure from classical electromagnetic theory was subsequently reinterpreted to the effect that each mode of the cavity has associated with it an energy E which satisfies the relation

$$E = nh\nu. \tag{1.1}$$

Here ν is the frequency of the mode, h is Planck's constant and n is a positive integer. (In more sophisticated treatments n is replaced by $n + \frac{1}{2}$ but it remains the case that the energy can only change by multiples of the quantity $h\nu$.) These units of energy, or quanta of radiant energy, are referred to as photons. The value of h, determined by fitting Planck's theoretical expression to the experimentally determined energy density, is found to be approximately 6.626×10^{-34} J s.

These ideas are reinforced by photoelectric phenomena. It is found that radiation incident on a metal surface gives rise to the emission of electrons whose energy is closely related to the frequency ν of the incident radiation, rather than to its intensity. The effect is usually observed in a vacuum photodiode, the photoelectrons being emitted from the cathode. Einstein showed that the effect is consistent with the idea that the radiation, of frequency ν, arrives as photons whose energies satisfy the relation

$$E_\gamma = h\nu \tag{1.2}$$

where h is again equal to Planck's constant. The energy of each photon is completely given up to an individual electron.

Although the frequency of visible radiation approaches 10^{15} Hz, Planck's constant is so small that a photon in this part of the spectrum carries an energy of only a few times 10^{-19} J. It can therefore be readily appreciated why the quantized nature of radiation eluded discovery for so long. Clearly the joule is an inconveniently large unit in this context and the electron volt ($1\ \text{eV} \simeq 1.6 \times 10^{-19}$ J) is more appropriate.

〉1.3 The special theory of relativity

Einstein published his theory of the photoelectric effect in 1905 in the journal *Annalen der Physik*. The same volume also contains his first publication on the special theory of relativity. The details of this will not concern us greatly since Schrödinger's wave mechanics is a non-relativistic theory. One result in particular is important to us here, however, and that is the relativistic relation between total energy, momentum and mass. Einstein showed that the mass of a body depended on its state of motion relative to the observer, and that mass and energy were aspects of the same thing. This can be summarized in the relation

$$E^2 = p^2c^2 + m^2c^4. \tag{1.3}$$

E is the total energy of the particle, p is its momentum, m is its mass measured in a frame in which it is at rest and c is the speed of light.

If we attempt to apply equation (1.3) to a photon rather than to a material particle we must set the rest mass m equal to zero. The justification for this is that there is no frame in which a photon is at rest; the theory of relativity hinges on the experimental observation that the speed of light in all frames is the same, namely c. Furthermore, the result of setting $m=0$ is that

$$E_\gamma^2 = p^2c^2. \tag{1.4}$$

This ratio between the energy and momentum carried by electromagnetic radiation agrees with both the classical theory and experiment. Equations (1.2) and (1.4) lead to the result that, for a photon of frequency ν and wavelength λ,

$$p = h\nu/c = h/\lambda. \tag{1.5}$$

Direct verification of this result is found in the phenomenon of Compton scattering. Here an energetic photon in the x-ray region of the spectrum is observed to collide with an electron. Analysis of the collision on the lines of a collision of billiard balls, using equation (1.5) to give the momentum of the photon in terms of its energy, leads to the conclusion that the scattered photon should be degraded in energy and momentum to account for the energy and momentum of the recoiling electron, and this is precisely what is found in practice.

〉1.4 Wave–particle duality

It is frequently convenient in what follows to introduce the angular frequency $\omega = 2\pi\nu$, the wavenumber $k = 2\pi/\lambda$ and the quantity $\hbar = h/2\pi$. In terms of these, the fundamental equations for a photon, (1.2) and (1.5), become

$$E = \hbar\omega \tag{1.6}$$

and

$$p = \hbar k. \tag{1.7}$$

These two equations encapsulate the dual nature of radiation. The left-hand sides of the equations refer to energy and momentum, both particle-like properties. To be sure, waves carry both energy and momentum but such quantities would have to refer to some specific volume of radiation-filled space whereas here they refer to single entities, namely photons. In contrast, the right-hand sides of the equations refer to frequency and wavenumber, both wave-like properties. Particle-like properties predominate in events involving the emission or absorption of radiation, as in the photoelectric effect. Wave-like properties manifest themselves in interference effects, as in the well-known double-slit experiments of Young, and also in x-ray diffraction. Before discussing the wave–particle duality of radiation, however, we will introduce another crucial input to quantum theory, originated by de Broglie.

In 1924 de Broglie, stimulated by the proposition that radiation consisted of particle-like photons, speculated that a similar dual nature was exhibited by objects hitherto considered to be particles, such as electrons or, indeed, any other material object. It followed that particles should manifest wave-like behaviour such as interference and diffraction, under suitable circumstances. Specifically, he proposed that the relations (1.6) and (1.7) should hold for all particles.

It is a general property of waves that interference and diffraction become evident when the waves interact with objects whose dimensions are comparable to the wavelength. The small size of Planck's constant means that the de Broglie wavelength of all macroscopic objects is too minute to give rise to observable effects. When asked how the waves might be detected, de Broglie suggested that diffraction effects should be observable when electrons were scattered by crystals. Since the wavelength of a 100 eV electron is 0.12 nm and the spacing of atoms in crystals is typically 0.1 nm we see that the condition for interference mentioned above is readily achieved at the microscopic level.

Just such effects were observed, initially by accident, in 1925 by Davisson and Germer. They were studying the scattering of electrons from polycrystalline nickel. In recovering from an accidental exposure of the hot nickel to air they heat-treated the sample in such a way that it recrystallized into larger crystals. Prominent peaks then appeared in the distribution of scattered electrons at certain scattering angles and for certain electron energies. Later experiments with single crystals showed that the observations were consistent with de Broglie's hypothesis that the electrons did indeed behave as if they had a wave nature, their wavelength was as prescribed by equation (1.6) and the whole process was exactly similar in nature to that of x-ray scattering.

〉1.5 The nature of matter waves

Rather than giving a detailed analysis of Davisson and Germer's experiment here, we will study the results of a conceptually simpler experiment, that of two-slit interference. This will highlight aspects of wave–particle duality and lead us to an interpretation of the nature of the waves associated with material particles.

Let us first consider an experiment in which objects which are quite unequivocally particle like in nature, marbles or bullets say, are projected in a stream towards a screen in which there are two parallel slits. The situation is shown in figure 1.1(*a*). On the other side of the screen the particles are collected in a movable receptacle so that the probability $P(x)$ of arrival at any particular point x on the target can be determined. The experiment is performed in three stages. First slit 1 only is open, then slit 2 only and, finally, both slits are open simultaneously. In each experiment the same number of particles is incident on the screen. The distribution at the target will approximate a geometrical shadow of the screen and its slits. However, we will allow for a small amount of scattering occurring when some particles strike the sides of the slits as they pass through, so that there is some overlap of the two distributions. The three patterns resulting from the three experiments are as shown in the diagram. The points to note are as follows.

(*a*) The receptacle registers the arrival of discrete particles.

(*b*) The probability distribution arising when both slits are open is simply the sum of the two probability distributions arising from the two slits alone:

$$P(x) = P_1(x) + P_2(x). \tag{1.8}$$

(*c*) We are at liberty to observe which of the slits each particle passes through, and of course each particle passes through one or the other but not both.

The experiment is now repeated but with the stream of particles replaced by a continuous wave, water waves for example. The same sequence is adopted with first one, then the other and, finally, both slits open. We choose the wavelength to be similar to the slit separation, and replace the bucket-shaped receptacle with a device which responds to the amplitude A of the wave at the point x (figure 1.1(*b*)). For water waves this could be a float with a pointer attached, which moves relative to a scale. There are a number of points to note.

(*a*) The detector registers a continually varying displacement; in other words, a wave!

(*b*) The intensity of the wave when both slits are open is not equal to the sum of the intensities observed with the slits open singly. Interference occurs between the waves which arrive at the screen via the two slits. The resulting amplitude depends upon the relative phase of the two waves,

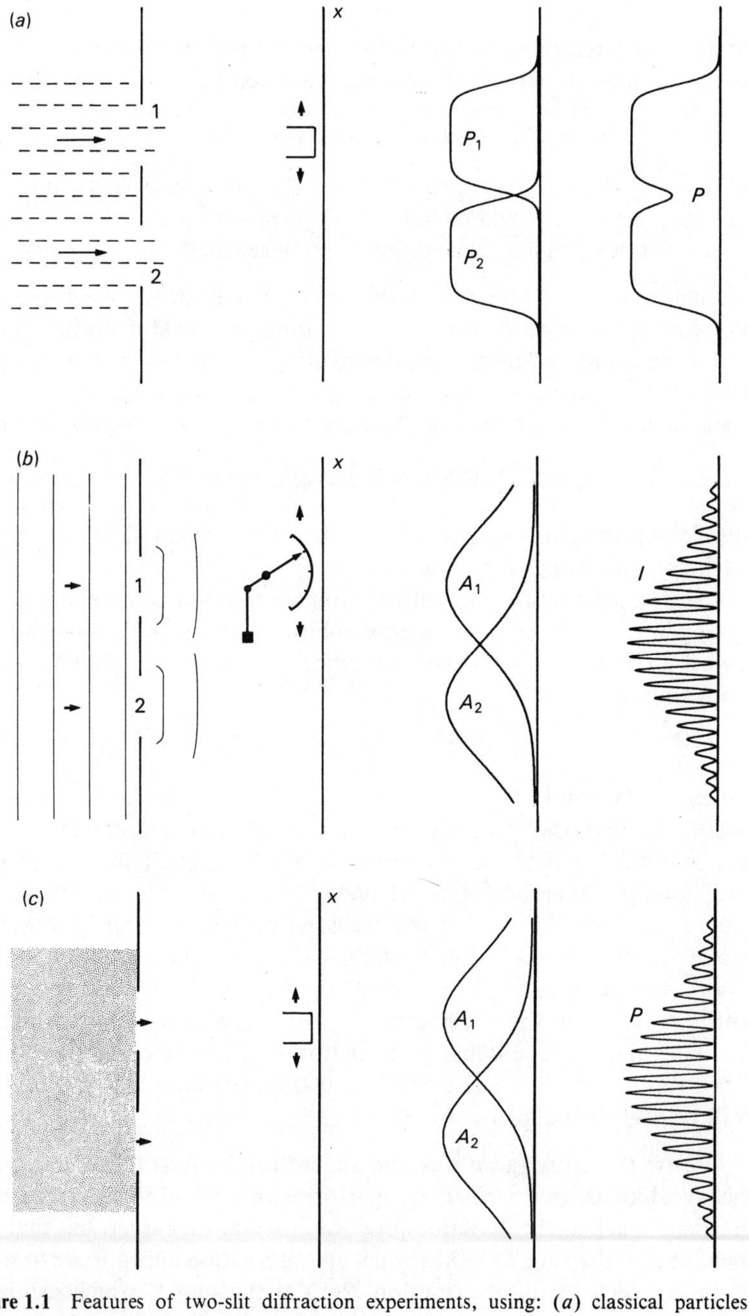

Figure 1.1 Features of two-slit diffraction experiments, using: (*a*) classical particles, (*b*) classical waves, and (*c*) electrons or photons.

and this is determined by the path difference between the two routes. In fact, the intensity with both slits open is given by

$$I(x) = (A_1(x) + A_2(x))^2. \tag{1.9}$$

$A_1(x)$ and $A_2(x)$ are the respective wave amplitudes at x when the slits are open separately, and include phase information.

(*c*) The waves pass through both slits when both slits are open.

Finally, the experiment is repeated with a beam of electrons. The detector at the target becomes either a photographic film or else a hollow electrode connected to a current measuring device with a short response time (figure 1.1(*c*)). As usual we perform three trials opening first one, then the other, then both slits. We now make the following observations.

(*a*) The detector observes discrete particles; the electronic device registers sharp pulses of current as each electron is collected. When developed, the photographic plate registers individual blackened grains, each corresponding to an electron.

(*b*) The probability $P(x)$ of receiving an electron at x when both slits are open is not the sum of the probabilities with the individual slits, but rather an interference pattern is observed. We therefore write, in analogy with equation (1.9),

$$P(x) = (A_1(x) + A_2(x))^2 \tag{1.10}$$

where $A_1(x)$ and $A_2(x)$ are interpreted as the amplitudes of waves associated with the electrons. The squares of these amplitudes give the probabilities of detecting electrons at x when slits 1 and 2 are open individually. When both slits are open we must sum the amplitudes and then square the sum to get the resulting probability and, just as with water waves, it is this process which leads to the interference pattern.

(*c*) Any attempt to observe which of the two slits the electron passes through results in the disappearance of the interference pattern. This is because of an unavoidable perturbation of the electron by the act of observation, and is a manifestation of the uncertainty principle, which will be discussed shortly.

We are therefore faced with the uncomfortable fact that the electrons behave like particles when we perform an operation appropriate to particles, such as the detection processes used at the target, but that they behave as if they are wave like when an observation appropriate to waves is made, such as the distribution $P(x)$ at the target, which exhibits a wave-like interference pattern. If we attempt to have it both ways and attempt simultaneously to observe electrons (i) passing through a specific slit (particle-like behaviour), and (ii) exhibiting an interference pattern at the target (wave-like behaviour) then the uncertainty principle frustrates the endeavour and we witness one or the other phenomenon but not both.

It is very important to realize that there is no suggestion that electrons are interfering with each other to create the interference pattern. The beam intensity can be reduced to the point where there is never more than a single electron in transit at any time, and the pattern remains the same. The quantities which are interfering are the probability amplitudes. There must be finite probabilities that the electron went through either slit. If we observe it passing through a particular one the probability of it passing through the other is zero and there is nothing left to provide interference.

The essential insight we gain from this discussion is the interpretation to be given to the wavefunction. The square of its value at any point x is the probability of the particle being detected at x. In succeeding chapters we will make this interpretation much more precise.

Computer exercise 1.1: electron diffraction and interference

The program 'Two-Slit Electron Diffraction' provides a computer simulation of the results of the third experiment described above: the arrival of electrons at a fluorescent screen following diffraction at a pair of slits as in figure 1.1(c). (The form of the distribution $P(x)$ in equation (1.10) which is being simulated here is identical to the one seen in optics in the well-known 'Young's slits' experiment and a physical optics text should be consulted if more information is required.) As in optics, the general characteristics of the pattern are completely determined by the ratio of the slit separation to the slit width, and any value can be chosen for this. The pattern consists of a series of parallel interference fringes whose spacing is determined by the slit separation, and which is modulated by a diffraction envelope whose scale is determined by the slit width. How much of the pattern you can see on the screen is set by the horizontal scale. If this is set to one then the interference fringes belonging to the central lobe of the diffraction pattern are displayed. Increase the scale to see more of the diffraction lobes.

Points relevant to our discussion of wave–particle duality to note are:

(a) the 'electrons' arrive one at a time;

(*b*) they are highly localized on detection at the screen; and

(*c*) the interference/diffraction pattern emerges as a non-random distribution of impacts as the events accumulate.

〉1.6 The uncertainty principle

We will now propose a specific form for the wavefunction whose nature we have been discussing. We will assume that the probability amplitude for a particle moving in one dimension along the x axis can be expressed as the wave

$$\Psi(x, t) = A \exp[i(kx - \omega t)] \tag{1.11}$$

where k and ω are related to the particle's momentum and energy as in equations (1.6) and (1.7). The use of the complex exponential form is quite deliberate. In other branches of physics its use is usually a matter of convenience, with the implication that it is the real part of the function which corresponds to physical reality. We will find, however, that the wavefunction is necessarily a complex quantity. This will cause no difficulty with the probability interpretation if we assume that it is the squared magnitude which gives the probability. Specifically, we assert that the probability of finding the particle between x and $x + dx$ is given by

$$P(x, t)\,dx = \Psi^*(x, t)\Psi(x, t)\,dx. \tag{1.12}$$

With this definition the wavefunction (1.11) leads to

$$P(x, t) = A^*A \tag{1.13}$$

which is independent of x; there is a uniform probability of finding the particle at any point on the x axis. The reason for this is that we have chosen for our wavefunction an expression which implies a definite value of momentum since, from equation (1.5), $p = \hbar k$.

We have been led to an extreme example of the uncertainty principle, first proposed by Heisenberg, and which may be expressed as follows:

$$\Delta p\,\Delta x \geqslant \tfrac{1}{2}\hbar. \tag{1.14}$$

The interpretation of this is that it is impossible to determine precisely both the position and the momentum of a particle simultaneously. The determination of one necessarily involves an unknown perturbation of the other. The more precise the determination of one quantity, the more the other is rendered uncertain. In our example above, $\Delta p = 0$ and so $\Delta x \to \infty$; all particle locations are equally likely.

The way in which the uncertainty principle operates has been illustrated by Heisenberg and others in various idealized thought or 'gedanken' experiments. In giving an example of a gedanken experiment we will take the opportunity to justify the statement made above in connection with the electron diffraction experiment, namely that any attempt to determine which slit the electron passed through destroys the interference pattern.

We imagine the apparatus set up with a microscope observing the region of the slits and with an intense light source, which serves to illuminate the electrons as they emerge from one slit or the other. If, as in figure 1.2(*a*), the slits are separated by a distance d and the screen is at a distance large compared with this, then the first minimum of the interference pattern will occur at an angle θ, where

$$\tfrac{1}{2}\lambda_e = d \sin\theta \tag{1.15}$$

where λ_e is the wavelength of the electron as defined in equation (1.5).

Now consider the effect of observing the electron. In order to minimize the disturbing effect of the illuminating light, we suppose a single photon only to be scattered by the electron into the optical system of the microscope, as in figure 1.2(*b*). Now an analysis of the resolving power of a microscope on the basis of wave optics leads to an expression for the resolving limit, r, the smallest separation for which two objects can be resolved, which is

$$r = 1.22\lambda_\gamma/2 \sin\alpha. \tag{1.16}$$

Here α is the angle subtended by the radius of the objective lens at the object and λ_γ is the wavelength of the light. Evidently we must arrange for $r \leqslant d$ so that the slits may be resolved. Hence

$$d \simeq \lambda_\gamma/2 \sin\alpha. \tag{1.17}$$

Since the scattered photon carries momentum the electron will be deflected from its original trajectory. This deflection would be calculable if we knew the details of the collision. Unfortunately, we have no way of knowing which part of the objective lens the photon passes through. We can thus only estimate the order of magnitude of the deflection angle to within approximately

$$\Delta\theta \simeq \Delta p_e/p_e \simeq (p_\gamma \sin\alpha)/p_e \tag{1.18}$$

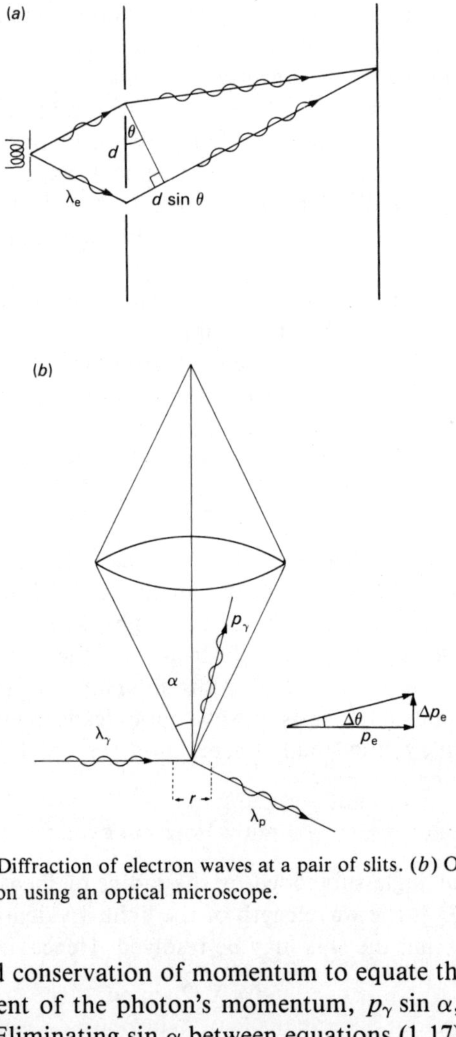

Figure 1.2 (*a*) Diffraction of electron waves at a pair of slits. (*b*) Observation of an electron using an optical microscope.

where we have used conservation of momentum to equate the unknown transverse component of the photon's momentum, $p_\gamma \sin\alpha$, to that of the electron, Δp_e. Eliminating $\sin\alpha$ between equations (1.17) and (1.18) leads to

$$\Delta\theta = \lambda_e/2d. \tag{1.19}$$

Comparing equations (1.15) and (1.19) and roughly equating $\sin\theta$ and θ since they have the same order of magnitude for $0 < \theta < \frac{1}{2}\pi$, we find that the act of observing the electron has introduced an angular uncertainty into its trajectory which is of the same magnitude as the angle of deflection itself! Clearly this will render the interference pattern unobservable.

Notice that it is the dual natures of the electron and the photon which have led to this result. In the collision particle-like behaviour is dominant, but it is the wave-like nature of radiation which requires the large aperture of the microscope in order to achieve the required resolving power, and this in turn leads to the unknown elements in the momenta of the photon and electron.

Setting $d = \Delta x$ in equation (1.19), and using $p_e = h/\lambda_e$, gives the result

$$\Delta p_e \, \Delta x = h \tag{1.20}$$

which is in accord with the uncertainty principle, equation (1.14). The act of locating the electron to within Δx has introduced an uncertainty in the corresponding component of momentum of $h/\Delta x$.

The probability density in equation (1.13) is independent of time as well as of position. This is related to the fact that the wavefunction (1.11) represents a state of definite energy, $E = \hbar\omega$. Since the wavetrain has an infinite length we could, in principle, determine the frequency with infinite precision, although this would require an infinite time to accomplish. These ideas lead to another expression of the uncertainty principle which relates energy and time:

$$\Delta E \, \Delta t \geqslant \tfrac{1}{2}\hbar. \tag{1.21}$$

Wavetrains of more restricted length, such as the wavepackets discussed in the next chapter, correspond to states in which the energy is less well defined, in accordance with equation (1.21). We interpret this equation as a statement of the minimum time required in order to measure the energy of a system with a precision of ΔE.

〉1.7 The Bohr atom

One of the major achievements of Schrödinger's wave mechanics is that it leads to an accurate description of the energy levels and spectrum of atomic hydrogen. Prior to the development of this 'new quantum theory', Bohr had obtained the same energy values on the basis of a semiclassical description (the 'old quantum theory'), in which the electron is described as being in one of certain prescribed circular orbits about the nucleus which by hypothesis do not radiate, as classical laws would require. Radiation is emitted only when the electron abruptly changes to an orbit of lower energy. One criterion for determining which orbits are allowed is that the circumference should equal a whole number of de Broglie waves in length, so that the wave reinforces itself on each orbit. It is readily shown that an equivalent statement is that the allowed orbits have angular momenta, L, given by

$$L = n\hbar \tag{1.22}$$

where n, the 'principal quantum number', is a positive non-zero integer. It is then a simple matter to show that the allowed energy levels of the hydrogen atom are given by

$$E = -me^4/2(4\pi\varepsilon_0)^2n^2 \tag{1.23}$$

where m and e are respectively the mass and charge of the electron. The energies of emitted photons are given by the expression

$$E_\gamma = hcR(1/n_f^2 - 1/n_i^2) \tag{1.24}$$

where the Rydberg constant, R, is

$$R = me^4/4\pi\hbar^3c(4\pi\varepsilon_0)^2 \tag{1.25}$$

and n_i and n_f are the quantum numbers of the initial and final states.

The radii of the allowed orbits are given by the expression

$$r = a_0n^2. \tag{1.26}$$

The radius of the 'first Bohr orbit', for which $n = 1$, is

$$a_0 = 4\pi\varepsilon_0\hbar^2/me^2. \tag{1.27}$$

The values

$$hcR/e = 13.6 \text{ eV} \tag{1.28}$$

and

$$a = 5.29 \times 10^{-11} \text{ m} \tag{1.29}$$

set the scale of atomic energies and dimensions, respectively.

〉 Problems

1.1 Calculate the energy in eV for photons typical of:

(a) A long-wave radio transmitter ($\lambda \simeq 1500$ m).

(b) The 2.7 K cosmic background radiation. (Assume $h\nu = kT$.)
(c) The Sun. (Assume $T \simeq 5800$ K.)

The transmitter of (a) radiates a power of 100 W. How many photons per second are emitted?

1.2 Calculate the wavelength and the frequency of the radiation from:

(a) The Sun. (See 1(c) above.)
(b) A synchrotron source of 1 keV x-rays.
(c) A radioactive source emitting 1 MeV γ-rays.

The power output of the synchrotron source in (b) after passing through a monochrometer is 0.1 W. How many photons per second are emitted?

The source in (c) has a strength of 1 Curie, which means that it emits 3.7×10^{10} particles in one second. Calculate the power output of the source in watts.

1.3 A photon from the cosmic background radiation (see 1(b) above) collides with a stationary electron and is directly back scattered. Calculate the change in energy of the scattered photon and the momentum of the recoiling electron.

1.4 In a two-slit interference experiment using 50 keV electrons the slits were separated by 2 μm, and the fringes were observed by means of an electron microscope at a distance of 35 cm. Calculate the fringe separation.

1.5 Use uncertainty principle arguments to find an expression for the minimum energy a particle of mass m must have, if it is to be confined within a one-dimensional box of length L. Ignoring the three-dimensional nature of the objects involved, estimate the energies, in eV, of:

(a) An electron in an atom, $L = 10^{-10}$ m.
(b) A nucleon in a nucleus, $L = 10^{-14}$ m.
(c) An electron in a nucleus.

Although radioactive nuclei emit electrons with energies of a few MeV, your answer to (c) should indicate that they can hardly have had an independent existence within the nucleus before emission.

1.6 Show that the wavelength of a photon whose energy is equal to the rest mass energy of a particle of mass m is $\lambda_c = h/mc$. λ_c is known as the Compton wavelength of the particle. Use a non-relativistic argument based on the uncertainty principle to calculate the minimum kinetic energy of a particle of mass m confined in a region of dimensions equal to its Compton wavelength. You should find that the particle is close to becoming relativistic.

1.7 Find expressions, on the basis of the Bohr theory of the hydrogen atom, for the frequency of the emitted radiation for a quantum jump in which $n \to n-1$, and the orbital frequency of the electron in the state n. Show that they approach the same value as $n \to \infty$.

Radio telescopes have observed radiation attributed to transitions between quantum states of hydrogen atoms with $n = 110$ and $n = 109$. Calculate, on the basis of Bohr theory, the radii of such atoms, the wavelength and frequency of the radiation and the orbital frequency of the electron.

1.8 Consider a particle of mass m in a circular orbit of radius r. Show that the uncertainty principle applied to this system can be expressed in the form

$$\Delta L \; \Delta\phi \geqslant \tfrac{1}{2}\hbar$$

where L is the angular momentum of the particle and ϕ denotes its orbital position. If the particle is actually an electron in one of the Bohr states of a hydrogen atom, what can be concluded regarding the possibility of determining the position of the electron in its orbit?

Suppose we agree that the classical concept of orbital motion of the particle requires that it be localized within $\Delta\phi \leqslant \frac{1}{10}$ radians, say. Find the uncertainty Δn in the quantum number of the atom for this to be the case. Deduce that a classical orbit, in which the position of the particle is precisely defined, requires the system to be in a highly excited state. This is one manifestation of the correspondence principle, which asserts that the behaviour of a system becomes classical in the limit of large quantum numbers.

Chapter 2

Schrödinger's Equation

2.1 The postulates of quantum mechanics

It is not possible to derive the theory of quantum mechanics *ab initio*, any more than Euclid could have formulated his geometry without introducing certain basic postulates. These are unprovable in themselves, but once accepted their logical consequences lead to a theory of great predictive power. In every case where the predictions have been tested they have been found to be in accord with reality and this leads us to believe in the validity of the postulates, and of the theory in general. The postulates can be reduced to some six or seven statements, and a particular formulation of them may be found in Appendix 1. We shall not have occasion to resort to all of them in this book, but nevertheless it is important, in a subject as counter-intuitive as quantum mechanics usually appears to be on first acquaintance, to be quite clear about what is being assumed and what follows logically.

We have already met one of the postulates in the introductory chapter. It concerns the existence and interpretation of the wavefunction, and can be expressed formally as follows.

Postulate 1. For a particle moving under the influence of an external potential there is an associated wavefunction. That wavefunction determines everything that can be known about the particle and is a single-valued function of the space coordinates and the time. In general, it is a complex function whose squared modulus, evaluated at a particular point and time, is proportional to the probability density of the particle at that point and time.

We have not previously mentioned the role of the potential, which is a way of introducing the effects of applied forces, since we have so far considered only free particles. This will be covered in detail in what follows, as will the point about the single valuedness of the wavefunction. The essential point for the moment is that we will assume the content of this postulate to be true, although all we were able to show in the earlier discussion was that there is a certain plausibility to the assumption.

〉2.2 The wave equation

In other branches of physics, where there is a wave there is a differential wave equation to be satisfied by the mathematical representation of the wave. This can usually be derived from the nature of the wave and the medium in which it propagates. In the case of the probability amplitude waves which we are considering, it is not possible to show that any particular wave equation is a logical necessity and, in fact, the validity of the one we choose, Schrödinger's equation, is the subject of another of the postulates of quantum mechanics. Nevertheless, it is possible to make the choice plausible and this is what will now be attempted.

We try first to find the wave equation satisfied by a free particle whose wavefunction we will assume to be as given by equation (1.11). If we substitute the de Broglie relations (1.5) and (1.6) into (1.11) we obtain for this wavefunction

$$\Psi(x, t) = A \exp[\mathrm{i}(px - Et)/\hbar] \tag{2.1}$$

thus making explicit the fact that we are dealing with a particle of momentum p and energy E. A natural choice of wave equation is the familiar one:

$$\frac{\partial^2 \Psi}{\partial x^2} = \frac{1}{v^2} \frac{\partial^2 \Psi}{\partial t^2}. \tag{2.2}$$

Substituting (2.1) into this results in

$$-\frac{p^2}{\hbar^2} \Psi = -\frac{1}{v^2} \frac{E^2}{\hbar^2} \Psi \tag{2.3}$$

or

$$E^2 = p^2 v^2. \tag{2.4}$$

However, as we saw in Chapter 1 this expression is only valid for a photon, when $v = c$. For a particle of finite rest mass m the more general relation

$$E^2 = p^2c^2 + m^2c^4 \tag{2.5}$$

applies. Clearly then the wave equation (2.2) is unsuitable to describe the waves associated with material particles.

Equation (2.5) is relativistically exact and if we persist in searching for a wave equation which was consistent with it we would be led to the Klein–Gordon equation, as indeed was Schrödinger originally, or perhaps the Dirac equation. (These equations are used in relativistic quantum mechanics to describe integral and half-integral spin particles, respectively.) However, in many cases of interest relativistic effects are very small, and a non-relativistic treatment is justified. For example, the Bohr theory of the hydrogen atom indicates that the speed of the electron in the lowest orbit is αc, where α is the fine structure constant and has the value $\alpha = e^2/\hbar c(4\pi\varepsilon_0) \simeq \frac{1}{137}$.

The non-relativistic expression relating momentum and energy which corresponds to equation (2.5) is

$$E = p^2/2m \tag{2.6}$$

where E is now the kinetic energy of the particle, rather than the total relativistic energy. Considering the wavefunction of equation (2.1), we note that each differentiation with respect to time t has the effect of multiplying by $-\mathrm{i}E/\hbar$, and each differentiation with respect to position x has the effect of multiplying by $\mathrm{i}p/\hbar$. Since equation (2.6) relates the first power of E to the second power of p we are prompted to try the wave equation

$$-\frac{\hbar}{\mathrm{i}}\frac{\partial \Psi}{\partial t} = -\frac{\hbar^2}{2m}\frac{\partial^2 \Psi}{\partial x^2} \tag{2.7}$$

on the grounds that substitution into it of equation (2.1) leads to equation (2.6). Equation (2.7) is in fact Schrödinger's equation for a free particle. In case we appear to be performing an empty exercise in going from a solution of a differential equation to the equation itself, it should be pointed out that equation (2.7) is not the only equation consistent with equation (2.6). We could, for example, have experimented with a term in $(\partial\Psi/\partial x)^2$ and arrived at a non-linear equation. The adoption of (2.7), which is a linear equation, carries with it the implication that we can superimpose different solutions of the equation, an operation crucial to the description of interference and many other phenomena in quantum mechanics.

It must nevertheless be admitted that equation (2.7) is very limited in scope since it applies only to a free particle. We must find some way of introducing the effects of forces on the particle. We can do this if we introduce a potential function $V(x, t)$ such that

$$F_x = -\frac{\partial}{\partial x} V(x, t). \tag{2.8}$$

The total energy of the particle is now given by

$$E = p^2/2m + V(x, t). \tag{2.9}$$

Adopting the convention that each term in the wave equation should correspond to a term of the energy–momentum equation multiplied by Ψ, as was true of equations (2.6) and (2.7), we propose

$$-\frac{\hbar}{\mathrm{i}}\frac{\partial\Psi}{\partial t} = -\frac{\hbar^2}{2m}\frac{\partial^2\Psi}{\partial\Psi^2} + V(x, t)\Psi \tag{2.10}$$

as a generalization of equation (2.7). Clearly, equation (2.10) reduces to equation (2.7) if the potential is uniform, because only potential differences have physical significance and a uniform potential can be set equal to zero. On the other hand, if $V(x, t)$ is actually a function of position or time the wavefunction (2.1) will not satisfy equation (2.10), and new solutions must be found, corresponding to the modified motion of the particle resulting from the applied force. Note that the introduction of $V(x, t)\Psi$ preserves the linearity of the equation.

Equation (2.10) is referred to as the time-dependent Schrödinger equation, which we will abbreviate to TDSE. We have not derived it; it is the subject of one of the postulates of quantum mechanics (Appendix 1) which may be expressed as follows.

Postulate 2. The wavefunction satisfies the time-dependent Schrödinger equation at all times.

A feature of the TDSE which does not usually occur in the differential equations of classical physics is that it contains the quantity i. This requires Ψ to be complex for the equation to balance. Any physically observable quantity must be real so the theory must be formulated in a way which takes account of this. We have already seen an example of this in equation (1.12), where $\Psi^*\Psi$ is used for the probability density.

〉2.3 The wave equation in three dimensions

So far we have been considering motion in one dimension only. However, equation (2.10) is rather easily generalized to three dimensions. Equation (2.9) becomes

$$E = (p_x^2 + p_y^2 + p_z^2)/2m + V(\boldsymbol{r}, t) \tag{2.11}$$

where $\boldsymbol{r}$ is the position vector of the particle, and so equation (2.10) becomes

$$-\frac{\hbar}{\mathrm{i}}\frac{\partial \Psi}{\partial t}(\boldsymbol{r}, t) = -\frac{\hbar^2}{2m}\left(\frac{\partial^2 \Psi}{\partial x^2} + \frac{\partial^2 \Psi}{\partial y^2} + \frac{\partial^2 \Psi}{\partial z^2}\right)\Psi(\boldsymbol{r}, t) + V(\boldsymbol{r}, t)\Psi(\boldsymbol{r}, t) \tag{2.12}$$

or

$$-\frac{\hbar}{\mathrm{i}}\frac{\partial \Psi}{\partial t}(\boldsymbol{r}, t) = -\frac{\hbar^2}{2m}\nabla^2\Psi(\boldsymbol{r}, t) + V(\boldsymbol{r}, t)\Psi(\boldsymbol{r}, t). \tag{2.13}$$

In general Ψ and V will each be functions of $\boldsymbol{r}$ and t.

〉2.4 The time-independent Schrödinger equation

Many important phenomena in physics are time dependent. For example, the absorption of radiation by an atom occurs when the atom responds to the fluctuating electromagnetic fields of the radiation. In such cases the potential function V in equation (2.13) would indeed be time dependent. On the other hand, an electron in an atom not subject to a radiation field experiences an electrostatic force due to the nuclear charge which can be described by a potential function which depends only on position. Solutions to the TDSE when V is a function of time, though very important, are beyond the scope of this book. We will concentrate on situations in which V is independent of the time. It is then possible to find solutions to the TDSE of the form

$$\Psi(\boldsymbol{r}, t) = u(\boldsymbol{r})T(t) \tag{2.14}$$

i.e. a product of a function of position only and a function of the time only. Such solutions have the character of a standing wave. To show that this factorization is possible we use the technique of variable separation. If equation (2.14) is substituted into the TDSE, equation (2.13), we find

$$-\frac{\hbar}{\mathrm{i}}u(\boldsymbol{r})\frac{\partial}{\partial t}T(t) = T(t)\left(-\frac{\hbar^2}{2m}\nabla^2 + V(\boldsymbol{r})\right)u(\boldsymbol{r}) \tag{2.15}$$

and dividing by $u(r)T(t)$ we obtain

$$-\frac{\hbar}{\mathrm{i}}\frac{1}{T(t)}\frac{\partial}{\partial t}T(t)=\frac{1}{u(r)}\left(-\frac{\hbar^2}{2m}\nabla^2+V(r)\right)u(r). \tag{2.16}$$

The equation is now arranged so that the left-hand side is a function only of time and the right-hand side is a function only of position. However, the equality must hold for all times and at all positions and these are independent variables. Clearly this can only be so if the two sides are independently equal to the same constant, which we will call K. We can therefore separate the equation into two, linked by this common constant K,

$$-\frac{\hbar}{\mathrm{i}}\frac{\partial}{\partial t}T(t)=KT(t) \tag{2.17}$$

$$\left(-\frac{\hbar^2}{2m}\nabla^2+V(r)\right)u(r)=Ku(r). \tag{2.18}$$

We can identify K by multiplying equation (2.17) by $u(r)$ and recalling that $u(r)T(t)=\Psi(r,t)$. Hence

$$-\frac{\hbar}{\mathrm{i}}\frac{\partial}{\partial t}\Psi(r,t)=K\Psi(r,t). \tag{2.19}$$

In our discussion of the individual terms of the TDSE, the left-hand expression in the above equation was identified with the total energy, and so it is clear that K is in fact just equal to E. Equations (2.17) and (2.18) can therefore be written as

$$-\frac{\hbar}{\mathrm{i}}\frac{\partial}{\partial t}T(t)=ET(t) \tag{2.20}$$

and

$$\left(-\frac{\hbar^2}{2m}\nabla^2+V(r)\right)u(r)=Eu(r). \tag{2.21}$$

Equation (2.21) is referred to as the time-independent Schrödinger equation, which we will abbreviate to TISE. Finding solutions of it for a variety of systems is our principal concern in the remainder of this book. Equation (2.20) can be solved immediately to give

$$T(t)=\exp(-\mathrm{i}Et/\hbar) \tag{2.22}$$

and, consequently, the complete wavefunction has the form

$$\Psi(r,t)=u(r)\exp(-\mathrm{i}Et/\hbar). \tag{2.23}$$

〉2.5 Stationary states

Equation (1.12) expresses the probability interpretation of the wavefunction as defined in postulate 1. In three dimensions this becomes

$$P(\boldsymbol{r}, t)\,\mathrm{d}\tau = \Psi^*(\boldsymbol{r}, t)\Psi(\boldsymbol{r}, t)\,\mathrm{d}\tau \tag{2.24}$$

where $\mathrm{d}\tau$ is an element of volume. Inserting the wavefunction of (2.23) shows that

$$P(\boldsymbol{r}, t) = u^*(\boldsymbol{r})u(\boldsymbol{r})\exp(\mathrm{i}Et/\hbar)\exp(-\mathrm{i}Et/\hbar). \tag{2.25}$$

Hence

$$P(\boldsymbol{r}, t) = u^*(\boldsymbol{r})u(\boldsymbol{r}) \tag{2.26}$$

which is time independent. Such states are therefore described as *stationary* and describe systems of definite energy. A more general solution to the TDSE would consist of a superposition of functions like equation (2.23):

$$\Psi(\boldsymbol{r}, t) = \sum_i A_i u_i(\boldsymbol{r})\exp(-\mathrm{i}E_i t/\hbar). \tag{2.27}$$

Clearly such a state is not stationary and describes a system for which the energy is not well defined. In future we will use the symbol 'u' exclusively to represent stationary states.

〉2.6 Normalization

The probability of finding a particle in the volume element $\mathrm{d}\tau$ at $\boldsymbol{r}$ at time t is given by equation (2.24). Now the probability of finding the particle *somewhere* must be unity, so it follows that

$$\int_{\substack{\text{All}\\\text{space}}} \Psi^*(\boldsymbol{r}, t)\Psi(\boldsymbol{r}, t)\,\mathrm{d}\tau = 1 \tag{2.28}$$

where the integral is over all space. For a stationary state this becomes

$$\int_{\substack{\text{All}\\\text{space}}} u^*(\boldsymbol{r})u(\boldsymbol{r})\,\mathrm{d}\tau = 1. \tag{2.29}$$

Now $\Psi(\boldsymbol{r}, t)$ and $u(\boldsymbol{r})$ are solutions of homogeneous second-order differential equations. Consequently, any solution we obtain may be multiplied by an arbitrary constant and remain a solution. We may exploit this freedom by choosing the multiplying constant so that equations (2.28) and (2.29) do in fact apply. This process is referred to as normalization.

〉2.7 Probability flux

The free-particle wavefunction (2.1) is an example of the stationary states described in section 2.5; the corresponding probability density is time independent. At first sight this is a little surprising since it represents a particle in motion with momentum p and the mental picture we have of a moving classical particle leads us to expect the probability density to be time dependent. Nevertheless, motion is implicit in the wavefunction, just as the motion of charge is implied by the presence of an electric current in a wire, regardless of whether the current is in a steady state or not. Below (section 2.9) we will discuss wavepackets. Such wavefunctions relate more closely to classical particles and do have time-dependent probability densities. In this section we will develop a formalism which quantifies the particle motion for all types of wavefunction.

First of all we will investigate the one-dimensional case. The probability that a particle whose wavefunction is $\Psi(x, t)$ will be found in a rectangular box of cross-sectional area $\mathscr{A}$, lying between the planes $x = a$ and $x = b$ (see figure 2.1(a)), is

$$\mathscr{A} \int_a^b P(x, t)\, \mathrm{d}x = \mathscr{A} \int_a^b \Psi^*(x, t)\Psi(x, t)\, \mathrm{d}x. \tag{2.30}$$

To see how this changes with time we differentiate both sides of the equation:

$$\mathscr{A} \frac{\partial}{\partial t} \int_a^b P(x, t)\, \mathrm{d}x = \mathscr{A} \int_a^b \left(\frac{\partial \Psi^*}{\partial t} \Psi + \Psi^* \frac{\partial \Psi}{\partial t} \right) \mathrm{d}x. \tag{2.31}$$

We can replace the time derivatives on the right by using the TDSE, equation (2.10), and its complex conjugate. The terms in $V(r, t)$ are real and so cancel, and we have

$$\begin{aligned} \mathscr{A} \frac{\partial}{\partial t} \int_a^b P(x, t)\, \mathrm{d}x &= -\mathscr{A} \frac{\mathrm{i}\hbar}{2m} \int_a^b \left(\Psi \frac{\partial^2 \Psi^*}{\partial x^2} - \Psi^* \frac{\partial^2 \Psi}{\partial x^2} \right) \mathrm{d}x \\ &= -\mathscr{A} \frac{\mathrm{i}\hbar}{2m} \int_a^b \frac{\partial}{\partial x} \left(\Psi \frac{\partial \Psi^*}{\partial x} - \Psi^* \frac{\partial \Psi}{\partial x} \right) \mathrm{d}x \end{aligned}$$

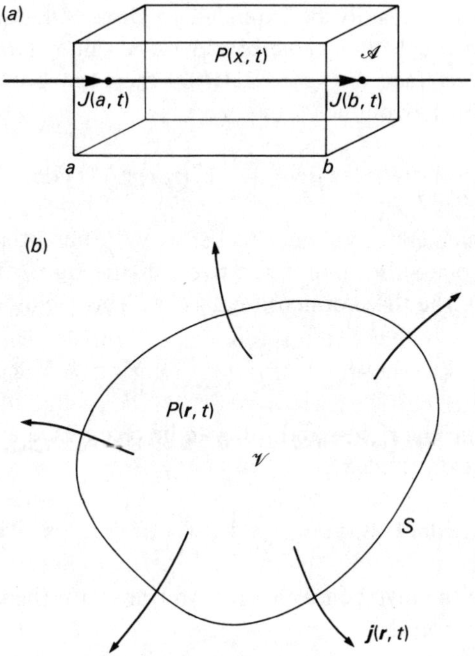

Figure 2.1 (*a*) One-dimensional particle current. If current in at *a* is not balanced by current out at *b*, the probability density inside the box must be changing. (*b*) Three-dimensional particle flux.

$$= -\mathscr{A}\frac{i\hbar}{2m}\left[\Psi\frac{\partial\Psi^*}{\partial x} - \Psi^*\frac{\partial\Psi}{\partial x}\right]_a^b$$
$$= J(a, t) - J(b, t). \tag{2.32}$$

Stated in words, the last equation tells us that the probability that the particle is to be found between a and b is changing at a rate equal to the difference between the values taken by a quantity $J(x, t)$ at $x = a$ and at $x = b$, where

$$J(x, t) = \mathscr{A}\frac{i\hbar}{2m}\left(\Psi(x, t)\frac{\partial\Psi^*(x, t)}{\partial x} - \Psi^*(x, t)\frac{\partial\Psi(x, t)}{\partial x}\right). \tag{2.33}$$

It is clear that the quantity $J(x, t)$ must be interpreted as a current. Consider the analogy, suggested above, of a wire carrying an electrical current. If the current entering at one end is not equal to the current leaving the other end the charge on the wire must be changing. Here, however, the current is one of *probability*!

The analysis can readily be extended to three dimensions using the methods of vector calculus. If we consider a certain volume of space $\mathscr{V}$, bounded by a surface S (figure 2.1(*b*)), then the probability of the particle being in the volume $\mathscr{V}$ at time t is

$$\int_{\mathscr{V}} P(r, t)\, \mathrm{d}\tau = \int_{\mathscr{V}} \Psi^*(r, t)\Psi(r, t)\, \mathrm{d}\tau \tag{2.34}$$

where $\mathrm{d}\tau$ is an element of volume. As before we differentiate with respect to time and replace the time derivatives on the right, this time using equation (2.13), the three-dimensional TDSE. This yields

$$\frac{\partial}{\partial t}\int_{\mathscr{V}} P(r, t)\, \mathrm{d}\tau = -\frac{\mathrm{i}\hbar}{2m}\int_{\mathscr{V}} (\Psi\nabla^2\Psi^* - \Psi^*\nabla^2\Psi)\, \mathrm{d}\tau. \tag{2.35}$$

The integrand on the right-hand side can be recognized as a divergence, since it can be expressed as

$$\frac{\partial}{\partial t}\int_{\mathscr{V}} P(r, t)\, \mathrm{d}\tau = -\frac{\mathrm{i}\hbar}{2m}\int_{\mathscr{V}} \nabla\cdot(\Psi\nabla\Psi^* - \Psi^*\nabla\Psi)\, \mathrm{d}\tau \tag{2.36}$$

and we can use the divergence theorem to transform the right-hand side into a surface integral:

$$\frac{\partial}{\partial t}\int_{\mathscr{V}} P(r, t)\, \mathrm{d}\tau = -\frac{\mathrm{i}\hbar}{2m}\int_{S} (\Psi\nabla\Psi^* - \Psi^*\nabla\Psi)\cdot \mathrm{d}S. \tag{2.37}$$

The expression on the left is the rate of change of the probability that the particle is in the volume, so we can interpret the right-hand side as a current of probability. The integrand therefore has the nature of a current *density* or flux density. We accordingly define the *probability flux density* as

$$\begin{aligned} j(r, t) &= \frac{\mathrm{i}\hbar}{2m}[\Psi(r, t)\nabla\Psi^*(r, t) - \Psi^*(r, t)\nabla\Psi(r, t)] \\ &= \mathrm{Re}\,\frac{\hbar}{\mathrm{i}m}\Psi^*(r, t)\nabla\Psi(r, t). \end{aligned} \tag{2.38}$$

Equating integrands, equation (2.35) now reads

$$\frac{\partial}{\partial t}P(r, t) = -\frac{\mathrm{i}\hbar}{2m}\nabla\cdot j(r, t). \tag{2.39}$$

This is directly analogous to the equation of continuity, or charge conservation, which appears in the theory of electromagnetism. In this case the equation is expressing the conservation of probability. The rate of change of the particle probability density at a point is equal to the divergence of the probability flux density at that point. Similarly, equation (2.37) is an integral statement of probability conservation, and refers to a finite volume.

If we allow the surface S to retreat to infinity, the integral on the left-hand side of equation (2.37) becomes the normalization integral since it is now to be taken over all space. Now the right-hand side of the equation will equal zero, provided Ψ falls to zero at infinity, which is necessary if the wavefunction is to be normalizable. From this we can conclude that the normalization integral is a constant; once normalized the wavefunction remains so. It would certainly be disquieting if this were not the case!

In one dimension, equation (2.38) reduces to

$$j(x, t) = \frac{\mathrm{i}\hbar}{2m}\left(\Psi \frac{\partial \Psi^*}{\partial x} - \Psi^* \frac{\partial \Psi}{\partial x}\right) \tag{2.40}$$

and if we apply this to the free-particle wavefunction, equation (2.1), we obtain

$$j(x, t) = \frac{\mathrm{i}\hbar}{2m}\left(-\frac{\mathrm{i}}{\hbar} p \,|A^2| - \frac{\mathrm{i}}{\hbar} p\, |A^2|\right) = \frac{p\,|A|^2}{m} = v\,|A|^2 \tag{2.41}$$

which we can interpret as the probability density in the wave multiplied by the particle speed v, a very reasonable result. We shall have occasion to make use of the flux density $\boldsymbol{j}(\boldsymbol{r}, t)$ when we study particle scattering at potential steps and barriers.

〉2.8 Physical constraints on the wavefunction: boundary conditions

The probability amplitude interpretation of the wavefunction places a number of constraints on the form of the wavefunction. One or two have been indicated already; we will assemble them all here.

(*a*) There cannot be more than one probability of a particle being found at a certain point, so the quantity $|\Psi|^2$ must be single valued. Ψ itself is not necessarily single valued, though it is often presumed to be so. We will meet an example of a double-valued wavefunction when we deal with electron spin in Chapter 6.

(*b*) The probability that the particle is in any given region of space cannot exceed unity! This requires that the wavefunction must be 'square integrable' (the left-hand side of equation (2.28) must be finite) so that it can be normalized. It is sometimes convenient to use functions which violate this condition; equation (2.1) is an example. When we do so it is to be understood that we are using an unphysical idealization of a properly behaved wavefunction. This will be discussed further in the next section.

(*c*) The wavefunction must be continuous everywhere. The probability density must be defined everywhere even if it is zero in some places.

(*d*) A more subtle requirement is that the wavefunction must be smooth. More precisely, we require the gradients $\partial\Psi/\partial x$, etc, to be continuous. A kink in the wavefunction implies a step in the first derivative and this in turn implies an infinite spike in the second derivative, as in figure 2.2.

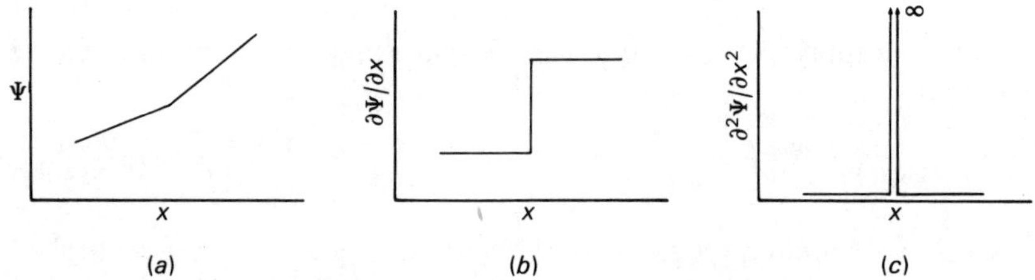

Figure 2.2 A slope discontinuity in Ψ (*a*) leads to an infinity in $\partial^2\Psi/\partial x^2$ (*c*). This is inadmissible except at infinite steps in potential energy.

However, from our discussion of the terms in the TDSE we know that the second derivative is related to the kinetic energy and so cannot be infinite. In fact, we occasionally consider cases in which the potential is infinite as an idealization of a very large potential, and in such cases there *will* be a kink in the wavefunction at the boundary, although it will still be continuous. Figure 2.3 illustrates some inadmissible wavefunctions, which violate one or more of conditions (*a*), (*b*), (*c*) and (*d*).

The conditions of continuity and smoothness are often referred to as boundary conditions since they must be satisfied even at boundaries where there may be (finite) discontinuities in the potential function. Thus if there is such a boundary between regions 1 and 2 at, say, $x = a$ (figure 2.4), we require that

$$\Psi_1(a, t) = \Psi_2(a, t) \tag{2.42}$$

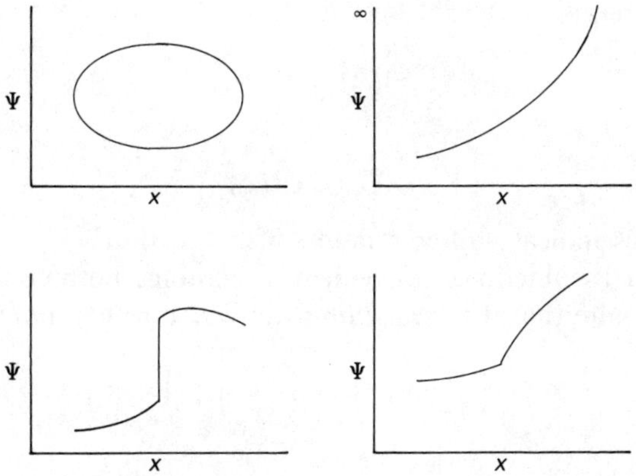

Figure 2.3 Some inadmissible wavefunctions.

and

$$\frac{\partial}{\partial x}\Psi_1(a,t)=\frac{\partial}{\partial x}\Psi_2(a,t). \tag{2.43}$$

We can avoid expressing the time dependence constantly by considering a specific time t_0 and letting $\Psi(x,t_0)\equiv\psi(x)$. The boundary conditions can

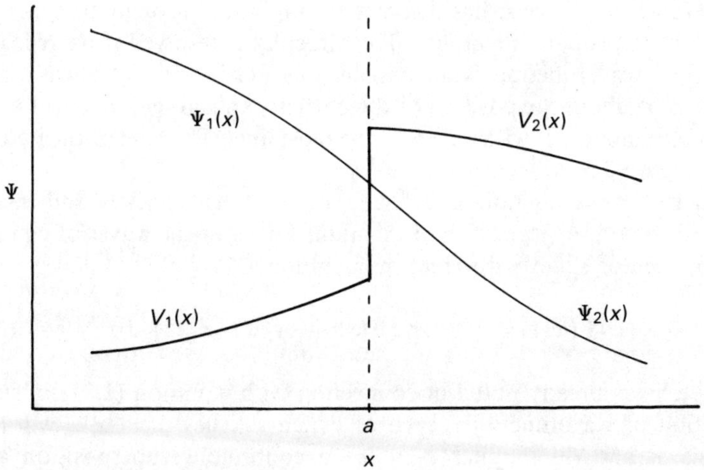

Figure 2.4 Even though the potential function may change discontinuously at a boundary, the wavefunction must be smooth.

then be written as

$$\psi_1(a) = \psi_2(a) \tag{2.44}$$

and

$$\psi_1'(a) = \psi_2'(a) \tag{2.45}$$

where primes indicate differentiation of a function with respect to its argument. It is sometimes convenient to combine both conditions into one, and require that the *logarithmic derivative* be continuous, i.e.

$$\left(\frac{\partial}{\partial x} \ln \psi_1\right)_a = \left(\frac{\partial}{\partial x} \ln \psi_2\right)_a$$

or

$$\left(\frac{\psi_1'}{\psi_1}\right)_a = \left(\frac{\psi_2'}{\psi_2}\right)_a \tag{2.46}$$

this being equivalent to the quotient of equations (2.44) and (2.45).

〉2.9 Wavepackets

It is unfortunate that the very first wavefunction we have to deal with (equation (2.1)) should violate one of the formal requirements of a wavefunction! Since it has a constant amplitude throughout space it is not, in fact, square integrable. The difficulty is resolved if we recognize that this wavefunction is an unrealizable ideal. We will always know something about the position of the particle, and so the wave will have a variable amplitude, with a peak at the most likely location of the particle, as in figure 2.5.

Such a wavefunction is referred to as a wavepacket and can be described as a superposition of an infinite array of the waves of equation (2.1), each of slightly different momentum p:

$$\Psi(x, t) = \int_{-\infty}^{\infty} A_p(p) \exp\left(\frac{\mathrm{i}}{\hbar}(px - E_p t)\right) \mathrm{d}p. \tag{2.47}$$

As we have already noted in connection with equation (2.27), a superposition of wavefunctions, each of which satisfies the TDSE, will itself satisfy it and this applies even for a continuous superposition as in equation (2.47). The factor $A_p(p)$ in equation (2.47) describes the distribution of wave amplitudes with momentum.

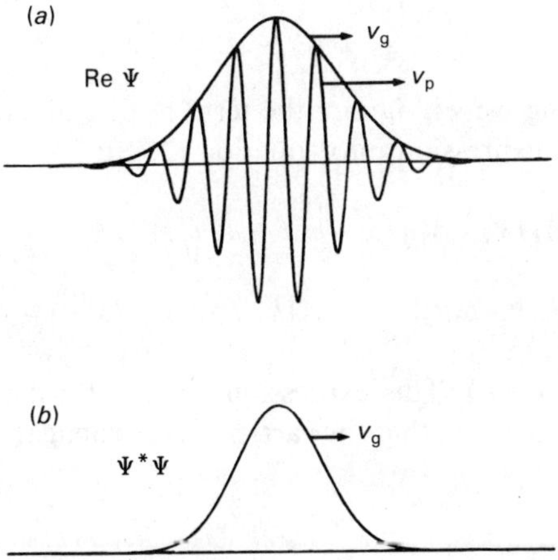

Figure 2.5 A Gaussian wavepacket. (*a*) The real part of the time-dependent function. (*b*) The probability density.

We can gain a great deal of insight into the wave nature of matter by studying the properties of such wavepackets. It is more convenient to work with ω and k than with E and p, and with these variables the wavepacket becomes

$$\Psi = \int_{-\infty}^{\infty} A_k(k) \exp[i(kx - \omega t)] \, dk. \tag{2.48}$$

We will now suppose that the range of k, though technically infinite, is principally concentrated within a small range centred at k_0. Now ω will normally be a function of k, $\omega(k)$ say, and the relation between these two variables is referred to as a *dispersion relation*. We can proceed further with the analysis without even knowing the explicit form of the dispersion relation, by making use of Taylor's theorem to expand $\omega(k)$ about the point k_0. This gives the result

$$\omega(k) = \omega(k_0) + (k - k_0)\left(\frac{d\omega}{dk}\right)_{k_0} + \tfrac{1}{2}(k - k_0)^2\left(\frac{d^2\omega}{dk^2}\right)_{k_0} + \cdots. \tag{2.49}$$

We will simplify the notation by means of the following substitutions:

$$\omega_0 = \omega(k_0) \qquad k_- = k - k_0 \qquad v_g = \left(\frac{d\omega}{dk}\right)_{k_0} \qquad \beta = \left(\frac{d^2\omega}{dk^2}\right)_{k_0} \tag{2.50}$$

so that

$$\omega(k) = \omega_0 + k_- v_g + \tfrac{1}{2} k_-^2 \beta + \cdots. \tag{2.51}$$

For the time being we will ignore the term in β, and substitute the first two terms of the expression into equation (2.48):

$$\begin{aligned} \Psi &\simeq \int_{-\infty}^{\infty} A_k(k) \exp[\mathrm{i}(kx - \omega_0 t - k_- v_g t)]\, \mathrm{d}k \\ &\simeq \exp[\mathrm{i}(k_0 x - \omega_0 t)] \int_{-\infty}^{\infty} A(k_-) \exp[\mathrm{i}k_-(x - v_g t)]\, \mathrm{d}k \end{aligned} \tag{2.52}$$

where $A_k(k) \equiv A(k_-)$. This expression has the form (carrier wave) × (envelope) and we can see that the carrier wave propagates with the phase velocity

$$v_p = \omega_0 / k_0 \tag{2.53}$$

whereas the envelope propagates with the group velocity v_g:

$$v_g = \left(\frac{\mathrm{d}\omega}{\mathrm{d}k}\right)_{k_0}. \tag{2.54}$$

We can now clear up a hitherto baffling aspect of the de Broglie hypothesis (equations (1.6) and (1.7)). For a free particle of energy $E = \hbar\omega$ and momentum $p = \hbar k$, the phase velocity is $v_p = \omega/k = E/p$. But for a free particle moving with velocity v in a constant potential V,

$$p = mv \qquad E = \tfrac{1}{2} mv^2 + V \tag{2.55}$$

so

$$v_p = \frac{\tfrac{1}{2} mv^2 + V}{mv}. \tag{2.56}$$

However, V is an *arbitrary* constant, so that both v_p and ω are arbitrary in the same way that the absolute value of potential energy is arbitrary. For a constant potential, a convenient choice would be to make $V = 0$, in which case $v_p = \tfrac{1}{2} v$, but any other choice would give a different value! We therefore conclude that the phase velocity has no physical significance.

In order to calculate the group velocity $\mathrm{d}\omega/\mathrm{d}k$ we must express ω in terms of k which we can do using (2.54) and the de Broglie relations:

$$\hbar\omega = \frac{\hbar^2 k^2}{2m} + V. \tag{2.57}$$

On differentiating we have

$$v_g = \frac{\mathrm{d}\omega}{\mathrm{d}k} = \frac{\hbar k}{m} = \frac{p}{m} = v \tag{2.58}$$

and we find that the group velocity is to be associated with the particle velocity. Since the group velocity is the velocity of the envelope and the latter relates to the position of the particle this is a very reasonable result.

〉2.10 Gaussian wavepackets and the uncertainty principle

As a specific example of a wavepacket we will examine one with a Gaussian profile. We will take the amplitude distribution function in equation (2.52) to be

$$A(k_-) = A_0 \exp(-\alpha k_-^2). \tag{2.59}$$

This is illustrated in figure 2.6(a). The parameter α determines the characteristic width of the distribution. Inserting $A(k_-)$ into equation (2.52) we obtain

$$\Psi = A_0 \exp[\mathrm{i}(k_0 x - \omega_0 t)] \int_{-\infty}^{\infty} \exp[-(\alpha k_-^2 + \mathrm{i}k_-(x - v_g t))]\, \mathrm{d}k_-. \tag{2.60}$$

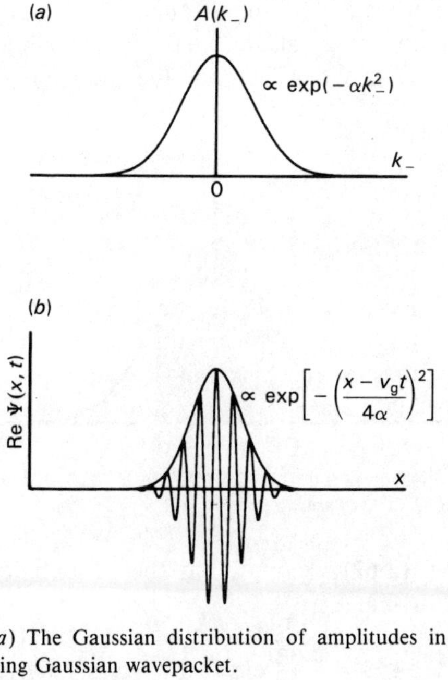

Figure 2.6 (a) The Gaussian distribution of amplitudes in $k_- = k - k_0$. (b) The resulting Gaussian wavepacket.

The integral has the form of a Fourier transform, and may be performed by writing the exponent as a perfect square. Thus it may be verified that

$$\Psi = A_0 \exp[i(k_0 x - \omega_0 t)] \exp[-(x - v_g t)^2/4\alpha] \times \int_{-\infty}^{\infty} \exp\{-\alpha[(k_- + i(x - v_g t))/2\alpha]^2\} \, dk_-. \quad (2.61)$$

Despite the complexity of the exponent, the integral is now simply that of a Gaussian and takes the value $(\pi/\alpha)^{1/2}$, so that the wavepacket becomes

$$\Psi = A_0(\pi/\alpha)^{1/2} \exp[i(k_0 x - \omega_0 t)] \exp[-(x - v_g t)^2/4\alpha]. \quad (2.62)$$

Hence the envelope of the wavepacket is Gaussian if the distribution function $A(k_-)$ is Gaussian (figure 2.6(b)).

Now that we have chosen an explicit form for the wavefunction we can adopt a precise definition for the uncertainties in position and momentum. To do this we recall that in probability theory the normal distribution is a Gaussian and takes the form

$$G(z) = \left(\frac{\sigma^2}{2\pi}\right)^{1/2} \exp(-z^2/2\sigma^2) \quad (2.63)$$

where σ^2 is the variance. The significance of σ is that if a variable z is distributed according to equation (2.63), the probability that $|z| \leqslant \sigma$ is 0.683 (see figure 2.7). We will adopt the quantity σ as a measure of uncertainty in z.

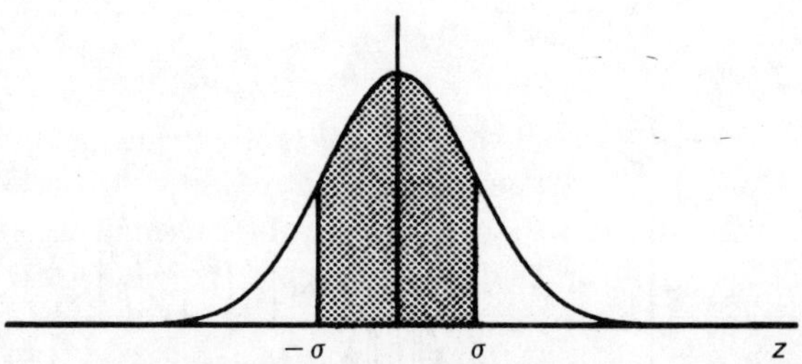

Figure 2.7 The standard normal distribution. The shaded area between $\pm\sigma$ represents 0.683 of the total.

From equation (2.62) we find that the probability density for the particle is the Gaussian

$$\Psi^*\Psi = \frac{\pi A_0^2}{\alpha} \exp[-(x - v_g t)^2/2\alpha]. \quad (2.64)$$

Normalization requires that

$$\frac{\pi A_0^2}{\alpha} = \left(\frac{1}{2\pi\alpha}\right)^{1/2} \tag{2.65}$$

which determines A_0. Hence

$$\Psi^*\Psi = \left(\frac{1}{2\pi\alpha}\right)^{1/2} \exp[-(x - v_g t)^2/2\alpha]. \tag{2.66}$$

Comparing this with equation (2.63) and identifying Δx with σ we find

$$\Delta x = \sqrt{\alpha}. \tag{2.67}$$

Similarly, from equation (2.59),

$$|A(k_-)|^2 = A_0^2 \exp(-2\alpha k_-^2) \tag{2.68}$$

and so

$$\Delta k_- = \frac{1}{2\sqrt{\alpha}}. \tag{2.69}$$

Hence the product of uncertainties is

$$\Delta x \, \Delta k_- = \tfrac{1}{2}. \tag{2.70}$$

Multiplying by $\hbar$ we have, for the Gaussian wavepacket,

$$\Delta x \, \Delta p_x = \tfrac{1}{2}\hbar. \tag{2.71}$$

It can be shown that the Gaussian profile leads to the minimum possible product, so that more generally

$$\Delta x \, \Delta p_x \geqslant \tfrac{1}{2}\hbar \tag{2.72}$$

which is the usual statement of the uncertainty principle.

We will now remedy an omission in the treatment of section 2.9, and examine the consequences of including the third term of the expansion of $\omega(k)$ (equation (2.51)) in the expression for the wavepacket, equation (2.48). Carrying through the analysis in a similar fashion leads to the following expression for the probability density:

$$\Psi^*\Psi = \left(\frac{1}{2\pi}\right)^{1/2} \left(\frac{\alpha}{\alpha^2 + \beta^2 t^2}\right)^{1/2} \exp\left(\frac{-\alpha(x - v_g t)^2}{2(\alpha^2 + \beta^2 t^2)}\right). \tag{2.73}$$

The profile remains Gaussian, but now we find that the amplitude decreases with time, and the width increases (figure 2.8). Physically this comes about because an initial uncertainty in momentum evolves into a future uncertainty in position.

In conclusion, we have shown that realistic normalizable particle wavefunctions can be constructed, exhibiting properties which link particle and wave behaviour in an elegant fashion. The price to be paid is that the momentum and energy of the particle become less well defined, a manifestation of the uncertainty principle. The other price to be paid is in the added complexity of the packet wavefunctions. We will find that for most purposes it is perfectly adequate to deal with functions like the one in equation (2.1).

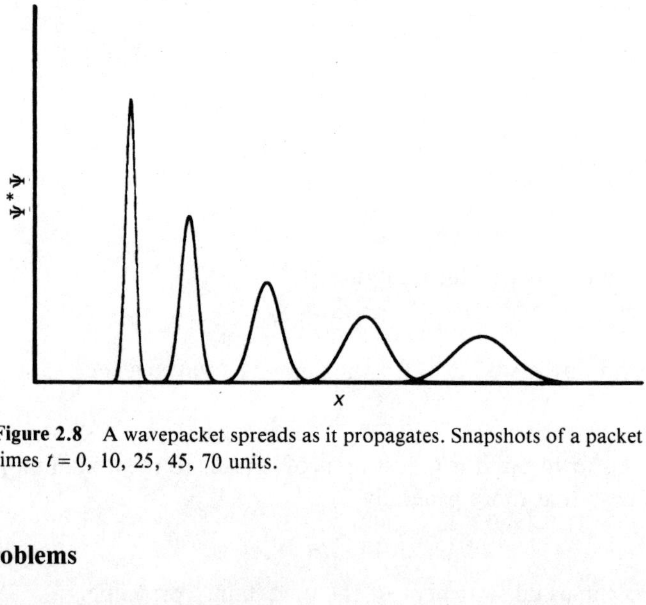

Figure 2.8 A wavepacket spreads as it propagates. Snapshots of a packet at times $t = 0$, 10, 25, 45, 70 units.

〉 Problems

2.1 A wavepacket is defined as in equation (2.48) with

$$A_k(k) = 0 \qquad |k| > K$$
$$A_k(k) = A \qquad |k| \leqslant K$$

where A is a constant. Find an expression for $\psi(x) = \Psi(x, 0)$ by setting $t = 0$ in equation (2.48), and performing the Fourier transform. Sketch the two functions $|A_k(k)|^2$ and $|\psi(x)|^2$. Choose simple definitions for the widths of these two distributions and show that the uncertainty principle is obeyed by this wavepacket.

2.2 Carry out the analysis leading to equation (2.73) for the propagation of a Gaussian wavepacket including the effect of the β term in equation (2.51).

2.3 Use equation (2.73) to find an expression for the ratio of the width of a free-particle Gaussian wavepacket at time $t = t'$ compared to that at time $t = 0$. Show that when this ratio is equal to 2 the particle has travelled a distance

$$x = (2\sqrt{3}\pi)(\Delta x/\lambda_0)\Delta x$$

where $\lambda_0 = 2\pi/k_0$ is the de Broglie wavelength of the particle and $\Delta x = \sqrt{\alpha}$ is the initial width of the packet. Expressed in words, this equation tells us that the spreading of the wavepacket becomes significant when it has travelled its original width multiplied by the number of wavelengths contained within that width.

2.4 Use the results of problem 3 to evaluate the following (assume Gaussian wavepackets in all cases):

(a) The time taken for an electron wavepacket of initial width $\Delta x = 0.1$ nm to double in size.
(b) The distance travelled by a 10 eV electron before its wavepacket doubles in width, if the initial width is $\Delta x = 1$ mm.
(c) The distance travelled by a particle of mass 1 g and speed $0.1\ \mathrm{m\,s^{-1}}$ before its wavepacket doubles in width, if the initial width is $\Delta x = 1$ cm.

2.5 (a) Show that the probability currents and flux densities of equations (2.34) and (2.40) are always real quantities.
(b) Show that Ψ must be inherently complex in order to have a non-zero flux (i.e. a *constant* complex overall phase factor is not enough!).
(c) Show that the probability flux density is zero everywhere for any stationary state.

2.6 The wavefunction of a particle at time $t = 0$ is

$$\Psi(x, 0) = f(x)\ \mathrm{e}^{ikx}$$

where $f(x)$ is a real function. Calculate the probability flux density for the particle, and interpret the result.

Chapter 3

One-Dimensional Potential Wells

3.1 Potential wells

There are many examples in nature where particles are confined to a certain region of space, but within which they have some freedom of motion. Examples are electrons in metals, electrons in atoms, and nucleons in nuclei. They are confined to these regions by the action of forces. In quantum mechanics these forces are represented by potential energy functions and for this reason we refer to the particles as being confined in potential wells. A schematic diagram of a potential well in one dimension is shown in figure 3.1.

The curve represents the potential energy of the particle, $V(x)$, as a function of position. Also shown is the total energy of the particle E, a constant. Since the kinetic energy is equal to $E - V$ and cannot be negative we see that, according to classical arguments, the particle is confined between the points marked by C, denoting the classical limits of motion. A useful model of a one-dimensional potential well is a wire bent into the shape of the curve in figure 3.1, with a bead sliding on the wire. If the wire sits in a vertical plane in the Earth's gravitational field, the potential energy of the bead will be determined by the height of the wire at any point. If the bead is given some kinetic energy it will oscillate back and forth, coming to rest momentarily at the points where its kinetic energy has been converted entirely to potential energy. These points are the classical limits of motion.

The restoring force which keeps the particle in the well is given by

$$F_x = -\frac{\mathrm{d}V(x)}{\mathrm{d}x}. \tag{3.1}$$

Thus the steeper the sides of the well, the greater the deceleration near the limit of motion. A well with vertical sides would reverse the motion instantaneously; the acceleration would be infinite. Clearly this is not a physically realizable situation but, nevertheless, for reasons of mathematical simplicity, such a well will be the first system we study. We will also assume that the potential well is infinitely deep, for the same reason. This system is often referred to as a 'particle in a box', because we can think of a particle bouncing back and forth without loss of energy in a box with absolutely rigid walls. It is also referred to as an 'infinite square well'.

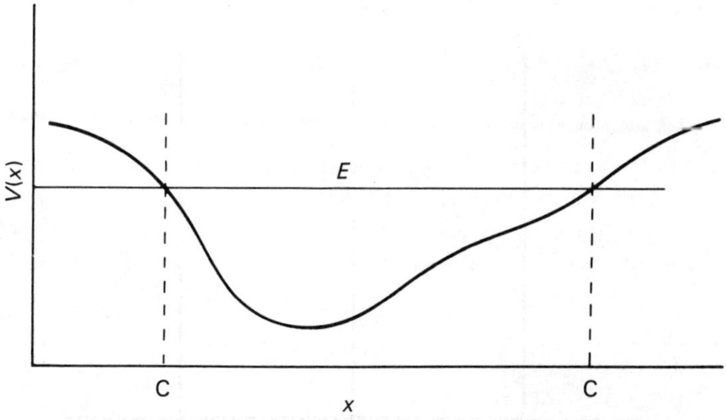

Figure 3.1 A potential well. The points C mark the limits of classical motion for a particle of energy *E*.

〉3.2 The one-dimensional infinite square well

The one-dimensional infinite square well (figure 3.2) is described in terms of a potential $V(x)$ such that

$$\begin{aligned} V(x) &= 0 \qquad |x| < a \\ V(x) &= \infty \qquad |x| \geqslant a. \end{aligned} \tag{3.2}$$

Consequently, the TISE takes the following two forms, corresponding to the regions inside and outside the well, respectively:

$$\frac{-\hbar^2}{2m}\frac{d^2u(x)}{dx^2} = Eu(x) \qquad |x| < a \tag{3.3}$$

$$\frac{-\hbar^2}{2m}\frac{d^2u(x)}{dx^2} = (E-\infty)u(x) \qquad |x| \geqslant a. \tag{3.4}$$

Considering equation (3.4) first, and remembering that the left-hand side of the equation represents the kinetic energy, which can hardly be infinite, we recognize that the only solution is

$$u(x) = 0 \qquad |x| \geqslant a. \tag{3.5}$$

Just as we would expect, there is zero probability of the particle being found outside the well.

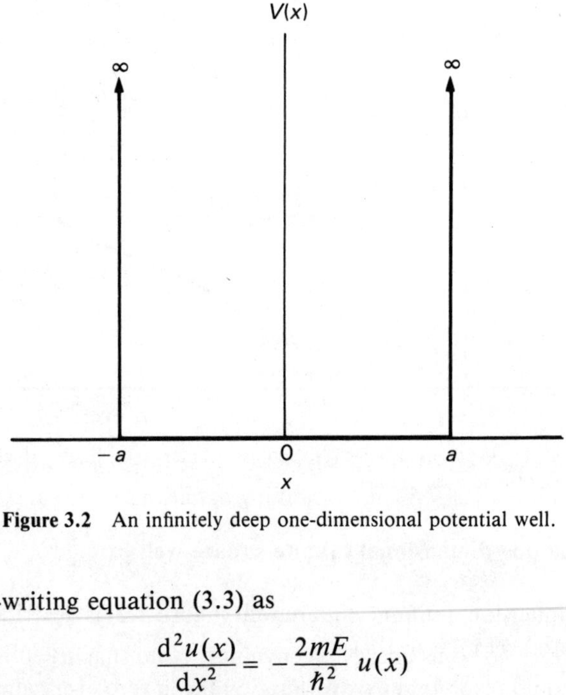

Figure 3.2 An infinitely deep one-dimensional potential well.

Now re-writing equation (3.3) as

$$\frac{d^2u(x)}{dx^2} = -\frac{2mE}{\hbar^2}\,u(x) \tag{3.6}$$

we recognize an equation with the same form as that of simple harmonic motion. The general solution can be written as

$$u(x) = A\cos kx + B\sin kx \tag{3.7}$$

where

$$k = (2mE)^{1/2}/\hbar. \tag{3.8}$$

The constants A and B must be chosen so that the solution satisfies the general requirements of a wavefunction as discussed in section 2.8. Since the wavefunction is zero at all points outside the well and must be continuous, the wavefunction inside must fall to zero at the boundary:

$$u(\pm a) = 0. \tag{3.9}$$

Hence

$$0 = A \cos ka + B \sin ka \tag{3.10}$$

and

$$0 = A \cos ka - B \sin ka. \tag{3.11}$$

Thus

$$A \cos ka = 0 \qquad \textit{and} \qquad B \sin ka = 0. \tag{3.12}$$

These conditions can be satisfied non-trivially if either

$$B = 0 \qquad \text{and} \qquad \cos ka = 0$$

so that

$$ka = (2mE)^{1/2} a/\hbar = \tfrac{1}{2} n\pi \qquad n \text{ odd} \tag{3.13}$$

or

$$A = 0 \qquad \text{and} \qquad \sin ka = 0$$

so that

$$ka = (2mE)^{1/2} a/\hbar = \tfrac{1}{2} n\pi \qquad n \text{ even.} \tag{3.14}$$

We have discovered that satisfactory wavefunctions exist only for particular values of E, which, combining equations (3.13) and (3.14), are given by

$$E = \frac{n^2\pi^2\hbar^2}{8ma^2} \qquad n = 1, 2, 3, 4, \ldots. \tag{3.15}$$

These are referred to as energy eigenvalues. Note that $n = 0$ is excluded since this would result in the wavefunction being zero everywhere, and so there is a finite minimum energy the particle must have if it is to be confined to the well.

The corresponding wavefunctions, referred to as energy eigenfunctions, are

$$\begin{aligned} n &= 1 \qquad & u_1 &= A_1 \cos \pi x/2a \\ n &= 2 \qquad & u_2 &= B_2 \sin 2\pi x/2a \\ n &= 3 \qquad & u_3 &= A_3 \cos 3\pi x/2a \\ n &= 4 \qquad & u_4 &= B_4 \sin 4\pi x/2a \end{aligned} \tag{3.16}$$

etc. These are shown in figure 3.3, together with their associated probability densities, $u_n^* u_n$.

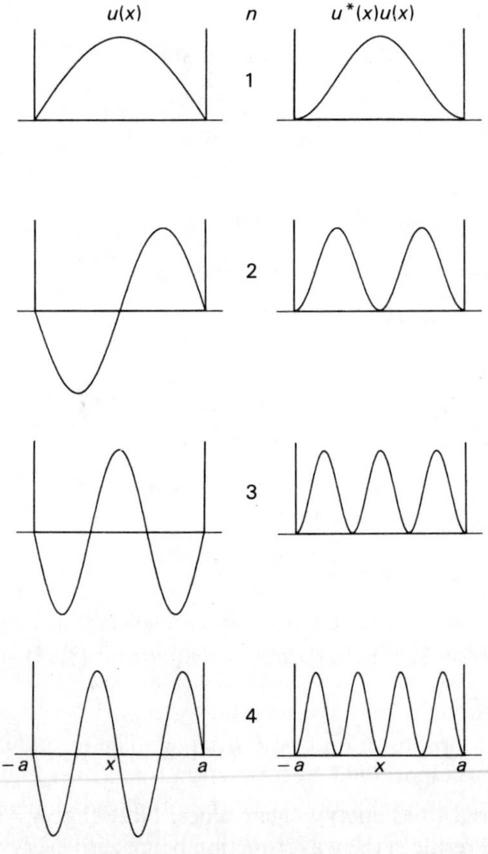

Figure 3.3 The lowest four eigenfunctions for the infinite potential well, with the corresponding probability density distributions on the right.

We can find the normalization constants A by requiring that

$$\int_{-a}^{a} A_n^2 \cos^2 kx \, \mathrm{d}x = 1$$

where

$$k = n\pi/2a. \tag{3.17}$$

Now

$$\int_{-a}^{a} A_n^2 \cos^2 kx \, \mathrm{d}x = \tfrac{1}{2} A_n^2 \int_{-a}^{a} (1 + \cos 2kx) \, \mathrm{d}x$$

$$= \tfrac{1}{2} A_n^2 \left[x + \frac{1}{2k} \sin 2kx \right]_{-a}^{a}$$

$$= A_n^2 a = 1. \tag{3.18}$$

Similarly

$$B_n^2 a = 1.$$

Hence

$$A_n = B_n = a^{-1/2} \tag{3.19}$$

and so the normalized wavefunctions can be written as

$$u_n = a^{-1/2} \begin{Bmatrix} \cos \\ \sin \end{Bmatrix} \frac{n\pi x}{2a} \qquad \begin{matrix} n \text{ odd} \\ n \text{ even.} \end{matrix} \tag{3.20}$$

〉3.3 Discussion of infinite square-well results

The solution of this very simple quantum system affords the opportunity to introduce several quantum mechanical concepts of more general application. We first notice two properties which prove to be related. The first is that, although the wavefunction $\Psi(x, t)$ is necessarily complex, the solutions we have found for $u(x)$ are real. The second is that there is only one linearly independent eigenfunction for each eigenvalue. We will first show that the second property is true of all one-dimensional bound systems. Let us suppose that two different functions, u_1 and u_2, each satisfy the TISE with the same eigenvalue, E. Then

$$\begin{aligned} u_1'' &= \frac{2m}{\hbar^2}(V - E)u_1 \\ u_2'' &= \frac{2m}{\hbar^2}(V - E)u_2. \end{aligned} \tag{3.21}$$

Hence, dividing these two equations and rearranging,

$$u_1'' u_2 - u_2'' u_1 = 0$$

or

$$(u_1' u_2)' - (u_2' u_1)' = 0. \tag{3.22}$$

Integrating, we find

$$u_1' u_2 - u_2' u_1 = \text{constant}.$$

However, for bound states u_1 and u_2 must go to zero at infinity and hence the constant must be zero. Then

$$\frac{u_1'}{u_1} = \frac{u_2'}{u_2} \tag{3.23}$$

and so, integrating again,

$$\ln u_1 = \ln u_2 + \ln c$$

or

$$u_1 = c u_2. \tag{3.24}$$

Hence the two solutions cannot be linearly independent, but are essentially identical.

We can now use this result to prove that the bound-state eigenfunctions can always be chosen to be real. Let us consider a complex solution of the form

$$u(x) = \text{Re } u(x) + \text{i Im } u(x) \tag{3.25}$$

with eigenvalue E. Now both the real and imaginary parts of $u(x)$ must separately satisfy the TISE with eigenvalue E, but there can be only one linearly independent solution. Hence the real and imaginary parts must be identical to within a constant multiplier and the wavefunction can be written in the form

$$u(x) = A e^{i\delta} \text{ Re } u(x) \tag{3.26}$$

where A is real. The complex constant coefficient $e^{i\delta}$ has no physical significance (it disappears, for example, in the physically significant quantity u^*u) and can be discarded. (This result, though not the proof, also applies in the three-dimensional case.)

We now note that there are two categories of wavefunction: those which are unchanged on reflection through the origin ($x \to -x$) and those which change sign. These are described as having even and odd parity, respectively. This is a general characteristic of the eigenfunctions of any potential well which satisfies

$$V(x) = V(-x) \tag{3.27}$$

i.e. which is itself unchanged on reflection through the origin. The physical background to this is simply that there is no reason for the particle to distinguish in any way between the two sides of the well if they are identical. We therefore expect that the probability density will be the same on the two sides, so that

$$P(x) = (u(x))^2 = (u(-x))^2. \tag{3.28}$$

($P(x)$ takes this form since $u(x)$ is real.) This in turn implies that

$$u(x) = \pm u(-x). \tag{3.29}$$

The plus and minus signs refer to even- and odd-parity eigenfunctions, respectively. We can show that the eigenfunctions will always have this property in the following way. $u(x)$ satisfies the TISE with $V(x)$ for all x, so

$$-\frac{\hbar^2}{2m}\frac{\mathrm{d}^2u(x)}{\mathrm{d}x^2} + V(x)u(x) = Eu(x). \tag{3.30}$$

On replacing x with $-x$ the equation must remain true, but using equation (3.27) we obtain

$$-\frac{\hbar^2}{2m}\frac{\mathrm{d}^2u}{\mathrm{d}x^2}(-x) + V(x)u(-x) = Eu(-x). \tag{3.31}$$

We find therefore that $u(x)$ and $u(-x)$ satisfy the same equation with the same energy eigenvalue E. They must therefore be identical apart from a multiplicative constant which we call α:

$$u(x) = \alpha u(-x). \tag{3.32}$$

Changing x to $-x$ and multiplying equation (3.32) by the resulting equation we find

$$\alpha^2 = 1$$

or

$$\alpha = \pm 1 \tag{3.33}$$

which proves the result. We will obtain the same result in Chapter 5 using a more elegant operator technique.

Another feature common to all particles confined in potential wells is that only discrete values of the energy are allowed, and that there is a minimum possible energy which is higher than the bottom of the well. In other words, the particle is not permitted to have zero kinetic energy if it is confined in space. This is a consequence of the uncertainty principle, equation (1.14). The particle being confined, its uncertainty in position is finite, and so the uncertainty in momentum must also be greater than zero. The possession of momentum (even though it fluctuates and must average to zero), implies the possession of kinetic energy.

The probability distribution which we have discovered for the particle is completely unexpected from the classical viewpoint. A particle bouncing back and forth between rigid walls would travel at constant velocity in the well, and so we should expect a uniform probability distribution, whereas we find a distribution with peaks and troughs. One of the guiding principles used in the development of quantum theory was Bohr's correspondence principle which indicates that quantum theory should give results which correspond to classical physics in the limit of large quantum numbers. Now it can readily be seen that, for large n, the distribution, if averaged over distances greater than a wavelength, approximates the classical one. However, the averaging requirement is not very satisfactory and a fuller explanation is that the states we have found are stationary states, whereas the bouncing-ball picture is in essence a dynamic one: the probability density changes from moment to moment. Quantum mechanics can represent this situation by means of a superposition of a large number of energy eigenfunctions to construct a wavepacket. However, if we confine ourselves to consideration of a single energy state a wavepacket cannot be constructed. This is a situation which we do not confront in everyday life; macroscopic systems have energy levels whose relative separations are infinitesimal, so we are very unlikely to find a tennis ball, say, in a pure energy eigenstate. Only with microscopic systems are the levels widely separated. We should not, therefore, be concerned to find unfamiliar results from our theory. When we do it is usually because we are considering situations outside our common experience.

The solutions which we have found for the infinite potential well have a discontinuity in gradient at the edges of the well. However, it was pointed out in section 2.8 that this can be expected to occur in a case of an infinite discontinuity in potential, which is the situation here. The more realistic case of a finite potential well will be considered in section 3.6 where it will be shown that the gradient discontinuity no longer occurs.

〉3.4 Eigenvalue equations

The equation which we have just solved, the time-independent Schrödinger equation, is referred to as an eigenvalue equation. The terms eigenvalue and eigenfunction are Anglicizations of the German *eigenwert* and *eigenfunktion*; eigen is equivalent to 'proper' or 'allowed'. The significance of this is that, although the TISE may be satisfied by many functions, only certain special ones are satisfactory from the physical point of view, and these require a particular choice of a parameter of the equation, namely E.

The TISE

$$-\frac{\hbar^2}{2m}\frac{d^2u}{dx^2} + V(x)u(x) = Eu(x) \tag{3.34}$$

can be written in the form

$$\hat{H}u(x) = Eu(x) \tag{3.35}$$

where the circumflex accent indicates that $\hat{H}$ is an operator, in this case a differential one, given by

$$\hat{H} \equiv \left(-\frac{\hbar^2}{2m}\frac{d^2}{dx^2} + V(x)\right). \tag{3.36}$$

$\hat{H}$, the Hamiltonian operator, represents the total energy of the system; the first term corresponds to the kinetic energy and the second to the potential energy. Later on we will find that all the dynamical variables in quantum mechanics are represented by operators and each has eigenfunctions and eigenvalues associated with it which satisfy equations similar to equation (3.35).

Examining equation (3.35), we can see that solving an eigenvalue problem amounts to finding those functions (eigenfunctions) which, when operated on by the relevant operator, reconstruct the original function unchanged except for a multiplying factor. The multiplying factor is the eigenvalue corresponding to the eigenfunction.

〉3.5 The finite square well

We now remove one of the unrealistic characteristics of the infinite square well and consider a square potential well of finite depth (figure

3.4). This well is described by a potential $V(x)$ such that

$$\begin{aligned} V(x) &= 0 \qquad |x| < a \\ V(x) &= V \qquad |x| \geqslant a. \end{aligned} \tag{3.37}$$

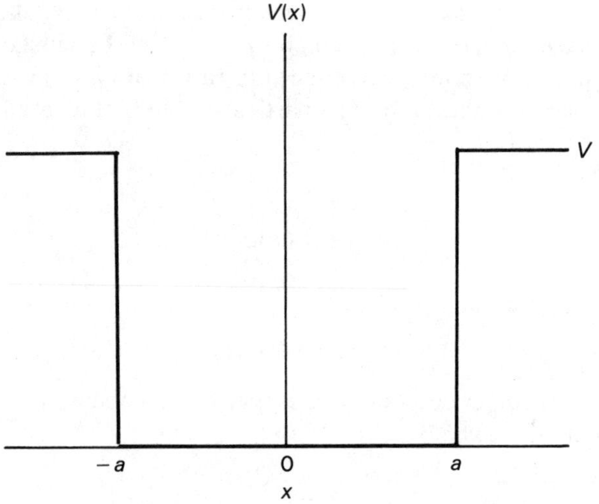

Figure 3.4 A one-dimensional potential well of finite depth.

Again the TISE takes two forms relating to regions inside and outside the well:

$$\frac{-\hbar^2}{2m}\frac{d^2u(x)}{dx^2} = Eu(x) \qquad |x| < a \tag{3.38}$$

$$\frac{-\hbar^2}{2m}\frac{d^2u(x)}{dx^2} = (E - V)u(x) \qquad |x| \geqslant a. \tag{3.39}$$

As before (equation (3.7)), the solution for the interior region is in terms of sines and cosines. We expect the solutions to have definite parities and so are led to try

$$u(x) = A \cos kx \tag{3.40}$$

and

$$u(x) = B \sin kx \tag{3.41}$$

where

$$k = (2mE)^{1/2}/\hbar. \tag{3.42}$$

Considering now equation (3.39) for the external region, there are two possibilities depending on the sign of $E - V$. If this is positive the solution is again in terms of oscillating trigonometrical functions, which in this case extend to infinity; the particle is not confined. If $E - V$ is negative the solutions must be in terms of real exponentials. We will treat this case first. Consider a wavefunction of the form

$$u(x) = C \exp(k'x) + D \exp(-k'x). \tag{3.43}$$

This satisfies equation (3.40) if

$$k' = [2m(V - E)]^{1/2}/\hbar. \tag{3.44}$$

However, equation (3.43) fails to meet the requirements of a wavefunction since it tends to infinity as x tends to positive or negative infinity. We therefore retain only decaying exponentials:

$$u(x) = C \exp(k'x) \qquad x \leqslant -a \tag{3.45}$$

$$u(x) = D \exp(-k'x). \qquad x \geqslant a. \tag{3.46}$$

Considering first the even-parity internal solution (3.40), it must be made to match (3.46) smoothly at the boundary $x = a$. Continuity of $u(x)$ requires

$$A \cos ka = D \exp(-k'a) \tag{3.47}$$

whilst continuity of $\mathrm{d}u(x)/\mathrm{d}x$ requires

$$-kA \sin ka = -k'D \exp(-k'a). \tag{3.48}$$

Dividing equation (3.48) by equation (3.47) yields

$$\tan ka = k'/k = (V/E - 1)^{1/2}. \tag{3.49}$$

The odd-parity solution, (3.41), leads to a similar equation:

$$\cot ka = -k'/k = -(V/E - 1)^{1/2}. \tag{3.50}$$

Matching solutions at the $x = -a$ boundary leads to the same two equations.

The two transcendental equations (3.49) and (3.50) require numerical or graphical techniques for their solution. The latter method is more instructive and is accomplished as follows. Starting with equation (3.49) we introduce a dummy variable z such that

$$z = \tan[(2mE)^{1/2}a/\hbar] = (V/E - 1)^{1/2}. \tag{3.51}$$

Making the substitutions

$$\theta = (2mE)^{1/2}a/\hbar \tag{3.52}$$

and

$$\gamma = (2mV)^{1/2}a/\hbar \tag{3.53}$$

equation (3.49) becomes

$$z = \tan\theta = (\gamma^2/\theta^2 - 1)^{1/2}. \tag{3.54}$$

Similarly, equation (3.50) leads to

$$z = -\cot\theta = (\gamma^2/\theta^2 - 1)^{1/2}. \tag{3.55}$$

Plotting the three expressions for z from equations (3.54) and (3.55) (the right-hand side expression being common to both) as functions of θ gives a result similar to the one shown in figure 3.5(*a*).

The curves for $\tan\theta$ and $-\cot\theta$ require no discussion. The exact form for the curve

$$z = (\gamma^2/\theta^2 - 1)^{1/2} \tag{3.56}$$

depends on the values of V and a, but in all cases the curve terminates on the θ axis at the point $\theta = \gamma$. Also, as $\theta \to 0$, $z \to \infty$.

Even and odd solutions to the equations occur where the curve (3.56) cuts the $\tan\theta$ and $-\cot\theta$ curves, respectively. The energy eigenvalues E are obtained from the values of θ at which the intersections occur using equation (3.52).

The number of solutions $n_{\max}$ is easily seen to be one greater than the number of $\frac{1}{2}\pi$ intervals contained in γ:

$$n_{\max} = \mathrm{INT}(2\gamma/\pi) + 1. \tag{3.57}$$

Because γ has this significance it is referred to as the *strength parameter* of the well. Examination of figure 3.5 and the analysis that leads to it reveals a number of important features of the eigenvalue spectrum of the finite square well, as follows.

(*a*) Energy levels are discrete and finite in number.

(*b*) There is always at least one level.

(*c*) The number of levels n increases with V and a (γ increases with both of these).

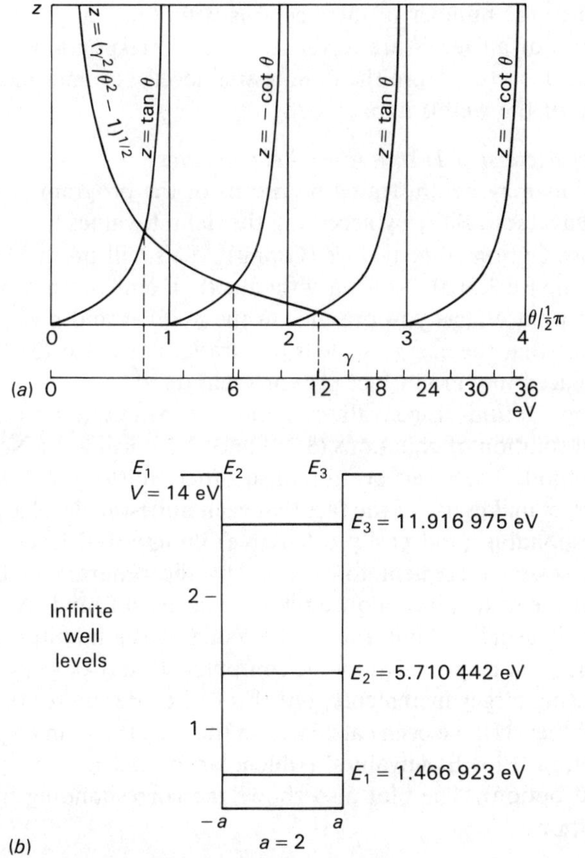

Figure 3.5 (*a*) Graphical solution of equations (3.54) and (3.55). Note the non-linear energy scale: $E \propto \theta^2$. (*b*) Linear plot of the resulting energy levels. The well is 14 eV deep and has a half-width of 0.4 nm.

(*d*) As V increases, the curve of equation (3.56) becomes flatter. The number of intersections $n_{\max}$ grows and they become more evenly spaced in θ. For V infinite, all the intersections occur on the asymptotes of the $\tan\theta$ and $-\cot\theta$ curves, are exactly evenly spaced and are infinite in number. The energies then become equal to the infinite-well energies, equation (3.15), and are proportional to n^2.

(*e*) As a and the number of intersections increase, the level separation decreases. When a becomes infinite we have an energy continuum; the particle is now free.

(*f*) When the number of intersections is finite (V and a both finite), the energies of all levels are lower than the corresponding infinite-well levels. The departure from the infinite-well levels becomes more marked as the top of the well is approached.

Computer exercise 3.1: square-well eigenvalues
These points may be illustrated by means of the program '1-D Square-Well Eigenvalues'. Start by accepting the default values in the program, and choose Option 'Eigenvalues (Graph)'. This will produce a diagram similar to figure 3.5(a). Now vary the depth V and width a of the well, observing what effects you produce in the number and positions of the levels. Note that the energy scale is quadratic, not linear (3.52) so levels equally spaced in θ are in fact proportional to n^2.

Selecting Option 'Eigenvalues (Num.)' provides a routine for the numerical solution of equations (3.49) and (3.50) using the Newton–Raphson method. The levels are found in order, starting with the lowest. The program makes use of the fact that each finite-well level is lower than the corresponding (and easily calculable) infinite-well level. You must specify a search increment to be used by the program in finding the vicinity of the next solution once the first has been found. A safe search increment is usually about 1% of the value of the infinite-well ground level, which will be provided by the computer. It may be possible to save time by using larger increments, but this can cause the routine to fail.

Once the levels have been calculated in this way they can be plotted out using Option 'Plot Eigenvalues' (which can be called directly from the numerical option). The plot also shows the corresponding infinite-well level positions.

〉3.6 Finite square-well eigenfunctions

The general form of the eigenfunctions for the finite square well can be seen in figure 3.6.

Although they resemble the infinite-well eigenfunctions in functional form within the well, the crucial difference is that there is now a finite probability of the particle being found within the classically forbidden region outside the well. Here the wavefunctions have the form of a decaying exponential. According to classical mechanics, the particle would require a negative kinetic energy to exist in this region. Quantum mechanics, which does permit the particle to be found in this region, also contrives to prevent the observation of a negative kinetic energy, which would be absurd, through the operation of the uncertainty principle. It can be shown that observation of the particle's position with sufficient precision to locate it in this region would disturb its momentum to such an extent that a negative energy would be unobservable.

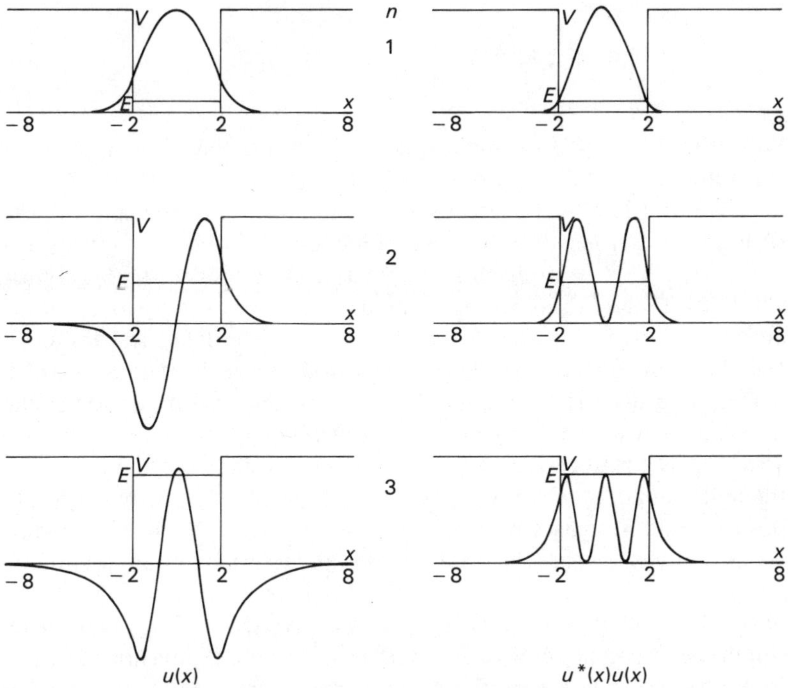

Figure 3.6 The three eigenfunctions of the finite square well of depth 14 eV and half-width 0.2 nm (x units are 0.1 nm).

The decay constant k' (equation (3.44)) of the exponential part of the wavefunction increases with V. As V tends to infinity the decay length tends to zero. We can therefore regard the infinite well as an extreme case of a finite well, and the discontinuities of wavefunction gradient which occur at its boundaries as limiting cases of exponential decay.

⟩ 3.7 The control of the wavefunction by the wave equation

We can gain valuable insight into the significance of eigenvalues and eigenfunctions by examining the way in which the TISE controls the behaviour of the wavefunction in regions of different potential energy and for different values of the total energy E. Recalling equation (3.34) and introducing the notation $u'(x) = \mathrm{d}u(x)/\mathrm{d}x$ for the slope of the wavefunction, we have

$$\frac{\mathrm{d}u'(x)}{\mathrm{d}x} = \frac{2m}{\hbar^2}(V(x) - E)u(x). \tag{3.58}$$

We see that the left-hand side equals the rate of change of slope of the wavefunction, which is proportional to its curvature. If $V(x) - E$ is negative the slope will decrease with x for $u(x)$ positive, and increase with x for $u(x)$ negative. The wavefunction is always concave towards the x axis. This is an essentially stable situation since, for any initial values of u and u', as we move along the x axis the wavefunction is deflected towards this axis. After it has crossed, it is deflected back again. We therefore find an oscillatory behaviour in such regions, with no tendency to diverge to infinity. This will be the case inside the square potential well where $V(x) = 0$. This contrasts with the behaviour when $V(x) - E$ is positive which it may be for example in the region external to the finite square potential well. Now the slope increases with x if $u(x)$ is positive, and decreases with x if $u(x)$ is negative. The wavefunction is always convex towards the x axis. This is an essentially unstable situation and the wavefunction in such regions tends to diverge to infinity. The only way in which this catastrophe can be averted is if $u(x)$ and $u'(x)$ simultaneously tend to zero, as was the case for the exponential decay we found for the square potential-well eigenfunctions. Clearly this is a very special circumstance which can only be engineered if the wavefunction is launched into this region with precisely the correct amplitude, slope and decay constant k'. Mathematically, it is a question of arranging conditions at the boundary so that only the decaying exponential in the solution (3.43) is excited; either C or D is zero, depending on the sign of x. Since k and k' are determined by E we begin to see why only certain values of the energy will work. These are the energy eigenvalues.

Although there exists the possibility of finding the particle in regions for which $E - V(x)$ is negative it is not possible to find a valid solution to the TISE if $E < V$ *everywhere*! In that case the wavefunction would be convex towards the x axis everywhere, it would never pass through a maximum or minimum and would therefore necessarily diverge.

〉3.8 Numerical integration of Schrödinger's equation

A computer can be programmed to calculate functions which satisfy the TISE. Since it is a second-order equation, we must specify initial values of the amplitude $u(x)$ and slope $u'(x)$ at some starting position x. In essence, the computer then makes use of equation (3.58) to calculate new values of u and u' at an adjacent position $x + \mathrm{d}x$, and so on. Any function so calculated will certainly satisfy the TISE but will not in general be a satisfactory wavefunction since it will usually violate one of the conditions specified in section 2.8. One reason for this may be that we have not correctly chosen initial values of u and u'. Choosing these is not as difficult as it might seem, especially if the symmetry of the well permits us to deduce that the wavefunction will have definite parity. We can then obtain an even parity function by starting at $x = 0$, with $u(0) = 1$ and $u'(0) = 0$. Clearly the slope must be zero in the middle for an even function and the choice of 1 for the amplitude is quite arbitrary, since this simply sets the overall scale of the function. If required it can be normalized later. Similarly an odd-parity function can be obtained by setting $u(0) = 0$ and $u'(0) = 1$ at $x = 0$. Again the choice of 1 for the slope simply determines the scale and has no physical significance apart from normalization.

Computer exercise 3.2: square-well eigenfunctions

The eigenfunctions for an electron in a finite square well can be examined with the help of the program '1-D Square Potential Well', which integrates the TISE in the manner described above.

The default values offered by the program will result in a well of half-width 0.2 nm and depth 10 eV. (Quantities of similar magnitude are commonly encountered in atomic and molecular physics.) The default energy of 1.3 eV is not quite equal to the lowest eigenenergy and the resulting wavefunction diverges, as will be seen on taking the option to plot the function. What has happened is that the curvature of the function within the well is insufficient to match a decaying exponential of the prescribed decay constant (equation (3.44)) in the external region. The external function is thus a mixture of positive and negative exponentials. Examination of equation (3.58) indicates that the magnitude of the curvature in the well will be increased by increasing the energy E, since $V(x) = 0$ there. Inserting the value 1.4 eV produces a function which diverges in the opposite sense; we have overshot.

Clearly the correct energy must be between 1.3 and 1.4 eV. 1.35 eV produces a better result and successive interpolations eventually lead to 1.356 629 eV. Plot the wavefunction with this energy and observe the satisfactory way in which it falls to zero outside the well. (Note that the limited precision of the computer will always result in divergence at sufficiently large distances from the well and this can be observed on increasing the maximum value $x = XF$ that the computer plots.)

This method of finding eigenvalues by interpolation is the one that is used in all but the handful of cases which permit an analytic solution. This is a time-consuming process; however, as we have seen above and in the previous program, eigenvalues for the finite square well can be found by solving equations (3.49) and (3.50) numerically and the routine for doing this is also included in this program. This option should now be called upon to find all the eigenvalues of the well. Thereafter eigenenergies can be input in the form $E(1)$, $E(2)$, etc.

The program should now be used to examine the eigenfunctions corresponding to each eigenvalue. The following points should be noted.

(*a*) The eigenfunctions alternate in parity. (Note that you must inform the computer what parity you expect so that it can choose the starting conditions appropriately.)

(*b*) The number of zero crossings of the wavefunctions increases by one with each successive eigenfunction. These zero crossings, where there is a zero probability of finding the particle, are referred to as nodes. The quantum number of the state is unambiguously defined as the number of nodes plus one.

(*c*) The wavefunction always curves towards the x axis within the well and away from the x axis outside the well.

(*d*) The penetration into the classically forbidden region increases with higher energy values. Plotting the probability density rather than the wavefunction gives a better impression of the likelihood of the particle being found outside the well.

(*e*) For the infinite well, each successive eigenfunction incorporates one extra half wave. With the finite well rather less than an extra half wave is added each time. The result is that the slope of the function at the boundary gets less, so that it matches an exponential of longer decay length. The fact that the wavelength (and hence the curvature) in the well is less than that of the equivalent infinite-well wavefunction corresponds to the fact that all the eigenenergies of the finite well are lower than those of the infinite well, the more so for the higher levels.

The program is not particularly well adapted to examining the eigenfunctions of an infinite square well; computers do not handle infinities well! Nevertheless, the infinite-well eigenfunctions can be displayed by using the following method. Set the potential V to any value larger than the eigenvalue of interest, and set XF equal to $a+h$. Enter the infinite-well eigenvalue, the correct parity and then plot the function.

Another facility provided by the program enables any wavefunction to be normalized. To do this you must select the tabulation option. At the same time as the function is being tabulated, the area of the probability density plot over the range tabulated (i.e. between $\pm XF$) is calculated. The program then works out the necessary scale factor for a plot normalized to any value you choose. (Choosing the value one may not always be a good idea since this may result in a plot of inconvenient scale.) Whenever quantitative comparisons of wavefunction amplitudes or probability densities are required it is essential that the functions are normalized to the same value, and this is readily accomplished using this facility.

As a simple exercise in normalization it may be verified that the infinite-well eigenfunction normalization factors are all $a^{-1/2}$.

〉3.9 Continuum solutions of the square-well potential

We have so far considered only the case where $E-V$ in equation (3.39) is negative, leading to real exponential solutions for the wavefunction. If $E-V$ is positive the solutions take the form of imaginary exponentials, or sines and cosines. These correspond to the fact that the particle is now free and can be found at arbitrary distances from the well. Although it is still necessary to match u and u' at the boundaries of the well where the potential is discontinuous, this can be achieved for any value of E. Even if the function approaches the boundary from the inside whilst increasing, since the function outside has an oscillatory nature, there is no possibility of divergence. Mathematically we have, for even parity internal and external solutions,

$$u(x)=A\cos kx \qquad |x|<a \tag{3.59}$$

$$u(x)=C\cos(k'x+\delta) \qquad x\geqslant a \tag{3.60}$$

$$u(x)=C\cos(k'x-\delta) \qquad x\leqslant -a \tag{3.61}$$

where

$$k = (2mE)^{1/2}/\hbar \tag{3.62}$$

and

$$k' = [2m(E - V)]^{1/2}/\hbar. \tag{3.63}$$

The forms (3.60) and (3.61) are equivalent to combinations of cosines and sines and are written in this form for convenience. Both can be present since neither diverges, unlike the earlier case of real exponentials in equation (3.43). Matching u and u' at $x = a$ yields the equations

$$A \cos ka = C \cos(k'a + \delta) \tag{3.64}$$

and

$$kA \sin ka = k'C \sin(k'a + \delta). \tag{3.65}$$

Dividing equation (3.65) by equation (3.64) we find

$$k \tan ka = k' \tan(k'a + \delta). \tag{3.66}$$

This equation can be satisfied for all real values of k and k' (i.e. all values of $E > V$), because of the freedom to choose δ. We find, therefore, that the energy eigenvalues form a continuum if $E > V$. The complete energy spectrum is made up of this unbound-particle continuum plus the discrete bound-particle levels discussed earlier.

Computer exercise 3.3: square-well continuum
Use the program '1-D Square Potential Well', inserting values of $E > V$ to investigate the nature of the unbound-particle wavefunctions. Vary E so as to obtain both positive and negative slopes at the $x = a$ boundary, verifying that the wavefunction in the external region is always oscillatory in nature. Show that odd-parity functions behave similarly in this respect.

〉3.10 Interpretation of unbound-particle wavefunctions

It is of interest to inquire into the physical significance of these solutions for the unbound particle. We deal with unbound particles in scattering experiments, when a physical system is probed by directing a stream of particles at it and observing the deflections they undergo. Since we are considering one dimension, the outcome in this case can only be either transmission or reflection. We now recall that we have restricted ourselves to finding eigenfunctions of definite parity. These are equivalent to standing waves which may be thought of as a combination of two equal-amplitude counter-propagating travelling waves. This is entirely appropriate to the case of the bound particle, but for the unbound case the eigenfunctions are not *required* to have definite parity. By so restricting ourselves we are considering a rather unlikely situation in which there is an equal probability of the particle approaching from the left or right and being transmitted or reflected at the potential well. A better representation of a one-dimensional scattering event, making use of travelling-wave rather than standing-wave solutions, will be discussed in Chapter 9. The essential result to be grasped at this point is the correlation of discrete and continuous energy spectra with bound- and unbound-particle states, respectively.

〉3.11 The double potential well

Our investigation of the finite square well indicated that there was a significant probability of finding the particle in the classically forbidden region within the barrier. X-ray scattering from electrons in atoms has shown that this does in fact occur. Another example of this which has important physical significance can be illustrated by analysing the double square well shown in figure 3.7.

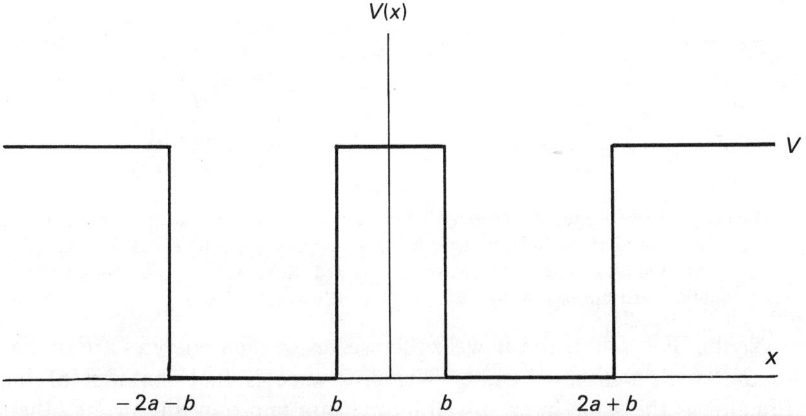

Figure 3.7 Two potential wells separated by a potential barrier. The wells communicate with each other through the mechanism of quantum mechanical tunnelling.

Finding the eigenvalues and eigenfunctions for this well follows very similar lines to the development of section 3.5. Wherever $E - V(x)$ is positive the wavefunction is oscillatory in nature, and wherever $E - V(x)$ is negative the wavefunction is exponential. At each boundary the different pieces of the wavefunction must join smoothly; $u(x)$ and $u'(x)$ must be continuous. A solution along these lines is straightforward but algebraically complex. We can deduce what to expect qualitatively quite easily, however, and then use the computer to obtain precise results. Considering the case where $E < V$, the wavefunction within each well and within the outer barriers will be similar to single-well eigenfunctions. The wavefunction in the central barrier must consist of real exponentials but there is no need to exclude increasing exponentials since the finite width of the barrier eliminates any possibility of the wavefunction diverging to infinity. Since the overall well is symmetric we expect to find eigenfunctions of definite parity, and this leads to an interesting new phenomenon. We find that there are two possible eigenfunctions corresponding to the single-well ground state, of opposite parity (figure 3.8).

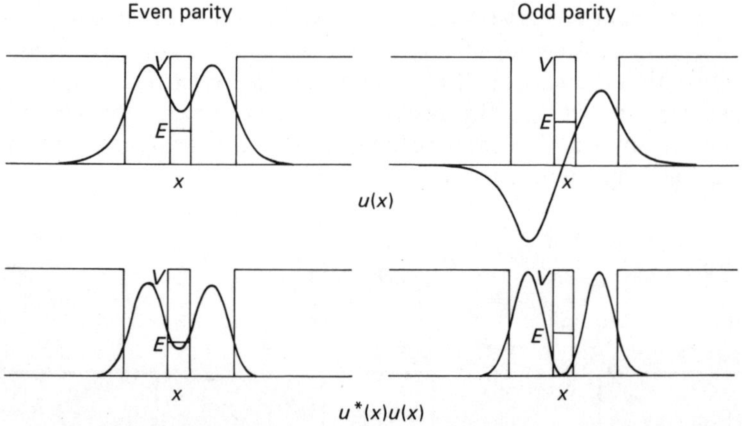

Figure 3.8 Even- and odd-parity wavefunctions for the double potential well, with their corresponding probability densities. The energies are similar, but that of the even function is a little lower. Note the marked difference in behaviour of the probability distributions in the region within the barrier.

Neither has a node in the wells themselves so their energies are similar to the single-well $n = 1$ state. The difference is that the sign of the function is the same in each well in one case and opposite in the other. Hitherto the overall sign (or phase) of the wavefunction has had no significance. However, if we regard the double well as being made up of two single wells with the overall wavefunction being the sum of the two individual ones then the relative sign or phase of these is all-important. We may imagine taking two widely separated square wells, with equal probability of the particle being in each, in the lowest energy state. On bringing the wells together the exponentially decaying tails begin to overlap. These will tend to reinforce or cancel depending upon their relative signs. Consequently, the even-parity eigenfunction has a significant probability density in the central barrier, whereas the odd-parity eigenfunction has a much lower density in this region.

The system we have described may be taken as a rudimentary one-dimensional model of a diatomic molecule. The simplest of all molecules is the hydrogen molecular ion consisting of two protons and one electron. The attractions between the protons and the electron provide the two potential wells. The covalent bond which holds the molecule together arises because of the electron probability density in the region between the two protons. This is much higher for the even-parity state, and in fact only this state is bound. In a real molecule the separation between the nuclei will adjust itself until there is an overall minimum in the total energy of the system, and this will include the electrostatic repulsion between the nuclei. Our model is defective in both these respects, but illustrates some of the basic ideas nevertheless.

Computer exercise 3.4: the double square potential well

Run the program '1-D Double Potential Well'. The default values define two wells each of width 0.2 nm and depth 10 eV, separated by 0.1 nm. The even-parity ground state is shown.

(*a*) Observe the cosine forms within the wells, the exponential decay in the external barriers and the large probability amplitude in the central barrier.

(*b*) Make a note of the energy and then attempt to find the energy eigenvalue for the odd-parity eigenfunction. Since we do not have an analytic equation for the eigenvalues you will need to use the eigenvalue-finding routine, the use of which is described in Appendix 4. When you have found the eigenvalue make a note of it and compare it with the even-parity value. Although it is significantly different, the ratio is much less than that between different quantum states of the single square well. Plot the eigenfunctions. Note that the wavefunctions in each individual well are all similar apart from their signs, with no nodes. The odd function does of course have a node but it is in the central barrier.

(*c*) Change to a display of the probability density and plot the two double-well eigenfunctions. Note the general similarity of the distributions apart from the region of the central barrier.

(*d*) Repeat (*b*) with well separations both greater and less than the default values. Run the program '1-D Square-Well Eigenvalues' to find the energy of a single well with the same dimensions as one of the wells of the double well. What is the general trend?

The ideas which we have developed for the double well can be applied and extended to the case of a long chain of potential wells, to serve as a model for a linear crystal. Discussion of this is deferred until Chapter 8.

〉3.12 The quantum bouncer

Consider a particle of mass m and energy E bouncing on a hard surface in a gravitational field. When it is a height y above the surface its potential energy is mgy, where g is the gravitational acceleration. Classical physics tells us that it will bounce to a height E/mg. The TISE for the system is

$$-\frac{\hbar^2}{2m}\frac{\mathrm{d}^2u(y)}{\mathrm{d}y^2}+V(y)u(y)=Eu(y) \tag{3.67}$$

where

$$\begin{aligned} V(y)&=\infty \qquad & y\leqslant 0 \\ V(y)&=mgy \qquad & y>0. \end{aligned} \tag{3.68}$$

Thus the particle is confined in an asymmetric V-shaped well, as in figure 3.9.

The equation can be written in a simpler form by changing the variable to z, where

$$y=E/mg+lz \tag{3.69}$$

$$l=(\hbar^2/2m^2g)^{1/3}. \tag{3.70}$$

The Schrödinger equation for $y>0$ then becomes

$$\frac{\mathrm{d}^2u(z)}{\mathrm{d}z^2}=zu(z) \tag{3.71}$$

which is known as Airy's equation. The new variable z is dimensionless and measures distance (upwards) from the classical peak of the motion in units of l, which is a very small unit indeed. For a mass of 1 g in the Earth's field it is about 10^{-21} m. For an electron it is about 1 mm.

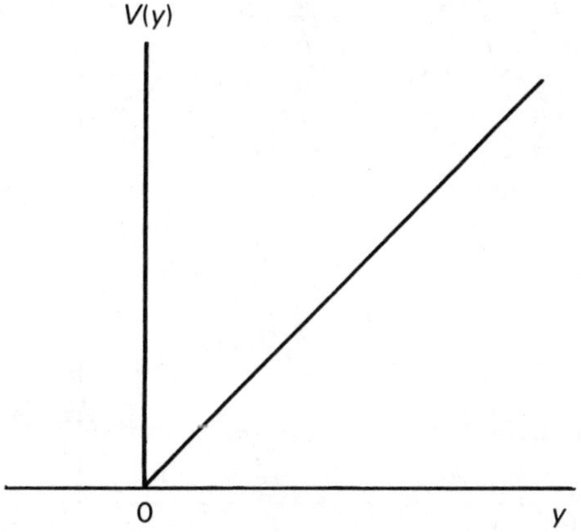

Figure 3.9 A linear potential well. This is the form of the gravitational (or electrostatic) potential energy in a uniform field extending in the y direction away from an impenetrable surface.

$u(y)$ must satisfy the boundary conditions

$$u(y=0) \equiv u(z= -E/mgl) = 0 \qquad (3.72)$$

$$u(y=\infty) \equiv u(z=\infty) = 0. \qquad (3.73)$$

The solutions of the Airy equation are well known and are the regular and irregular Airy functions, Ai(z) and Bi(z), shown in figure 3.10.

The method of solution is given in Appendix 2. The function Bi(z) can be rejected immediately since it diverges for large positive z. The function Ai(z) satisfies equation (3.73) and will satisfy (3.72) if

$$\mathrm{Ai}(-E/mgl) = 0. \qquad (3.74)$$

The eigenvalues E are therefore determined by the zeros of the Airy function, and successive eigenfunctions consist of an increasing number of half cycles of the Airy function itself, starting with the final lobe which includes the asymptotic tail. At large positive values of z this tail decays as

$$u(z) \sim z^{-1/4} \exp(-\tfrac{2}{3} z^{2/3})$$

so that there is a finite probability of finding the particle at a greater

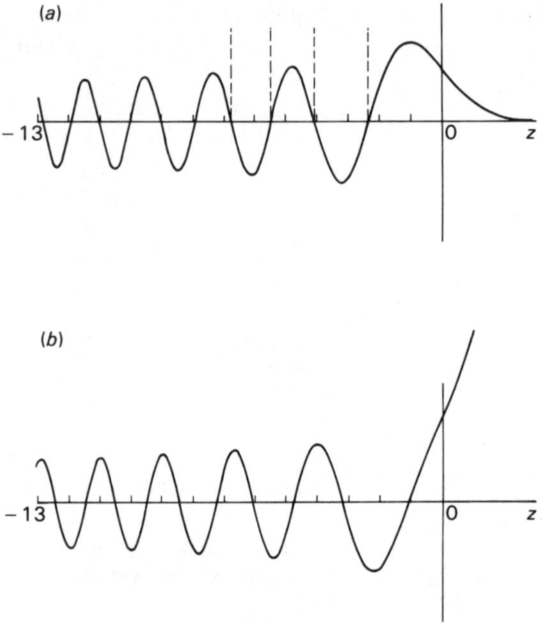

Figure 3.10 The regular and irregular Airy functions, (*a*) Ai(z) and (*b*) Bi(z). For the quantum bouncer z is a scaled vertical coordinate, with its origin at the classical limit of motion (i.e. the classical bounce height). All the quantum states have the same wavefunction, Ai(z), but the bouncing surface is at a different node for each, as indicated by the broken lines.

height than the classical turning point, $z = 0$. For each successive eigenfunction the bouncing surface is located one further node to the left in the diagram. An interesting property of this system is that the distance from $z = 0$ to the node at the surface is proportional to the eigenvalue, as can be seen from equation (3.72). This just corresponds to the classical result that the height of bounce is proportional to the energy. From figure 3.10 we can see that the eigenvalues get progressively more crowded, and will eventually approximate a continuum. The number of bound states is infinite. For large negative values of z, implying large values of the energy, the Airy function can be approximated by

$$\mathrm{Ai}(-z) \approx \pi^{-1/2} z^{-1/4} \sin(\tfrac{2}{3} z^{3/2} + \tfrac{1}{4}\pi).$$

This has nodes when

$$(\tfrac{2}{3} z^{3/2} + \tfrac{1}{4}\pi) = n'\pi \qquad n' = 1, 2, 3, \ldots \tag{3.75}$$

so that the eigenvalues are approximately

$$E \approx (\tfrac{9}{8} mg^2\hbar^2\pi^2)(n' - \tfrac{1}{4})^{2/3} \qquad n' \text{ integral}, \gg 1. \tag{3.76}$$

This is the first potential well which we have considered which is not symmetric so that the eigenfunctions do not have definite parity. In this case, however, we can recognize that it actually consists of one half of the symmetric well shown in figure 3.11.

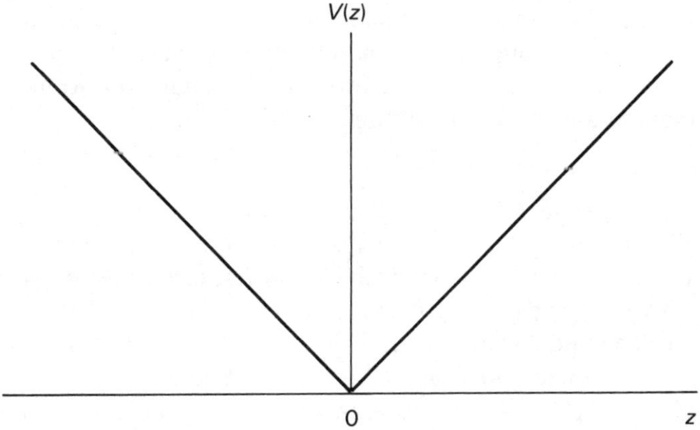

Figure 3.11 Symmetrical, bilinear potential well. The quantum bouncer well can be regarded as half of this well.

The eigenfunctions of the half well will be identical, for $y > 0$, with those which have a node at the centre of the symmetric well, namely the odd-parity solutions. In addition, the symmetric well will have an interleaved spectrum of even-parity solutions. The even-parity solutions will have an extremum in the centre of the well and so will satisfy the boundary condition

$$\mathrm{Ai}'(-E/mgl) = 0 \tag{3.77}$$

where the prime indicates differentiation of the Airy function with respect to its argument. For z large and negative this has the form

$$\mathrm{Ai}'(-z) \approx -z \cos(\tfrac{2}{3} z^{3/2} + \tfrac{1}{4}\pi). \tag{3.78}$$

Hence the even-parity eigenvalues are approximately as given in equation (3.76) with n' large and half integral. Letting $(n' - \tfrac{1}{4}) \to \tfrac{1}{2}(n + \tfrac{1}{2})$,

$n = 0, 1, 2, 3, \ldots$, the combined odd and even energy spectrum can be expressed as

$$E \approx (\tfrac{9}{64} m g^2 \hbar^2 \pi^2)^{1/3} (n + \tfrac{1}{2})^{2/3} \qquad n \text{ integral}, \gg 1. \qquad (3.79)$$

(Although the even-parity eigenfunctions are not relevant to the quantum bouncer, a charged particle in an electric field behaves similarly (see section 3.13) and it is easy to devise a system in which the field reverses abruptly on passing through a plane, leading to an even-parity potential well. For example, two parallel plates at the same positive potential with an intervening grid at a negative potential provide a system in which an electron would oscillate back and forth. Its eigenfunctions would include both the odd and the even solutions described above.)

Computer exercise 3.5: the quantum bouncer

Load the program '1-D General Potential (Sym.)'. The default values lead to the symmetric well, of which the lowest eigenfunction is plotted. Since this is of even parity it is not an eigenfunction of the quantum bouncer. The option 'atomic units' is selected. The effect of this is that the constants m, g and $\hbar$ are all set to unity and the potential is just entered as 'X'. (X in the computer program is equivalent to z except that its zero is shifted to the bouncing surface, i.e. the centre of the symmetrical well.) The energy unit is $mgl = 2^{-1/3}$ (equation (3.70)).

(*a*) Find the first few odd-parity eigenfunctions, using the computer routine to find the eigenvalues. The right-hand half of the function (say) represents the quantum-bouncer wavefunction. Compare your eigenvalues with the tabulated zeros of the Airy function, $\mathrm{Ai}(z)$, reproduced in Appendix 2. (Remember to apply the factor $2^{-1/3}$.)

(*b*) Find the first few even-parity functions, comparing your eigenvalues with the zeros of $\mathrm{Ai}'(z)$.

(*c*) Plot the eigenfunction for $n' = 10$ ($E = 10.182\,206\,85$ au) in order to examine a large section of the Airy function. Note the crowding together of the zeros towards the centre. Recall that these zeros also graph the energy levels, which become closer at higher levels. Also note that the peaks of the function are larger towards the edge of the well. This corresponds to the peak of the motion, where the particle momentum is least. Classically, this is where we would be most likely to find the particle on a random sampling.

⟩ 3.13 Electric fields and the tilted well

The Airy functions also provide solutions to the problem of the quantum bouncer when the particle bounces between two rigid surfaces. The potential well is then bounded on either side by infinitely high potential barriers, and has a sloping bottom. This problem is referred to as the infinite tilted well. It may be thought that these bouncer problems have no practical application, but the analysis applies equally well to the case of an electron in an electric field. If the electron in a finite square well is subjected to a uniform electric field $\mathscr{E}$ the result is a finite tilted well, with the potential

$$
\begin{aligned}
V(x) &= V_1 \qquad && x \leqslant -a \\
V(x) &= e\mathscr{E}x \qquad && |x| < a \\
V(x) &= V_2 \qquad && x \geqslant a.
\end{aligned}
\tag{3.80}
$$

This is shown in figure 3.12.

Such a situation might arise, for example, for an electron in a semiconducting layer which lies between two conducting regions with different potentials, V_1 and V_2. Quantum wells of this nature find many applications in semiconductor devices.

The TISE for this system now takes three forms corresponding to regions 1, 2 and 3 in figure 3.12:

$$-\frac{\hbar^2}{2m}\frac{\mathrm{d}^2u_1(x)}{\mathrm{d}x^2} = (E - V_1)u_1(x) \qquad x \leqslant a \tag{3.81}$$

$$-\frac{\hbar^2}{2m}\frac{\mathrm{d}^2u_3(x)}{\mathrm{d}x^2} = (E - e\mathscr{E}x)u_3(x) \qquad |x| < a \tag{3.82}$$

$$-\frac{\hbar^2}{2m}\frac{\mathrm{d}^2u_2(x)}{\mathrm{d}x^2} = (E - V_2)u_2(x) \qquad x \geqslant a. \tag{3.83}$$

Assuming a bound system, $E < V_1, V_2$, the solutions of equations (3.81) and (3.83) are decaying exponentials as for the finite square well:

$$u_1(x) = A_1 \exp(k_1 x) \tag{3.84}$$

$$u_2(x) = A_2 \exp(-k_2 x) \tag{3.85}$$

where

$$k_1 = [2m(V_1 - E)]^{1/2}/\hbar \tag{3.86}$$

$$k_2 = [2m(V_2 - E)]^{1/2}/\hbar. \tag{3.87}$$

Solution of equation (3.82) follows similar lines to the treatment of the quantum bouncer above. First we rescale the variable x:

$$z = E/e\mathscr{E} + lx. \tag{3.88}$$

The scale length is now

$$l = (\hbar^2/2me\mathscr{E})^{1/3}. \tag{3.89}$$

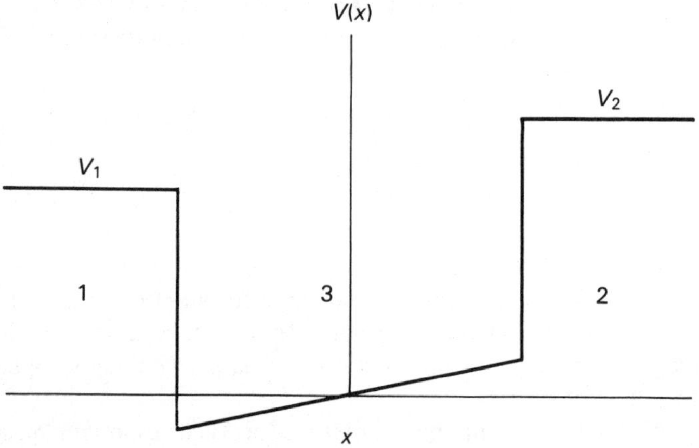

Figure 3.12 A tilted potential well. This is the potential experienced by an electron between two conductors held at potentials V_1 and V_2.

Again, the Schrödinger equation reduces to Airy's equation, (3.71). Within the well the wavefunction takes the form

$$u_3(x) = A_3\mathrm{Ai}(z) + B_3\mathrm{Bi}(z) \tag{3.90}$$

where $\mathrm{Ai}(z)$ and $\mathrm{Bi}(z)$ are the Airy functions. Both functions are now needed in order to provide a match to the exponential decays on the two sides of the well. The ratios of the four constants A_1, A_2, A_3 and B_3 are obtained by equating $u(x)$ and $u'(x)$ at the two boundaries of the well. This can only be achieved for certain values of E. Finding these eigenvalues by solving the resulting transcendental equation numerically is straightforward but not particularly illuminating and will not be pursued here. The interested reader may refer to Aguilera-Navarro *et al* (1981). However, the system may easily be studied with the aid of the computer.

Computer exercise 3.6: the tilted well

The finite tilted well is not a symmetric well, nor can it usefully be regarded as half of a symmetric well. This causes a difficulty in the computer integration routine, in that it is not easy to choose starting conditions since we do not expect the eigenfunctions to have a definite parity. The procedure we adopt is to start the integration a long way to the left of the well where the wavefunction should have decayed almost to zero, approximating this by setting $u(x) = 0$ and setting $u'(x)$ equal to a very small value. This value will determine the overall amplitude of the wavefunction but otherwise has no physical significance.

As a preliminary, run the program '1-D Square Well Eigenvalues' with half-width $a = 0.4$ nm and $V = 10$ eV. Make a note of the eigenvalues. Run the program '1-D General Potential Well (Asym)'. Accepting the default values will result in a tilted well like that of figure 3.12. This well is tilted because of the presence of an electric field. A potential difference of 2 volts exists across the 0.8 nm wide well. In other respects the well is identical to the finite well for which you have just determined the eigenvalues.

(*a*) Examine the ground-state wavefunction. It is significantly distorted and the peak lies on the deeper side of the well. Is this what you would expect intuitively? Notice the different decay rates on the two sides of the well.

(*b*) Now find the $n = 2$ eigenvalue (one node) with the help of the 'Find Eigenvalue' option. The untilted-well eigenvalue will give you some idea of where to start. When you have found it, plot the eigenfunction or, better, the probability density. Although the peaks are displaced towards the deeper side of the well, the one on the shallower side is bigger. The overall result is that the electron is more likely to be found on the shallow side.

(*c*) Plot the other eigenfunctions and verify that the trend noted in (*b*) continues. Compare the eigenvalues with those of the untilted well.

Again we observe the effect first noted with the quantum bouncer: the peaks in the wavefunction are larger where the well is shallower and the kinetic energy is less. This is not a very rigorous argument since the speed of a particle cannot properly be defined for a particle in a pure energy eigenstate. From another point of view, the result is rather startling since it indicates that in all but the ground state the electron will most probably be found on the side of the well opposite the one to which it is attracted by the electric field!

〉3.14 The harmonic oscillator: introduction

Infinite forces do not occur in nature and neither do discontinuities in potential functions! As a step in the direction of reality we now consider some systems described by potential wells with smoothly varying profiles. Of these, by far the most important is the harmonic oscillator which is at least as significant in quantum mechanics as it is in classical mechanics, and arguably more so. The reason is that almost any potential well can be modelled to some degree of approximation by the harmonic oscillator potential. Consider, for example, the potential-energy curve which describes the forces between the atoms of a diatomic molecule. The existence of the molecule as a stable entity indicates the action of an attractive force. On the other hand, the atoms are known to adopt a characteristic (non-zero) separation and this implies the existence of a repulsive force if the atoms get too close to one another. We also know that molecules dissociate if the atoms become separated too widely. Hence, without knowing anything about the origins of the forces, we can deduce that the potential curve of the molecule must look something like figure 3.13.

The coordinate x represents the interatomic distance and in classical equilibrium we would expect this to be equal to x_0, the position of the well minimum. The interatomic force is given by

$$F_x = -\frac{\mathrm{d}V(x)}{\mathrm{d}x} \tag{3.91}$$

which is equal to zero at $x = x_0$.

Expanding $V(x)$ about x_0 in powers of $(x - x_0)$ using Taylor's theorem we find

$$V(x) = V(x_0) + (x - x_0)\left(\frac{\mathrm{d}V(x)}{\mathrm{d}x}\right)_{x_0} + \tfrac{1}{2}(x - x_0)^2\left(\frac{\mathrm{d}^2V(x)}{\mathrm{d}x^2}\right)_{x_0} + \cdots. \tag{3.92}$$

The first term is a constant and can be ignored, the second is equal to zero so the third term is the first to have physical significance. If we move the origin of x to x_0 and let

$$K = \left(\frac{\mathrm{d}^2V(x)}{\mathrm{d}x^2}\right)_{x_0} + \cdots \tag{3.93}$$

then

$$V(x) \approx \tfrac{1}{2}Kx^2 \tag{3.94}$$

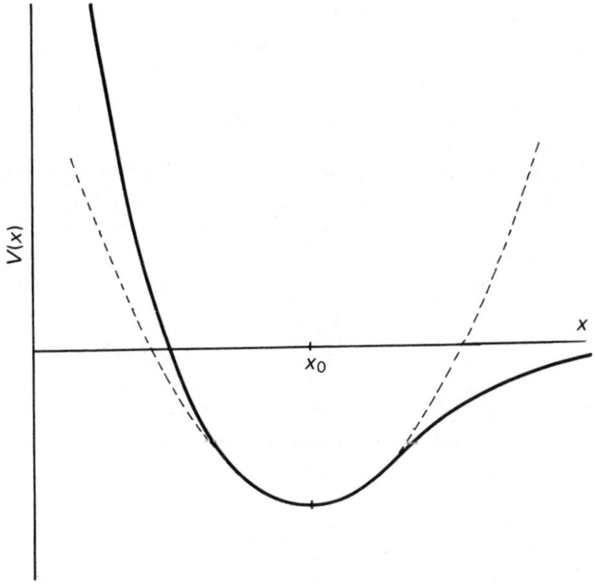

Figure 3.13 Schematic of the potential energy of a one-dimensional diatomic molecule, as a function of interatomic separation. x_0 is the equilibrium separation.

and

$$F_x = -\frac{d}{dx}\left(\tfrac{1}{2}Kx^2\right) = -Kx. \tag{3.95}$$

Approximating the potential by the first significant term in the expansion is equivalent to replacing the actual potential curve by the dotted parabola in figure 3.13. A similar procedure may be adopted for any potential-energy curve which exhibits a minimum. A particle subject to the potental of equation (3.94) satisfies the equation of motion

$$m\frac{d^2x}{dt^2} = -Kx \tag{3.96}$$

and so executes simple harmonic motion. Equation (3.96) could represent a simple mass attached to a spring of spring constant K, as in figure 3.14(a), with x representing the displacement of the mass from its equilibrium position. This is an example of a 'one-body problem'. Alternatively, we might consider the model of a diatomic molecule illustrated in figure 3.14(b).

Although this is now a two-body problem it may be shown (see Appendix 3) that equation (3.96) still applies, provided that we replace m by μ, the reduced mass, where

$$\mu = m_1 m_2/(m_1 + m_2) \tag{3.97}$$

and let x become the displacement from its equilibrium value of the *relative* coordinate (i.e. the separation) of the particles.

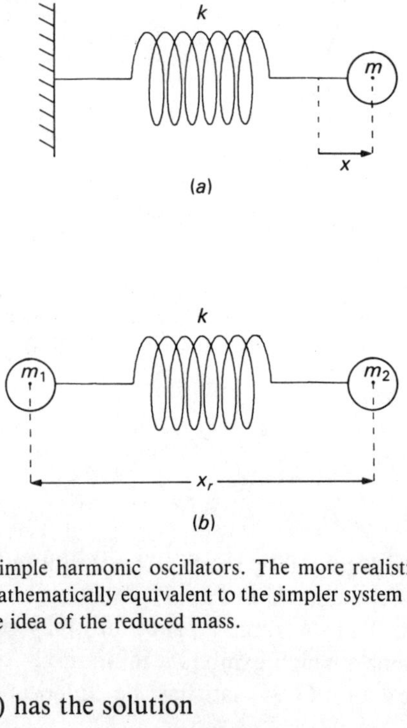

Figure 3.14 Simple harmonic oscillators. The more realistic model in (*b*) can be made mathematically equivalent to the simpler system shown in (*a*) by introducing the idea of the reduced mass.

Equation (3.96) has the solution

$$x = A\sin(\omega t + \delta) \tag{3.98}$$

where A and δ are determined by the initial conditions, and

$$\omega = (K/m)^{1/2}. \tag{3.99}$$

Hence we can express the potential in terms of the natural frequency of oscillation:

$$V(x) = \tfrac{1}{2}m\omega^2 x^2. \tag{3.100}$$

〉3.15 The quantum oscillator

We now begin our quantum mechanical treatment of the simple harmonic oscillator by writing the TISE with the potential of equation (3.100):

$$\left(-\frac{\hbar^2}{2m}\frac{\mathrm{d}^2}{\mathrm{d}x^2}+\tfrac{1}{2}m\omega^2x^2\right)u(x)=Eu(x). \tag{3.101}$$

To avoid carrying the constants through the analysis we first rearrange:

$$\left(\frac{\hbar}{m\omega}\frac{\mathrm{d}^2}{\mathrm{d}x^2}+\frac{2E}{\hbar\omega}-\frac{m\omega}{\hbar}x^2\right)u(x)=0 \tag{3.102}$$

and then substitute

$$y=(m\omega/\hbar)^{1/2}x=\alpha x \qquad \text{and} \qquad \lambda=2E/\hbar\omega \tag{3.103}$$

giving

$$\frac{\mathrm{d}^2u(y)}{\mathrm{d}y^2}+(\lambda-y^2)u(y)=0. \tag{3.104}$$

Note that y and λ are dimensionless; y is the displacement measured in units of $\alpha^{-1}=(\hbar/m\omega)^{1/2}$, and λ is the energy eigenvalue, measured in units of $\frac{1}{2}\hbar\omega$.

The solution of equation (3.104) follows a procedure known as the Frobenius method. There are several differential equations to be solved in this book using the technique so it will be described in some detail on this occasion.

As usual, it will turn out that most solutions of the equation are unsatisfactory as wavefunctions since they diverge at large values of y. We therefore start by investigating the asymptotic behaviour of the solutions as $y\to\infty$. If y is very large we may neglect λ in equation (3.104), in which case the equation is approximately satisfied by

$$u(y)=\mathrm{e}^{\pm y^2/2}. \tag{3.105}$$

We can see this by differentiating it twice, but first we reject the divergent solution with the positive exponent. Then

$$\frac{\mathrm{d}^2u(y)}{\mathrm{d}y^2}=\mathrm{e}^{-y^2/2}(y^2-1). \tag{3.106}$$

Equations (3.104) and (3.106) concur if λ is neglected in the former and 1 in the latter, so (3.105) does correctly describe the asymptotic behaviour. It fact, (3.105) satisfies (3.104) exactly if the two neglected parts are equal, i.e. if

$$\lambda = 1 \tag{3.107}$$

or

$$E = \tfrac{1}{2}\hbar\omega. \tag{3.108}$$

We have been lucky and stumbled upon the first exact eigenfunction, equation (3.105), and first eigenvalue, equation (3.108), with very little effort.

We were led to the asymptotic form $\exp(-\frac{1}{2}y^2)$ by neglecting the eigenvalue λ in equation (3.104). It is therefore expected to apply to all the eigenfunctions, so more generally we look for solutions of the form

$$y = H(y)\,\mathrm{e}^{-y^2/2}. \tag{3.109}$$

Since $\exp(-\frac{1}{2}y^2)$ decays very rapidly as y increases it will control any function $H(y)$ which increases less rapidly than itself. We can find the differential equation which $H(y)$ must satisfy by inserting equation (3.109) into equation (3.104) and performing the differentiations:

$$\frac{\mathrm{d}^2 H(y)}{\mathrm{d}y^2} - 2y\frac{\mathrm{d}H(y)}{\mathrm{d}y} + (\lambda - 1)H(y) = 0. \tag{3.110}$$

This is known as Hermite's equation. We now assume that $H(y)$ can be expressed as a series:

$$H(y) = a_0 + a_1 y + a_2 y^2 + \cdots + a_s y^s + \cdots \tag{3.111}$$

and insert it into equation (3.110). Performing the differentiations we find that the coefficient of a typical term, the one in y^s say, is

$$(s+2)(s+1)a_{s+2} - 2sa_s + (\lambda - 1)a_s. \tag{3.112}$$

Since the equation must be satisfied for all values of y the coefficient of every power of y must separately equal zero, so

$$a_{s+2} = \frac{2s + 1 - \lambda}{(s+2)(s+1)}\,a_s. \tag{3.113}$$

This recurrence equation relates alternate terms of the expansion (3.111). We can learn something about the behaviour of $H(y)$ by examining the ratio of alternate terms in equation (3.111) as s becomes large. Letting $s \to \infty$ in equation (3.113), we find

$$\frac{a_{s+2}}{a_s} = \frac{2}{s}. \tag{3.114}$$

Now consider the expansion of $\exp(y^2)$:

$$e^{y^2} = 1 + y^2 + \frac{y^4}{2!} + \cdots + \frac{y^s}{(\frac{1}{2}s)!} + \frac{y^{s+2}}{[\frac{1}{2}(s+2)]!} + \cdots. \qquad (3.115)$$

The ratio of adjacent terms is the same as in equation (3.114), namely $2/s$, implying that $H(y)$ must behave like $\exp(y^2)$ for large values of y. This in turn means that the combined solution $H(y)\exp(-\frac{1}{2}y^2)$ would be proportional to $\exp(\frac{1}{2}y^2)$ for large y, which diverges. It would therefore seem that no further physically acceptable solutions of the form (3.109) are to be found. However, if the series for $H(y)$ were to terminate, and become a polynomial rather than an infinite series, divergence with y could be avoided. Examining equation (3.113) we see that a_n will be the last term in the series if

$$2n + 1 - \lambda = 0$$

i.e. if

$$E = (n + \tfrac{1}{2})\hbar\omega \qquad n = 0, 1, 2, 3, \ldots. \qquad (3.116)$$

n is necessarily a positive integer or zero since it is an index in the expansion of $H(y)$. As in the case of the finite square well, the requirement that the wavefunction be well behaved at large distances has led to a restriction on the allowed values of the energy. Equation (3.116) therefore specifies the energy eigenvalues for the harmonic oscillator.

The eigenfunctions are obtained by using the recursion relation (3.113) to find the coefficients in the polynomial (3.111). Since this relates alternate terms it appears that there should be independent odd and even series but that equation (3.113) will only ensure the termination of one of them. However, a second-order differential equation permits the arbitrary choice of two constants, which we will take to be the coefficients a_0 and a_1, thereby providing a seed for each series. By setting one or the other of these equal to zero we can suppress either the odd terms or the even terms altogether, as we wish. We then set the other coefficient equal to 1. This simply sets the scale of the wavefunction, which can be normalized later.

As an example, consider the case $n = 3$. Then $\lambda = 2n + 1 = 7$. Let $a_0 = 0$ and $a_1 = 1$. From equation (3.113) we have, for $s = 1$ and $\lambda = 7$,

$$a_3 = -\tfrac{2}{3}. \qquad (3.117)$$

For $s = 3$ and $\lambda = 7$ we find

$$a_5 = 0 \qquad (3.118)$$

and the series terminates after two terms. Hence

$$H_3(y) = y - \tfrac{2}{3}y^3 \tag{3.119}$$

and the complete eigenfunction is

$$u_3(y) = (y - \tfrac{2}{3}y^3)\mathrm{e}^{-y^2/2}. \tag{3.120}$$

That this satisfies equation (3.104) can be verified by direct substitution. The solutions $H_n(y)$ of equation (3.110) which we have found are referred to as the Hermite polynomials and table 3.1 lists the first few members of the sequence, using the conventional normalization. Once the first two have been found they can all be generated by means of the recurrence relation

$$H_n = 2yH_{n-1} - 2(n-1)H_{n-2}. \tag{3.121}$$

Table 3.1 The first few Hermite polynomials. For the harmonic oscillator discussed in the text $y = \alpha x = (m\omega/\hbar)^{1/2}x$.

$H_0 = 1$
$H_1 = 2y$
$H_2 = 4y^2 - 2$
$H_3 = 8y^3 - 12y$
$H_4 = 16y^4 - 48y^2 + 12$
$H_5 = 32y^5 - 160y^3 + 120y$
$H_6 = 64y^6 - 480y^4 + 720y^2 - 120$

The complete, normalized wavefunctions are the Hermite–Gaussian functions:

$$u_n(x) = N_n \exp(-\tfrac{1}{2}\alpha^2x^2)H_n(\alpha x) \tag{3.122}$$

where

$$N_n = [\alpha/(\pi^{1/2}2^n n!)]^{1/2} \tag{3.123}$$

$$\alpha = (m\omega/\hbar)^{1/2} \tag{3.124}$$

and

$$y = \alpha x. \tag{3.125}$$

〉3.16 Discussion of harmonic oscillator eigenvalues and eigenfunctions

The harmonic oscillator potential is infinitely deep and, as with the infinite square well, there are infinitely many energy levels. As before there is a minimum energy, referred to as the zero-point energy. The reason for this designation is as follows. Amongst the many examples of systems which behave like harmonic oscillators are molecular vibrations, crystal lattice vibrations and the cavity modes of an electromagnetic radiation field briefly discussed in Chapter 1. Considering an assembly of oscillators in thermal equilibrium, the higher the temperature the higher will be the mean excitation level (i.e. the mean value of n) of the oscillators. The term zero-point energy refers to the fact that, even at absolute zero, no oscillator will have zero energy. In fact, from equation (3.116) we can see that the lowest energy that an oscillator of frequency ω can have is $\frac{1}{2}\hbar\omega$. A quantum oscillator is never at rest. This irreducible energy content is particularly interesting in the case of the electromagnetic field and has observable consequences in the detailed structure of atomic energy levels and elsewhere, which can only be fully explored with the aid of the theory of quantum electrodynamics.

The energy eigenvalues themselves increase linearly with n, in contrast to the infinite well where they were found to be proportional to n^2. Consequently, the energy separation of adjacent levels is constant at $\hbar\omega$, and the energy level scheme is as shown in figure 3.15.

Reverting to the simple model of a diatomic molecule, shown in figure 3.14(b), it is clear that simple harmonic vibrational motion will result in the emission of electromagnetic radiation if the individual atoms are charged, as is the case for a polar molecule, since we then have an oscillating electric dipole. Classically, the frequency of the radiation equals that of the oscillation, namely ω. Quantum mechanically, the radiation will occur when the quantum state of the system changes. Clearly the emitted frequency will be ω if the system jumps between any pair of adjacent levels so that the quantum number n changes by 1; $\Delta n = -1$. Transitions with $\Delta n = -2, \ -3$, etc, would lead to the emission of radiation with frequency 2ω, 3ω, etc. This should not occur if the quantum system is to behave as the classical system for large quantum numbers as the Bohr correspondence principle requires. (In this case, exceptionally, the correspondence extends even to small quantum numbers!) We therefore expect that a selection rule will apply:

$$\Delta n = \pm 1. \tag{3.126}$$

The plus sign in equation (3.126) refers to absorption of radiation and the minus sign to emission.

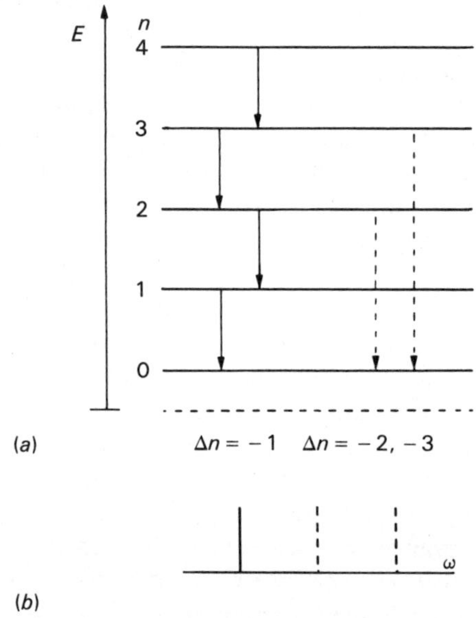

Figure 3.15 (*a*) Energy level diagram for a harmonic oscillator. (*b*) The resulting spectrum. The transitions shown broken occur only in the presence of anharmonicity.

Experimentally, it is found that diatomic molecules do behave rather like this but the situation is complicated by the fact that the actual potential is only approximately a simple harmonic one. The neglected terms in the expansion of equation (3.92) mean that the oscillator is actually anharmonic, and frequencies of approximately 2ω, 3ω, etc, *will* be emitted. Classically, this corresponds to a Fourier decomposition of the anharmonic motion. Quantum mechanically, it corresponds to quantum jumps for which $\Delta n = -1, -2$, etc, which violate the selection rule of equation (3.126). A more realistic analysis of a diatomic molecule allowing for motion in three dimensions, so that rotation can also be included, will be discussed in Chapter 7.

We now turn our attention to the harmonic oscillator eigenfunctions. The lowest state, $n = 0$, for which the wavefunction is a Gaussian, shows a bell-shaped distribution centred at $x=0$ (figure 3.16). The classical oscillator with zero energy would have a distribution which was a sharp spike at $x=0$. As we have found, the quantum oscillator must have a zero-point energy of $\frac{1}{2}\hbar\omega$; it is never at rest and the Gaussian probability distribution reflects this fact. Equations (3.122) and (3.124) show that $\alpha^{-1}=(\hbar/m\omega)^{1/2}$ is a measure of the width of the Gaussian wavefunction. If $\hbar$ were to be reduced to zero the curve would shrink to zero width and the zero-point energy would also approach zero; classical physics would be restored! The classical oscillator with finite energy would give

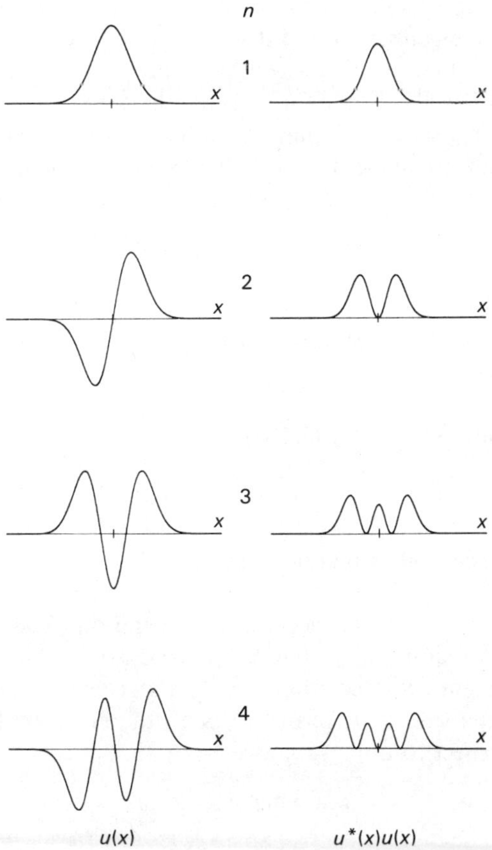

Figure 3.16 The lowest four eigenfunctions and corresponding probability densities for a harmonic oscillator.

maxima at the extremes of motion where the momentum is least. Examining the functions $H_n(y)$, we see that each is a polynomial of degree n, and consequently has n zeros. The eigenfunctions u_n (equation (3.122)) will therefore have n nodes, since the Gaussian exponential factor has no zeros itself. Each function $H_n(y)$ consists entirely of either even or odd powers of y and has even or odd parity accordingly. The Gaussian factor in all the eigenfunctions is of even parity so the wavefunctions have the parity of $H_n(x)$. A definite parity for the wavefunctions was to be expected in view of the symmetry of the well. These considerations are enough to enable us to make qualitative plots of the eigenfunctions. The first four are shown in figure 3.16, together with their associated probability densities. The Gaussian factor common to all the wavefunctions ensures rapid decay to zero as x extends outside the classical limit of motion. That the decay is more rapid than in the case of the finite square well is due to the fact that the potential increases as x^2 in the forbidden region rather than remaining constant. The classical limit is easily calculated by setting the kinetic energy equal to zero in the energy equation

$$E = \tfrac{1}{2} m\omega^2 x^2 + p^2/2m. \tag{3.127}$$

Hence

$$x_{\max} = (2E/m\omega^2)^{1/2} \tag{3.128}$$

$$= (2n+1)^{1/2}/\alpha \tag{3.129}$$

using equations (3.116) and (3.124).

〉3.17 The classical probability density

A classical oscillator moves according to equation (3.98). If we sample the system at random the probability $P(x)\,\mathrm{d}x$ of finding the particle in any small element $\mathrm{d}x$ at position x will be inversely proportional to the speed of the particle as it passes through the element. Differentiating (3.98) with respect to time we find

$$\dot{x} = \omega x_0 \cos(\omega t + \delta). \tag{3.130}$$

Now

$$P(x) \propto \dot{x}^{-1} \tag{3.131}$$

and we can express this in terms of x by noting that

$$\dot{x}/\omega = (x_0^2 - x^2)^{1/2} \tag{3.132}$$

so that

$$P(x) \propto (x_0^2 - x^2)^{-1/2}. \tag{3.133}$$

This curve is sketched in figure 3.17.

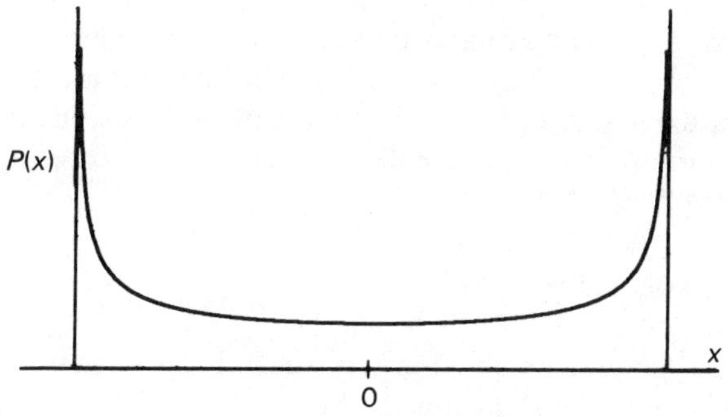

Figure 3.17 The probability density distribution for a classical harmonic oscillator.

As expected, the probability density is least in the centre of oscillation where the particle is moving fastest and becomes very large near the extremes of motion. At the limits themselves the particle is instantaneously at rest and $P(x)$ becomes infinite but this should not concern us since the probability $P(x)\,\mathrm{d}x$ remains finite; the particle is at rest at a single point only.

Computer exercise 3.7: one-dimensional harmonic oscillator
Run the program '1-D Harmonic Oscillator'. It differs from the earlier ones in that the potential well is not plotted since it is a simple parabola in all cases. The vertical lines correspond to the classical limits of motion. The program uses dimensionless quantities, i.e. lengths and energies are measured in the units of equation (3.103), so that the results can be adapted to apply to any oscillator given the effective mass m (or where appropriate the reduced mass μ), and the frequency ω. Plot the eigenfunctions and probability densities for $n = 0, 1, 2, 3, 4$ and verify the

following points.

(*a*) The parity is even if n is even and odd if n is odd.
(*b*) The peaks increase in amplitude towards the edge of the well.
(*c*) The penetration into the classically forbidden region becomes less significant as the quantum number increases.
(*d*) Non-integral values of n result in unsatisfactory wavefunctions, which diverge at large distances.

Plot the probability density function of a highly excited state. Try $n = 55$, with an x increment of $h = 0.01$ and $xF = 12$. Observe how the effects noted in (*b*) and (*c*) above have become more exaggerated. We are moving more towards a classical system; the particle is more confined to the classically allowed region and if the rapidly fluctuating probability density is smoothed, we approach the classical distribution of figure 3.17.

⟩ 3.18 More potential wells

The V-shaped potential studied in section 3.12, the harmonic oscillator and the square well are all members of a family of potentials which have the form

$$V(x) = C|x/a|^r \tag{3.134}$$

with $r = 1, 2$ and ∞, respectively. As r increases the walls of the well steepen until for $r \to \infty$ they become vertical. r can also be negative, with negative C for an attractive potential.

Computer exercise 3.8: general symmetrical one-dimensional potential wells
Run the program '1-D General Potential (sym)'. This permits the study of almost any symmetrical non-singular potential well on typing in the appropriate formula, which must be expressed in BASIC, using X as the variable. Three different functions can be employed for three zones which are separated by two 'break-points'. If the potential can be represented by a continuous function no break points are required and an arbitrary large value ($> xF$) should be entered for them. Nothing need then be entered for the function in the outer regions. Possible systems to investigate are potentials of the form of equation (3.134) with $r = 4$ and 8.

⟩ Problems

3.1 An electron is in a square potential well defined by

$$V(x) = 0 \qquad |x| \leqslant 10^{-10}\,\text{m}$$
$$V(x) = 20\ \text{eV} \qquad |x| > 10^{-10}\,\text{m}.$$

How many bound states are there? Find, by graphical or numerical means, the ground and first excited state energies, and sketch the eigenfunctions. Check your results using the '1-Dimensional Square Well' computer program.

The system is now reduced to a 'half potential well' by inserting an infinite barrier for $x < 0$, so that

$$V(x) = \infty \qquad -\infty < x < 0$$
$$V(x) = 0 \qquad 0 \leqslant x \leqslant 10^{-10}\,\text{m}$$
$$V(x) = 20\ \text{eV} \qquad x > 10^{-10}\,\text{m}.$$

State the ground-state energy and sketch the eigenfunction. (Hint: no further computation is required. Think about the boundary condition at $x = 0$ and the functions you have already considered for the first potential well.)

3.2 Consider a symmetrical square potential well of finite depth. The well is such that there is only one energy eigenstate, which is *just* bound. Sketch this wavefunction. Now consider a 'half' square well, where the potential is infinite on one side and finite on the other, similar to the one defined in question 1. Again the well is such that there is only one energy eigenfunction which is also *just* bound. Briefly argue the case, on the basis of your sketches, that there is invariably at least one bound state for a symmetrical square well, however narrow or shallow it may be, but not necessarily one for a 'half' square well. (Hint: in section 3.7 it is pointed out that $\mathrm{d}^2u/\mathrm{d}x^2$ is related to both the curvature of the wavefunction and the kinetic energy of the particle. Consider the lower limits to the overall curvatures/kinetic energies in the two cases, for a normalizable wavefunction.)

3.3 Write down the energy eigenvalues for the 'half harmonic oscillator' defined by

$$V(x) = \infty \qquad -\infty < x < 0$$
$$V(x) = \tfrac{1}{2} m\omega^2 x^2 \qquad 0 \leqslant x \leqslant \infty.$$

(Hint: see the hint for question 1 above.)

3.4 Use the program '1-Dimensional Square Well Eigenvalues' to find the eigenvalues of the symmetrical well defined in question 1. Now design other wells with different widths and depths which have eigenvalues (*a*) 4 times, and (*b*) $\frac{1}{4}$ as big as the original. Check your results using the program. (Hint: in section 3.5 the 'strength parameter' for a potential well is defined as $\gamma = (2mV)^{1/2} a/\hbar$.)

3.5 A one-dimensional harmonic oscillator is in its lowest energy eigenstate. Calculate the probability that the particle may be found outside its classical range of motion, defined in equation (3.128). (Hint: the integral you need to evaluate cannot be expressed in closed form. You should be able to convert it into the form

$$\frac{1}{\sqrt{2\pi}} \int_0^z e^{-z^2/2}\, dz$$

values of which are tabulated in tables of probability.)

3.6 The eigenfunctions and eigenvalues for a one-dimensional infinite square well positioned between $x = \pm a$ are given in equations (3.15) and (3.20). The well is now relocated so that it lies between $x = 0$ and $x = L$ $(= 2a)$. What are the new eigenvalues and eigenfunctions? What can now be said of the parity of the eigenfunctions?

3.7 A particle of mass m moves along the x axis subject to a potential

$$V(x) = -K/x \qquad x \geqslant 0$$
$$V(x) = \infty \qquad x < 0.$$

Show that a bound state exists with a wavefunction of the form

$$u(x) = Nx^n\, e^{-x/a} \qquad x > 0.$$

Find n and a in terms of K. Hence show that the energy of the state is $E = -mK^2/2\hbar^2$ and that the normalization constant is $N = 2a^{-3/2}$. (Hint: do not attempt to solve the TISE directly, but substitute the suggested solution into it. The result must be an identity, satisfied by all values of x, so the coefficients of all powers of x must vanish identically.) (NB: the one-dimensional computer programs are not usable with a potential of this form since it is singular at the origin.)

3.8 Consider the possibility of observing the quantum behaviour of a minimal 'macroscopic' particle in a box. A cubical dust particle 1 μm in size of density 10^3 kg m^{-3} moving with speed 10^{-3} m s^{-1} is trapped in a box of dimensions $a = 10^{-4}$ m. Obtain a rough value for the quantum number of the particle. Is quantum behaviour likely to be evident?

3.9 A particle moves in a double potential well defined by

$$\begin{aligned} V(x) &= \infty \qquad & |x| > 2a + b \\ V(x) &= V \qquad & |x| < b \\ V(x) &= 0 \qquad & b < |x| < 2a + b. \end{aligned}$$

(This is similar to figure 3.7 except that the potential is infinite on either side.) Find the form of the energy eigenfunctions ($E < V$) and show that the energy eigenvalues are given by the solutions of the equations:

$$\tan 2k_2 a = (k_0^2/k_2^2 - 1)^{1/2} \begin{Bmatrix} \tanh \\ \coth \end{Bmatrix} (k_0^2 - k_2^2)^{1/2} \qquad \begin{matrix} \text{even parity} \\ \text{odd parity} \end{matrix}$$

where

$$k_1 = [2m(V - E)]^{1/2}/\hbar \qquad k_2 = (2mE)^{1/2}/\hbar \qquad k_0 = (2mV)^{1/2}/\hbar.$$

Hence or otherwise show that:

(a) The even-parity functions always have lower energy than the corresponding odd-parity functions.

(b) When $V \to \infty$ the energies reduce to the single infinite square-well values for a well of width $2a$.

(c) when $b \to 0$ the energies approach those of a single infinite square well of width $4a$.

(Hint: parity considerations dictate the general form of the wave-functions; in particular they must be either cosh or sinh within the barrier. Once you have the required equations draw a schematic graphical solution, first omitting the tanh and coth factors, and then taking them into account.)

Chapter 4

Schrödinger's Equation in Three Dimensions

4.1 Introduction

Schrödinger's equation in three dimensions was introduced in section 2.3. The form taken by the time-independent equation (TISE) is

$$-\frac{\hbar^2}{2m}\nabla^2 u(\boldsymbol{r}) + V(\boldsymbol{r})u(\boldsymbol{r}) = Eu(\boldsymbol{r}). \tag{4.1}$$

The physics of the specific problem to be considered enters the equation only through the term $V(\boldsymbol{r})$. However, the physics also dictates the boundary conditions which the solutions must satisfy. Virtually all analytic solutions of the equation are obtained by the technique of separation of variables, by which means the problem is reduced to a set of three one-dimensional equations. The nature of $V(\boldsymbol{r})$ will determine a suitable choice of coordinate system and there are eleven systems in which the equation is separable. We will illustrate the method using the simplest of these, cartesian coordinates. A condition for separability in any system of coordinates is that it must be possible to write the potential as a sum of terms, each of which depends on one coordinate only. We will therefore assume that we can express $V(\boldsymbol{r})$ as

$$V(\boldsymbol{r}) = V_x(x) + V_y(y) + V_z(z) \tag{4.2}$$

in which case it will be possible to find solutions of the form

$$u(\boldsymbol{r}) = X(x)\,Y(y)Z(z). \tag{4.3}$$

This can be shown by inserting equations (4.3) and (4.2) in (4.1) and expressing the result in cartesian form:

$$-\frac{\hbar^2}{2m}\left(\frac{\partial^2}{\partial x^2}+\frac{\partial^2}{\partial y^2}+\frac{\partial^2}{\partial z^2}\right)XYZ+(V_x+V_y+V_z)XYZ=EXYZ. \quad (4.4)$$

Dividing by XYZ and rearranging we find

$$\left(-\frac{\hbar^2}{2m}\frac{1}{X}\frac{\partial^2 X}{\partial x^2}+V_x\right)+\left(-\frac{\hbar^2}{2m}\frac{1}{Y}\frac{\partial^2 Y}{\partial y^2}+V_y\right)$$
$$+\left(-\frac{\hbar^2}{2m}\frac{1}{Z}\frac{\partial^2 Z}{\partial z^2}+V_z\right)=E. \quad (4.5)$$

Each bracketed term in equation (4.5) is a function of a single coordinate. Since the coordinates are independent variables and E is a constant, the equation can only be true if each bracket on the left is independently constant, so that

$$\left(-\frac{\hbar^2}{2m}\frac{1}{X}\frac{\partial^2 X}{\partial x^2}+V_x\right)=E_x \qquad \text{etc.} \quad (4.6)$$

We can rearrange these as

$$\left(-\frac{\hbar^2}{2m}\frac{\partial^2}{\partial x^2}+V_x\right)X=E_xX$$
$$\left(-\frac{\hbar^2}{2m}\frac{\partial^2}{\partial y^2}+V_y\right)Y=E_yY \quad (4.7)$$
$$\left(-\frac{\hbar^2}{2m}\frac{\partial^2}{\partial z^2}+V_z\right)Z=E_zZ$$

where

$$E_x+E_y+E_z=E. \quad (4.8)$$

Note that the eigenfunctions are products, and the eigenvalues sums, of terms relating to each of the coordinates x, y, z.

〉4.2 Particle in a three-dimensional box

As an example of such a system we will take a rectangular box with sides of length, $2a$, $2b$ and $2c$, centred on the origin. The potential which describes the system is

$$
\begin{array}{llll}
V_x(x) = 0 & -a < x < a & V_x(x) = \infty & |x| \geqslant a \\
V_y(y) = 0 & -b < y < b & V_y(y) = \infty & |y| \geqslant b \\
V_z(z) = 0 & -c < z < c & V_z(z) = \infty & |z| \geqslant c.
\end{array}
\tag{4.9}
$$

We can immediately write down the solutions since the equations and boundary conditions to be satisfied are identical to those for the one-dimensional infinite square well. Thus

$$
\begin{array}{ll}
X(x) = a^{-1/2}\begin{Bmatrix}\cos\\ \sin\end{Bmatrix}\dfrac{n_x\pi}{2a}x & E_x = \dfrac{\pi^2\hbar^2}{8ma^2}n_x^2 \\[2ex]
Y(y) = a^{-1/2}\begin{Bmatrix}\cos\\ \sin\end{Bmatrix}\dfrac{n_y\pi}{2b}y & E_y = \dfrac{\pi^2\hbar^2}{8mb^2}n_y^2 \\[2ex]
Z(z) = a^{-1/2}\begin{Bmatrix}\cos\\ \sin\end{Bmatrix}\dfrac{n_z\pi}{2c}z & E_z = \dfrac{\pi^2\hbar^2}{8mc^2}n_z^2
\end{array}
\tag{4.10}
$$

where n_x, n_y, $n_z = 1, 2, 3$, etc, and the cosine and sine solutions relate to n odd and even, respectively. The complete eigenfunctions and eigenvalues for the system are therefore

$$u_{n_x,n_y,n_z}(x, y, z) = X_{n_x}(x)\,Y_{n_y}(y)Z_{n_z}(z) \tag{4.11}$$

and

$$E = \frac{\hbar^2\pi^2}{8m}\left(\frac{n_x^2}{a^2} + \frac{n_y^2}{b^2} + \frac{n_z^2}{c^2}\right). \tag{4.12}$$

We can visualize the probability density distribution as follows. If $n_x = n_y = n_z = 1$, so that the particle is in its lowest energy eigenstate, the probability density peaks at the origin and surfaces of constant density have the shape of somewhat rounded-off rectangular boxes (figure 4.1(a)). The density falls to zero at the walls. If $n_x = 2$, there will be a nodal plane parallel to the yz plane cutting the box in half, in which the probability density is zero. In each half of the box the density will be similar to the ground-state distribution (figure 4.1(b)). Each increment in one of the n inserts an extra nodal plane with the appropriate orientation, dividing the box into more and more subsidiary boxes.

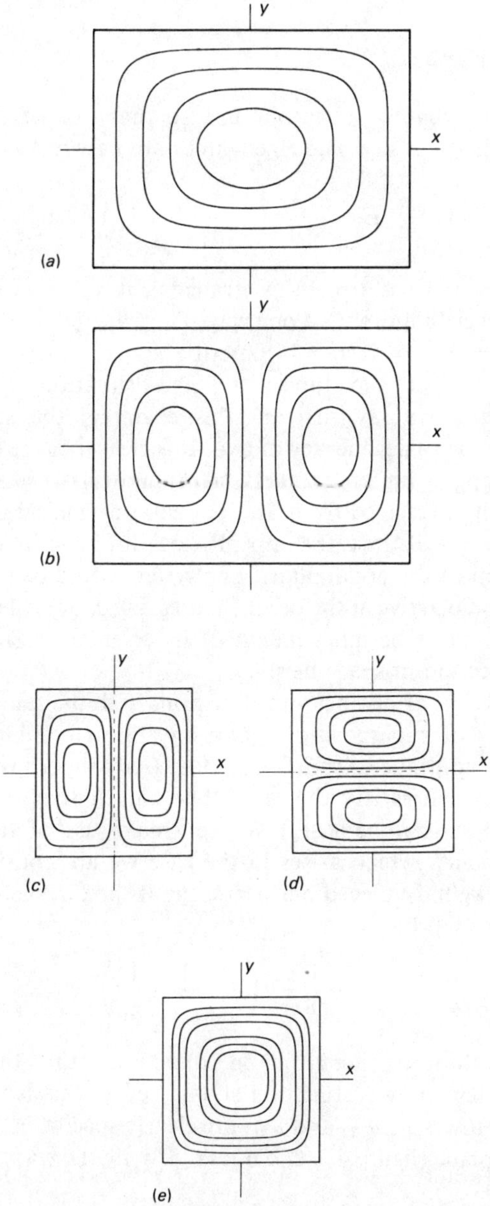

Figure 4.1 Particle in a box. (*a*) and (*b*) Contour plots of the probability densities for the states u_{111}, and u_{211}, for a rectangular box. (*c*) and (*d*) Contour plots for the degenerate eigenstates u_{211} and u_{121}, for a square box. (*e*) Contour plot for an incoherent superposition of u_{211} and u_{121}, which reflects the symmetry of the square box.

〉4.3 Degeneracy

Let us suppose that we alter our box so that two of the dimensions become equal; $a = b$ say. The eigenvalues are now

$$E = \frac{\hbar^2\pi^2}{8m}\left(\frac{n_x^2 + n_y^2}{a^2} + \frac{n_z^2}{c^2}\right). \tag{4.13}$$

The interesting new feature of this arrangement is that we can now arrive at a given energy in more than one way. Thus u_{121} ($n_x = 1$, $n_y = 2$, $n_z = 1$) and u_{211} ($n_x = 2$, $n_y = 1$, $n_z = 1$) describe states which have the same energy but distinct wavefunctions. This situation is described as degeneracy, and we say that (in this example) the state is doubly degenerate, or has a degeneracy of two. It is clear that the degeneracy is a reflection of the symmetry in the system, and this is nearly always the case. If not, it is referred to as an accidental degeneracy. Although the wavefunctions are distinct, the only physical difference in the two cases is the orientation of the nodal plane; one system could be rotated through $\frac{1}{2}\pi$ and we would arrive at the other (figures 4.1(c, d)). Clearly the energy of the system must be independent of its orientation since there is no preferred direction in space here.

The availability of distinct wavefunctions with the same energy opens the possibility of superposition, since any linear combination of such functions will continue to satisfy the TISE. (Any superposition of energy eigenfunctions will satisfy the TDSE, but only if they have a common energy will it satisfy the TISE.) We can make use of this freedom to resolve a difficulty which arises in the case we are considering. Let us suppose that we have placed our particle in its box in such a way that we know its energy to be

$$E = \frac{\hbar^2\pi^2}{8m}\left(\frac{1^2 + 2^2}{a^2} + \frac{1^2}{c^2}\right) \tag{4.14}$$

but we do nothing to single out one direction rather than another. It would seem that the wavefunction should reflect this lack of particular orientation. However, u_{121} and u_{211} are clearly specific as to the direction of the nodal plane (figure 4.1(c, d)). Let us write the wavefunction as the superposition

$$u = u_{121} + e^{i\delta}u_{211} \tag{4.15}$$

which satisfies the TISE. $e^{i\delta}$ is an arbitrary and unknown relative phase factor, while u_{121} and u_{211} are real. The probability density is then

$$|u|^2 = u_{121}^2 + u_{211}^2 + 2u_{121}u_{211}\cos\delta. \tag{4.16}$$

Averaging over the random phase factor eliminates the last term in the equation and the mean probability distribution becomes

$$|u|^2 = u_{121}^2 + u_{211}^2 \tag{4.17}$$

which reflects the symmetry of the system as it should (figure 4.1(*e*)). The justification for the averaging process will become clearer when we have examined the quantum mechanical theory of measurement in Chapter 5.

〉4.4 Schrödinger's equation in spherical polar coordinates: the hydrogen atom

Spherical polar coordinates are the natural choice in physical systems which exhibit a degree of spherical symmetry, such as the hydrogen atom. Placing the origin at the nucleus, the position of the electron is given by $\boldsymbol{r}$ and the potential function is

$$V(\boldsymbol{r}) = V(r) = -\frac{1}{4\pi\varepsilon_0}\frac{Ze^2}{r} \tag{4.18}$$

where r is the radial polar coordinate (see figure 4.2). Z, the atomic number, is included so that all hydrogen-like ions can be dealt with in the same analysis. Since the potential depends only upon r the force acting on the electron is a central one:

$$\boldsymbol{F} = -\nabla V(r) = -\frac{1}{4\pi\varepsilon_0}\frac{Ze^2}{r^2}\hat{\boldsymbol{r}}. \tag{4.19}$$

Classical mechanics tells us that angular momentum will be conserved in such a system and we will find that this quantity plays an equally crucial role in quantum mechanics.

The TISE for the hydrogen atom is

$$-\frac{\hbar^2}{2\mu}\nabla^2 u(\boldsymbol{r}) - \frac{1}{4\pi\varepsilon_0}\frac{Ze^2}{r}u(\boldsymbol{r}) = Eu(\boldsymbol{r}) \tag{4.20}$$

where

$$\nabla^2 = \frac{1}{r^2}\frac{\partial}{\partial r}\left(r^2\frac{\partial}{\partial r}\right) + \frac{1}{r^2\sin\theta}\frac{\partial}{\partial\theta}\left(\sin\theta\frac{\partial}{\partial\theta}\right) + \frac{1}{r^2\sin^2\theta}\frac{\partial^2}{\partial\phi^2}. \tag{4.21}$$

The reduced mass $\mu = mM/(m + M)$ (M is the mass of the nucleus) appears in (4.20) in place of m to take account of the nuclear motion in the usual way (see Appendix 3).

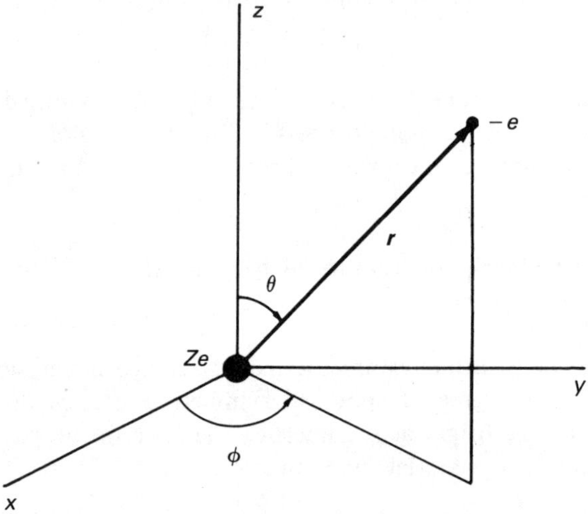

Figure 4.2 Spherical polar coordinate system for the electron in a hydrogenic ion.

This rather formidable equation can be greatly simplified if we restrict ourselves to a search for solutions which exhibit spherical symmetry. As was demonstrated in the treatment of the particle in the rectangular box, the existence of symmetry in the potential does not automatically result in a corresponding symmetry in the eigenfunctions. The same is true of the classical particle motion in an inverse square law force field. There is nothing spherically symmetric about an orbit. In fact, the restriction to spherically symmetric solutions is equivalent to searching for states of zero angular momentum, which are referred to as s states. These are excluded from consideration in Bohr's theory. Sommerfeld's generalization of Bohr's theory allows elliptical orbits, the eccentricity of which determines the angular momentum. An 'orbit' of zero angular momentum would consist of an ellipse of unit eccentricity, i.e. a linear oscillation. However, these so-called pendulum orbits were excluded on the reasonable grounds that they required the electron to pass through the nucleus! The existence of zero angular momentum states is a novel result of quantum mechanics and marks a significant departure from the old quantum theory.

For a solution with spherical symmetry the eigenfunctions $u(r)$ will be functions only of r. All the terms in θ and ϕ in equation (4.21) can therefore be neglected and the equation we must solve is

$$-\frac{\hbar^2}{2\mu}\frac{1}{r^2}\frac{\mathrm{d}}{\mathrm{d}r}\left(r^2\frac{\mathrm{d}}{\mathrm{d}r}\right)u(r)-\frac{1}{4\pi\varepsilon_0}\frac{Ze^2}{r}u(r)=Eu(r). \tag{4.22}$$

This can be written more neatly in dimensionless form by making the substitutions

$$(4\pi)^{1/2}u(r)=R(r)=\chi(r)/r \tag{4.23}$$

$$\rho=\frac{(8\mu\,|\,E\,|)^{1/2}}{\hbar}\,r \tag{4.24}$$

$$\beta=\left(\frac{\mu}{2\,|\,E\,|}\right)^{1/2}\frac{Ze^2}{4\pi\varepsilon_0\hbar}. \tag{4.25}$$

(The *radial wavefunction* $R(r)$, which differs only by the constant factor $(4\pi)^{1/2}$ from $u(r)$, may appear to be redundant. It is introduced here for two reasons. It provides continuity with our later and fuller treatment of the hydrogen atom including the effects of angular momentum (Chapter 7), when we will express $u(r)$ as $R(r)\Theta(\theta)\Phi(\phi)$ and, secondly, the computer program which computes the hydrogenic wavefunctions actually displays $R(r)$ rather than $u(r)$.)

Note that since the potential energy is chosen to be zero at infinity and is negative elsewhere, bound-state energies E in equations (4.24) and (4.25) will be negative. With these substitutions the TISE becomes

$$\frac{\mathrm{d}^2\chi}{\mathrm{d}\rho^2}+\left(\frac{\beta}{\rho}-\frac{1}{4}\right)\chi=0. \tag{4.26}$$

Solution of this equation follows the same procedure used for the harmonic oscillator in section 3.14. The asymptotic behaviour is found by neglecting the β/ρ term in equation (4.26) for large ρ. Then

$$\chi(\rho)\approx \mathrm{e}^{-\rho/2}\qquad \rho\to\infty \tag{4.27}$$

and we expect this behaviour for any eigenvalue β, because the asymptotic equation is independent of β, the energy eigenvalue term.

Next we look for functions $L(\rho)$ such that

$$\chi(\rho)=\rho L(\rho)\mathrm{e}^{-\rho/2} \tag{4.28}$$

is a solution of equation (4.26). We will assume that $L(\rho)$ can be expressed as a power series in ρ:

$$L(\rho) = \sum_{p=0}^{\infty} a_p \rho^p. \tag{4.29}$$

The factor ρ is included in (4.28) because of a boundary condition which must be satisfied at the origin. From equation (4.23) we find that if $u(0)$ is to remain finite at the origin, $\chi(0)$ must be zero. When equations (4.28) and (4.29) are substituted into (4.26) the following recurrence relation is found:

$$\frac{a_{p+1}}{a_p} = \frac{p-\beta}{p(p+1)}. \tag{4.30}$$

However, a power series in ρ with coefficients in this ratio behaves like e^{ρ} for large values of p. This diverges too rapidly with ρ to be controlled by $e^{-\rho/2}$ and acceptable wavefunctions are obtained only if the series terminates. This it will do if for some $p = n$ for example, if

$$n = \beta \qquad n = 1, 2, 3, \ldots. \tag{4.31}$$

Hence, from equation (4.25),

$$E_n = -\frac{Z^2 e^4 \mu}{2(4\pi\varepsilon_0)^2 \hbar^2} \frac{1}{n^2} \qquad n = 1, 2, 3, \ldots \tag{4.32}$$

and these are the energy eigenvalues for the s states of the hydrogen atom. It will be shown later that even when we drop the spherical symmetry requirement and include the effects of angular momentum no new energies will emerge. New states will be found but they are degenerate with these. The energies of equation (4.32) are precisely the same as the ones obtained in the Bohr theory and agree very well with experiment. There are infinitely many of them, but they are contained in a finite range with a maximum at $E = 0$. They become more crowded as n increases, as a result of the way the $-1/r$ potential well widens as $V \to 0$ (figure 4.3).

The functions $L(\rho)$ which are determined by equations (4.29)–(4.31) are a subset of the associated Laguerre polynomials, denoted $L_n^1(\rho)$, and can be found from the recurrence relation (4.30) with $\beta = n$. $a_0 = 1$ is chosen arbitrarily, and n set to the relevant positive integer. As an example consider the case $n = 2$. From equation (4.30) we find $a_1 = -\frac{1}{2}$. On the next iteration we find $a_2 = 0$, and the series terminates. Hence $L_2^1(\rho) = (1 - \frac{1}{2}\rho)$. Table 4.1 lists the first few s-state radial wavefunctions. Equation (4.23) has been used to obtain $R_n(r)$ from $\chi_n(r)$, but the functions are written in terms of ρ for the sake of brevity. The constants N_n are normalizing factors.

Using equations (4.25) and (4.32) we find that

$$\frac{\rho}{2} = \frac{(2\mu\,|\,E\,|)^{1/2}}{\hbar}\,r = \frac{Zr}{na_0} \tag{4.33}$$

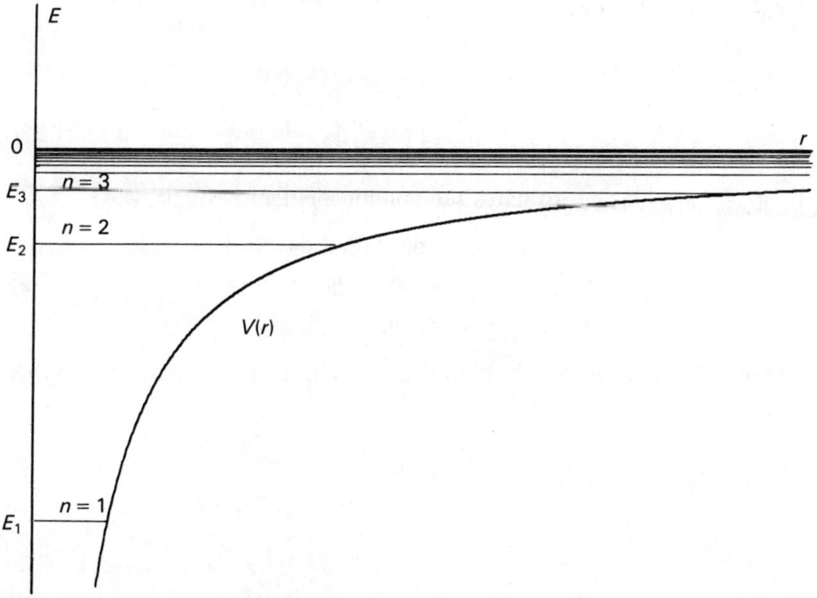

Figure 4.3 Energy level scheme for a hydrogenic ion, superimposed upon the potential-energy function. There is an infinite number of levels, confined within a finite range.

Table 4.1 Some s-state radial wavefunctions for hydrogen-like ions, $R_n(\rho) = N_n L_n^1(\rho) e^{-\rho/2}$, in terms of the variable $\rho = 2(Zr/na_0)$.

$R_1(\rho) = N_1 e^{-\rho/2}$
$R_2(\rho) = N_2(2-\rho)e^{-\rho/2}$
$R_3(\rho) = N_3(6-6\rho+\rho^2)e^{-\rho/2}$
$R_4(\rho) = N_4(24-36\rho+12\rho^2-\rho^3)e^{-\rho/2}$
$R_5(\rho) = N_5(120-240\rho+120\rho^2-20\rho^3+\rho^4)e^{-\rho/2}$
$R_6(\rho) = N_6(720-1800\rho+1200\rho^2-300\rho^3+30\rho^4-\rho^5)e^{-\rho/2}$

where

$$a_0 = \frac{\hbar^2(4\pi\varepsilon_0)}{\mu e^2} = 52.9177 \text{ pm} \tag{4.34}$$

and is the radius of the first Bohr orbit in hydrogen.

〉4.5 Normalization

Normalization of the eigenfunctions requires that

$$\int_{\text{All space}} |u(r)|^2 \, \mathrm{d}\tau = 1. \tag{4.35}$$

In spherical polar coordinates the volume element is (figure 4.4)

$$\begin{aligned} \mathrm{d}\tau &= \mathrm{d}r \; r \, \mathrm{d}\theta \; r \sin\theta \, \mathrm{d}\phi \\ &= r^2 \sin\theta \, \mathrm{d}r \, \mathrm{d}\theta \, \mathrm{d}\phi \\ &= r^2 \, \mathrm{d}r \, \mathrm{d}\Omega \end{aligned} \tag{4.36}$$

where $\mathrm{d}\Omega$ is the element of solid angle lying in the direction (θ, ϕ). In this

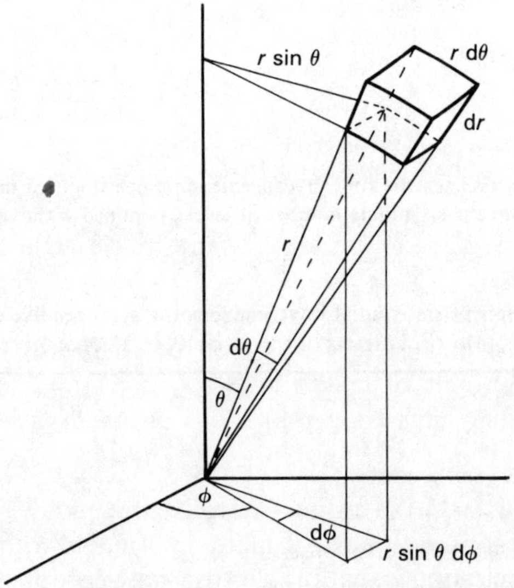

Figure 4.4 The volume element in spherical polar coordinates.

case u is independent of θ and ϕ and so on integrating over them the normalization requirement becomes

$$\int_0^\infty r^2 |u(r)|^2 \, dr \int_0^\pi \sin\theta \, d\theta \int_0^{2\pi} d\phi = 1$$

or

$$4\pi \int_0^\infty r^2 |u(r)|^2 \, dr = 1. \tag{4.37}$$

The quantity $4\pi r^2 \, dr$ is the volume of a spherical shell of thickness dr situated a distance r from the origin. Hence the quantity

$$4\pi r^2 |u(r)|^2 \, dr = D(r) \, dr \tag{4.38}$$

is the probability, integrated over all angles, of finding the electron at a distance between r and $r + dr$ from the origin. $D(r)$ is called the *radial distribution function.*

The normalized s-state eigenfunctions for $n = 1, 2, 3$ are

$$\begin{aligned} u_1(r) &= \pi^{-1/2}(Z/a_0)^{3/2} \exp(-Zr/a_0) \\ u_2(r) &= \pi^{-1/2}(Z/2a_0)^{3/2}(1 - Zr/2a_0)\exp(-Zr/2a_0) \\ u_3(r) &= \pi^{-1/2}(Z/3a_0)^{3/2}[1 - 2(Zr/3a_0) + \tfrac{2}{3}(Zr/3a_0)^2]\exp(-Zr/3a_0). \end{aligned} \tag{4.39}$$

Figure 4.5 shows the eigenfunctions for $n = 1, 2, 3$. The functions $\chi_n(r)$, $R_n(r)$ and $D_n(r)$ are plotted in each case. (Recall that $R(r) = (4\pi)^{1/2}u(r)$.)

Computer exercise 4.1: hydrogen-like ions

Run the program 'Hydrogen-like Ions'. This program deals with both zero and non-zero angular momentum cases. At present we are considering only the former, so $l = 0$ should be entered throughout the following demonstrations.

The program makes use of 'atomic units'. e, μ, $\hbar$ and $4\pi\varepsilon_0$ are set equal to 1. The unit of length is then $a_0 = 5.29 \times 10^{-11}$ m (equation (4.34)), and the unit of energy is $\mu e^4/\hbar^2(4\pi\varepsilon_0)^2 = 27.211\,65$ eV. The energy eigenvalues of hydrogen-like ions are then given by $-Z^2/2n^2$, $n = 1, 2, 3, \ldots$.

(*a*) Have a look at the ground-state eigenfunction, $n = 1$. Plot $R(r)$ (set $K = 1$) and observe that it is finite, and indeed a maximum at the origin. The implication is that the electron can penetrate the nucleus. Of course, we have not allowed for the finite size and structure of the

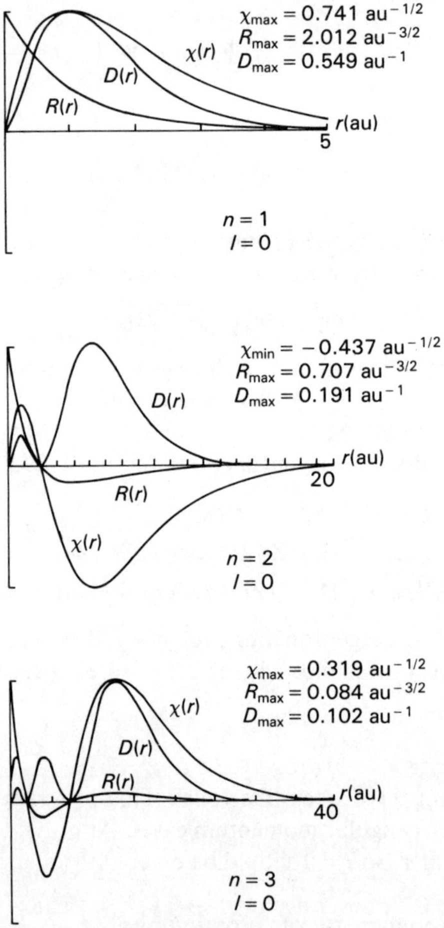

Figure 4.5 The first three spherically symmetric eigenfunctions of hydrogen, showing the relations between the radial wavefunction $R(r)$, the function $\chi(r)$ and the radial probability density $D(r)$. Note that the dimensions of these are all different, and so cannot be shown with the same ordinate calibration. The curves have been scaled so that in each drawing the three functions show the same maximum excursions. The amplitudes at the extrema are listed in the diagrams.

nucleus which would modify the potential and hence the wavefunction in this region, nevertheless the effect is real and is of considerable significance since electrons in such states interact especially strongly with the nucleus, giving rise to several observable effects. An example is hyperfine structure, a splitting and displacement of spectral lines resulting from an interaction between the nuclear and electron magnetic moments, which is particularly large in s states. Another is the phenomenon of K capture, an alternative to β^+ decay, in which the nucleus (of a heavy atom) swallows up an s electron, converting one of the nuclear protons into a neutron in the process.

(*b*) Plot the wavefunctions for $n = 2, 3, 4$. Observe that there are $n - 1$ nodes in each wavefunction. These indicate that there are nodal spheres situated at these radii for which the probability density is zero. Looking at the wavefunctions in table 4.1, we can see that the function for a given n has a polynomial factor of degree $n - 1$ which supplies the $n - 1$ zeros.

(*c*) In all cases the eigenfunctions decay at large distances as $\exp(-Zr/na_0)$. Observe in your plotting how the range of the function rapidly increases with n, since the exponential decay length is na_0/Z.

(*d*) Pick a value of n and plot the wavefunction $R(r)$ for various values of Z, corresponding to a sequence of hydrogen-like ions. Observe how the increase in nuclear charge shrinks the size of the structure, again because of the na_0/Z decay length.

(*e*) Plot $\chi(r) = rR(r)$ and $D(r) = r^2 \,|\, R(r)\,|^2$ $(K = 0, K = 2)$ for $n = 1$. The peak at the origin has become a zero, because of the r and r^2 factors, and we now have a maximum in the probability distribution which in this case occurs at a_0, the radius of the first Bohr orbit! Whereas the neighbourhood of the origin is the most favoured volume element, a_0 is the most favoured distance. The distinction arises because many volume elements lie at a distance r from the origin; all those which combine to make a spherical shell between r and $r + \mathrm{d}r$ of volume $4\pi r^2\,\mathrm{d}r$.

(*f*) Repeat (*e*) for several more values of n. The electron density is distributed amongst n spherical shells, with the probability increasing with r. The inner shells are always present, however, and give rise to the effects discussed in (*a*). Notice that these curves all have a point of inflection where $(V - E) = 0$. This point can be found on the screen because V is plotted as a function of r, and E is indicated. This point marks the classical limit, beyond which the exponential decay takes over.

(*g*) Verify that acceptable wavefunctions are only obtained for integral values of n.

A fuller treatment of the hydrogen eigenfunctions and eigenvalues requires an understanding of angular momentum in quantum mechanics, and for this we need to carry the formal development of quantum mechanics further than we have so far. It is to this that we now turn our attention.

〉 Problems

4.1 Consider a three-dimensional harmonic oscillator. A mechanical model for such a device would be a mass constrained by springs in three mutually perpendicular directions. For small oscillations in the x direction say, the lengths of the springs in the y and z directions would be essentially constant, so the three oscillators are substantially independent. The potential energy of the system is thus

$$V(x, y, z) = \tfrac{1}{2} m(\omega_x^2 x^2 + \omega_y^2 y^2 + \omega_z^2 z^2).$$

Deduce, on the basis of your knowledge of the one-dimensional oscillator, the energy eigenvalues and the general form of the eigenfunctions.

If the springs are all identical the oscillator is said to be isotropic, and the possibility of degeneracy arises. What is the degree of degeneracy of the level with energy $E = 7\hbar\omega/2$? Two quite different classes of eigenfunction occur with this energy. Write down an example of each.

4.2 A particle of mass μ moves in the xy plane under the influence of a potential $V(r)$ which is a function of the radial coordinate r only. Write down the TISE for the system in plane polar coordinates (r, ϕ). Try a solution of the form $u(r, \phi) = R(r)\Phi(\phi)$ and hence show that the equation can be separated. In plane polar coordinates

$$\nabla^2 = \frac{1}{r}\frac{\partial}{\partial r}\left(r\frac{\partial}{\partial r}\right) + \frac{1}{r^2}\frac{\partial^2}{\partial \phi^2}.$$

Show that the ϕ equation has solutions of the form

$$\Phi(\phi) = A\mathrm{e}^{\mathrm{i}m\phi}$$

where m is an integer. (Hint: the fact that m needs to be an integer comes from the requirement that the wavefunction be single valued, and that ϕ is cyclic, so that $\phi \equiv \phi + 2\pi$. This will be dealt with further in Chapter 6.)

4.3 A particle of mass μ moves in the xy plane under the influence of a potential $V(r)$ which is defined by

$$V(r) = 0 \qquad r \leqslant a$$
$$V(r) = \infty \qquad r > a.$$

Write down the TISE for the system in plane polar coordinates (r, ϕ) and adapt it to the case of circular symmetric solutions, i.e. $u(r, \phi) \to R(r)$. In plane polar coordinates

$$\nabla^2 = \frac{1}{r}\frac{\partial}{\partial r}\left(r\frac{\partial}{\partial r}\right) + \frac{1}{r^2}\frac{\partial^2}{\partial \phi^2}.$$

Make the substitution $r = (\hbar^2/2\mu E)^{1/2}\rho$ and show that the equation is satisfied by

$$R(\rho) = \sum_{k=0}^{\infty} a_k \rho^k \quad \text{where} \quad a_{k+2} = \frac{-a_k}{(k+2)^2} \quad \text{and} \quad a_1 = 0.$$

This is the Bessel function $J_0(\rho)$. It is an oscillating function, the first three zeros of which occur at $\rho = 2.4048$, 5.5201 and 8.6537. What are the energies of the lowest three circularly symmetrical states of the system, if the particle is an electron and $a = 0.1$ nm?

4.4 A particle of mass μ moves in three dimensions under the influence of a potential defined by

$$V(r) = 0 \qquad r \leqslant a$$
$$V(r) = V \qquad r > a.$$

Write down the TISE and simplify it by (a) considering only spherically symmetric solutions, $u(r, \theta, \phi) \to R(r)$ and (b) making the substitution $R(r) = \chi(r)/r$. Hence find the form of the eigenfunctions and sketch them.

Find an equation for the eigenvalues and find the two lowest eigenvalues for the case $a = 0.2$ nm and $V = 30$ eV.

Deduce that a spherical square well has no spherically symmetric bound state if $V < \pi^2\hbar^2/8\mu a^2$.

(Hint: after performing simplifications (a) and (b) you should find that you have arrived at an equation identical to the one-dimensional TISE for a finite square well, but remember that in this case $\chi(0) = 0$ in order that $R(r)$ remain finite at the origin. Review problem 1 of Chapter 3. Note that you can find the energies requested by judicious use of the *one-dimensional* square potential well eigenvalue computer program.)

4.5 The wavefunction for the ground state of hydrogen is

$$u_1 = Ae^{-r/a_0}.$$

Calculate

(a) The normalization constant A.
(b) The most probable value of r.
(c) The mean value of r.
(d) The probability that the electron will be found within a distance αa_0 of the origin.
(e) Approximate your result from (d) for the case $\alpha \ll 1$. Use this to estimate the probability of finding the electron *inside* the nucleus: assume the latter is spherical with $\alpha = 3 \times 10^{-5}$, and that the wavefunction retains its form inside the nucleus.
(f) Show that the same result for (e) is obtained more directly by assuming that the value of the wavefunction inside the nucleus is uniform with the value $u_1(0) = A$.

Comment on the validity of calculations (e) and (f) and the assumptions made therein. (Hint: in (c) if z occurs with (normalized) probability $P(z)$ the mean value of z is given by $\int zP(z)\,dz$, with the integral being taken over all z.)

Chapter 5

Formal Development of Quantum Mechanics

5.1 Eigenvalue equations, operators and measurements

It has already been noted in section 3.4 that the TISE,

$$\hat{H}u(r) = Eu(r) \tag{5.1}$$

has the form of an eigenvalue equation. The Hamiltonian operator $\hat{H}$, which represents the total energy of the system, takes the form

$$\hat{H} = -\frac{\hbar^2}{2m}\nabla^2 + V(r). \tag{5.2}$$

It is found that any measurement of E which is undertaken yields a result which is one of the eigenvalues of $\hat{H}$. This is not the same thing as saying that the system must have one of the eigenfunctions of $\hat{H}$ as its wavefunction. We have illustrated the possibility of wavefunctions which are superpositions of energy eigenfunctions in connection with wavepackets, for example, in section 2.9. However, if we make a single measurement of E we get a definite result rather than a range of values, and the result will be the eigenvalue belonging to one of the eigenfunctions which make up the superposition. (We ignore experimental limitations arising from the restricted precision of the apparatus.) The question which now arises is what to expect if we immediately re-measure the energy. We can answer this by recalling that the wavefunction is postulated to represent everything we can know about the system. Before we make the first measurement we only know that the energy lies in the range spanned by the wavepacket. After the measurement there is no longer any uncertainty in the energy, and the wavefunction should reflect the fact. We now have our answer: after the measurement the wavefunction of the system must be the eigenfunction corresponding to the measured eigenvalue, and the repeated measurement will give the same result as the first.

It is one of the postulates of quantum mechanics (number 3, Appendix 1) that all dynamical variables, which include for example position, momentum and angular momentum, as well as energy, can be represented by operators, that the results of measurements of all of these variables are eigenvalues of the corresponding operator, and that the system is left in the appropriate eigenfunction after the measurement.

Before we can examine the implications of this postulate further we need to know what the relevant operators are. This is the subject of another postulate (number 4, Appendix 1), which asserts that the operators representing position and momentum are

$$\hat{\boldsymbol{r}} = \boldsymbol{r} \tag{5.3}$$

and

$$\hat{\boldsymbol{P}} = -\mathrm{i}\hbar\nabla \tag{5.4}$$

respectively. (Although in general we will denote operators with circumflex accents, following normal practice this will be omitted in the case of $\boldsymbol{r}$ because the variable and its operator are identical.) Operators representing other variables are related to these in the same way that the classical quantities are related to the classical position and momentum variables. Examining equations (5.3) and (5.4) we can see that we have tacitly accepted these from the beginning in that they form components of the Hamiltonian operator, $\hat{\boldsymbol{P}}^2/2m + V(\boldsymbol{r}) = (-\hbar^2/2m)\nabla^2 + V(\boldsymbol{r})$.

〉5.2 The momentum operator, eigenvalues and eigenfunctions

The eigenvalue equation for $\hat{P}_x$ is

$$\hat{P}_x\phi = p_x\phi$$

i.e.

$$-\mathrm{i}\hbar\frac{\mathrm{d}\phi}{\mathrm{d}x} = p_x\phi. \tag{5.5}$$

Here we have taken the x component of equation (5.4) and also adopted a new notation. From now on we will use ϕ to denote an eigenfunction of any operator but the Hamiltonian, for which we will reserve the symbol u. For a general wavefunction or one whose precise character is not known or is not significant we will use ψ.

Equation (5.5) has the solutions

$$\phi = A\,\exp(\mathrm{i}p_x x/\hbar) \tag{5.6}$$

where p_x can take any value. These momentum eigenfunctions are simply the free-particle wavefunctions. Solutions of equation (5.5) with the form of (5.6) can be found for *any* value of p_x and so the momentum eigenvalue spectrum is continuous. For a free particle $\hat{H} = \hat{P}^2/2m$ so that the momentum eigenfunctions are also energy eigenfunctions. More generally, $\hat{H} = \hat{P}^2/2m + V(x)$ and this is *not* the case.

〉5.3 The position operator and its eigenfunctions

It is postulated that the position operator is simply the variable itself. The eigenvalue equation x is therefore

$$\hat{x}\phi(x) = x_0\phi(x) \tag{5.7}$$

where $\hat{x} \equiv x$ is a variable and x_0 is a particular value of x, as might result from a position measurement. This requires $\phi(x)$ to be a rather peculiar function, since it must produce a constant when it multiplies a variable! In fact, it is not a normal function at all, but a Dirac δ function, $\delta(x - x_0)$. Loosely speaking this is a function which is zero everywhere except at the point (x_0 in this case) where its argument is zero, and there it is infinite. It is best thought of as a limiting case of a sharply peaked function which becomes taller and narrower in such a way that its area remains constant. This property ensures that the function is square integrable, as required of a wavefunction. After a measurement of position the wavefunction becomes a position eigenfunction and must reflect the fact that x is now known precisely. Clearly the δ function has the necessary attributes.

An important property of a δ function is the effect it has on an integral when it appears as a factor in the integrand. It has the effect of evaluating the integrand at the location of the δ function, so for example

$$\int_{-\infty}^{\infty} \delta(x - x_0) f(x)\,\mathrm{d}x = f(x_0). \tag{5.8}$$

Mathematicians would claim that this is the only respectable way in which to employ a δ function!

We can gain further insight into momentum and position eigenfunctions if we consider a packet wavefunction of the type introduced in section 2.9:

$$\psi(x)=\frac{1}{(2\pi)^{1/2}}\int_{-\infty}^{\infty} a(k)\mathrm{e}^{\mathrm{i}kx}\,\mathrm{d}k. \tag{5.9}$$

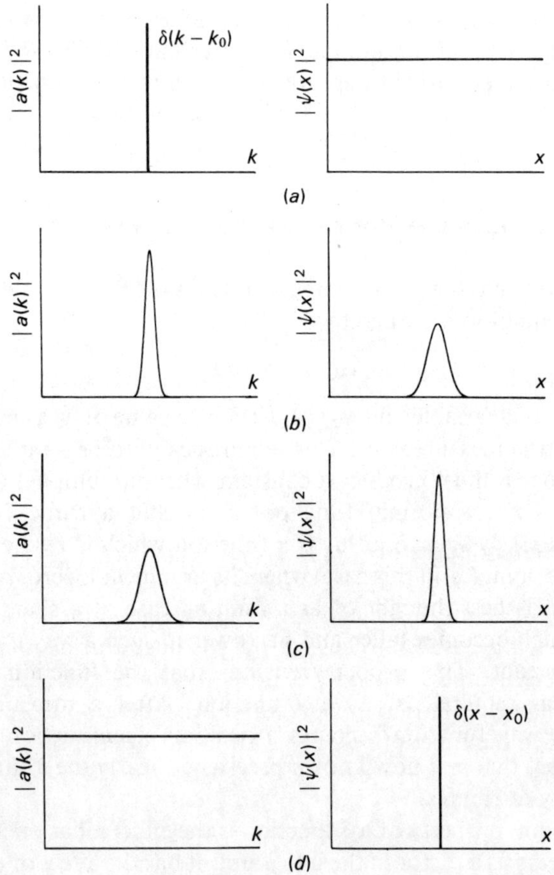

Figure 5.1 A progression from a momentum eigenstate (a), to a position eigenstate (d), via wavepackets of various widths ((b) and (c)). The probability density in real space is shown on the right and in k (effectively momentum) space on the left.

This is the time-independent wavefunction; it is equivalent to a snapshot taken at time $t=0$. The form of this equation shows that $\psi(x)$ and $a(k)$ are a Fourier transform pair, so

$$a(k)=\frac{1}{(2\pi)^{1/2}}\int_{-\infty}^{\infty}\psi(x)\mathrm{e}^{-\mathrm{i}kx}\,\mathrm{d}x. \tag{5.10}$$

Now consider the effect of allowing the k distribution $a(k)$ to become extremely narrow, so that it approximates a δ function in k, $a(k_0)\delta(k-k_0)$ say. The integral in equation (5.9) will take on the value of the integrand at $k=k_0$, and the result is a momentum eigenfunction with $p=\hbar k_0$:

$$\psi(x)=\frac{a(k_0)}{(2\pi)^{1/2}}\,\mathrm{e}^{\mathrm{i}k_0x}. \tag{5.11}$$

If, on the other hand, we allow $a(k)$ to become very broad $\psi(x)$ will become very narrow. For example, if $a(k)=a\exp(-\mathrm{i}kx_0)$, a function of constant amplitude, then from equation (5.9) we have

$$\psi(x)=\frac{1}{(2\pi)^{1/2}}\int_{-\infty}^{\infty}a\exp[\mathrm{i}k(x-x_0)]\,\mathrm{d}k\propto\delta(x-x_0) \tag{5.12}$$

a δ function in x which is a position eigenfunction. This follows because the integrand fluctuates with k and integrates to zero everywhere except at $x=x_0$. Hence momentum and position eigenfunctions are two extreme cases of packet wavefunctions. The transition from one to the other is shown in figure 5.1.

〉5.4 Hermitian operators

If the results of all measurements are to be the eigenvalues of appropriate operators it will be as well to ensure that the operators chosen do in fact have real eigenvalues. This is a property of a particular class of operators, namely Hermitian operators. An operator $\hat{Q}$ is Hermitian with respect to the functions ψ_1 and ψ_2 if it satisfies the rather surprising condition that

$$\int_{\substack{\text{All}\\\text{space}}}\psi_1^*\hat{Q}\psi_2\,\mathrm{d}\tau=\int_{\substack{\text{All}\\\text{space}}}\psi_2\hat{Q}^*\psi_1^*\,\mathrm{d}\tau. \tag{5.13}$$

In this equation ψ_1 and ψ_2 can be quite arbitrary as long as they are well enough behaved to be possible wavefunctions. $\mathrm{d}\tau$ is an element of volume. Integrals of this nature occur frequently in what follows and unless otherwise indicated will always be assumed to extend over all relevant space.

That a Hermitian operator has real eigenvalues can be seen as follows. Choose $\psi_1 = \psi_2 = \phi_n$ where ϕ_n is an eigenfunction of $\hat{Q}$. Then

$$\hat{Q}\phi_n = q_n\phi_n \tag{5.14}$$

and

$$\hat{Q}^*\phi_n^* = q_n^*\phi_n^* \tag{5.15}$$

where q_n is the eigenvalue corresponding to ϕ_n, and equation (5.15) is simply the complex conjugate of equation (5.14). Multiplying these equations on the left by ϕ_n^* and ϕ_n respectively and integrating we find

$$\int \phi_n^*\hat{Q}\phi_n\, \mathrm{d}\tau = q_n \int \phi_n^*\phi_n\, \mathrm{d}\tau. \tag{5.16}$$

$$\int \phi_n\hat{Q}^*\phi_n^*\, \mathrm{d}\tau = q_n^* \int \phi_n\phi_n^*\, \mathrm{d}\tau. \tag{5.17}$$

Since $\hat{Q}$ is Hermitian the left-hand sides of equations (5.16) and (5.17) are equal and hence so also are the right-hand sides. Therefore

$$q_n = q_n^* \tag{5.18}$$

and the eigenvalue is real.

Of the operators we have met, $\hat{x}$ (and $\hat{\boldsymbol{r}}$) are clearly Hermitian since the position of an algebraic factor in an integrand is irrelevant. That the momentum operator is Hermitian is less obvious since it is a differential operator. The proof is as follows. Consider

$$\int_{-\infty}^{\infty} \psi_1^*\hat{P}_x\psi_2\, \mathrm{d}x = \int_{-\infty}^{\infty} \psi_1^*\left(-\mathrm{i}\hbar\frac{\mathrm{d}}{\mathrm{d}x}\right)\psi_2\, \mathrm{d}x. \tag{5.19}$$

Integrating by parts we find

$$\int_{-\infty}^{\infty} \psi_1^*\hat{P}_x\psi_2\, \mathrm{d}x = -\mathrm{i}\hbar[\psi_1^*\psi_2]_{-\infty}^{\infty} + \mathrm{i}\hbar\int_{-\infty}^{\infty} \psi_2\frac{\mathrm{d}}{\mathrm{d}x}\psi_1^*\, \mathrm{d}x. \tag{5.20}$$

The first term on the right in equation (5.20) vanishes simply because any acceptable wavefunction will vanish at infinity. Hence

$$\int_{-\infty}^{\infty} \psi_1^*\hat{P}_x\psi_2\, \mathrm{d}\tau = \mathrm{i}\hbar\int_{-\infty}^{\infty} \psi_2\frac{\mathrm{d}}{\mathrm{d}x}\psi_1^*\, \mathrm{d}x = \int_{-\infty}^{\infty} \psi_2\hat{P}_x^*\psi_1^*\, \mathrm{d}\tau \tag{5.21}$$

which proves that $\hat{P}_x$ is Hermitian. It follows that $\hat{P}_y$, $\hat{P}_z$ and hence $\hat{\boldsymbol{P}}$ are also Hermitian. The Hermiticity of $\hat{\boldsymbol{P}}$ and $\boldsymbol{r}$ ensures the Hermiticity of $\hat{H}$.

〉5.5 Orthogonality of eigenfunctions

A useful property of the eigenfunctions of a Hermitian operator is that provided they are not degenerate they must be orthogonal. Two functions ϕ_n and ϕ_m are orthogonal if

$$\int \phi_n^* \phi_m \, \mathrm{d}\tau = 0 \tag{5.22}$$

where again the integral extends over all space. The proof proceeds by considering ϕ_n and ϕ_m to be two eigenfunctions of an operator $\hat{Q}$ with eigenvalues q_n and q_m, respectively. Then

$$\int \phi_n^* \hat{Q} \phi_m \, \mathrm{d}\tau = q_m \int \phi_n^* \phi_m \, \mathrm{d}\tau \tag{5.23}$$

$$\int \phi_m \hat{Q}^* \phi_n^* \, \mathrm{d}\tau = q_n \int \phi_m \phi_n^* \, \mathrm{d}\tau. \tag{5.24}$$

The left-hand sides of these two equations are equal because $\hat{Q}$ is Hermitian, and equating the right-hand sides we find

$$(q_m - q_n) \int \phi_n^* \phi_m \, \mathrm{d}\tau = 0. \tag{5.25}$$

Consequently, unless $q_m = q_n$, in which case the eigenfunctions are degenerate,

$$\int \phi_n^* \phi_m \, \mathrm{d}\tau = 0 \tag{5.26}$$

and the functions are orthogonal. In the event of degeneracy we can construct superposition states from the degenerate eigenfunctions which are also eigenfunctions and which can be so constructed as to be orthogonal to each other.

Functions which are both normalized and orthogonal are referred to as orthonormal, and satisfy the relation

$$\int \phi_n^* \phi_m \, \mathrm{d}\tau = \delta_{nm}. \tag{5.27}$$

The symbol δ_{nm} is the Kronecker δ function which has the properties

$$\begin{aligned} \delta_{nm} &= 0 \qquad m \neq \mathrm{n} \\ \delta_{nm} &= 1 \qquad m = n. \end{aligned} \tag{5.28}$$

It may be verified that the eigenfunctions which we have found for the particle in the infinite square well are all orthogonal, as are those for the harmonic oscillator and for the hydrogen-atom s states.

⟩ 5.6 The expansion postulate

The orthonormal properties which we have just discussed are similar to those possessed by the cartesian unit vectors $\hat{\imath}$, $\hat{\jmath}$, $\hat{k}$, since

$$\hat{\imath}\cdot\hat{\imath}=1, \text{ etc} \qquad \text{and} \qquad \hat{\imath}\cdot\hat{\jmath}=0, \text{ etc.} \tag{5.29}$$

The unit vectors are analogous to eigenfunctions and the scalar product has taken the place of the integration over all space of the product of functions. In fact this analogy can be taken much further. It is a feature of unit vectors that an arbitrary vector can be expressed in terms of a suitably weighted combination of them. Precisely the same thing can be done with a set of eigenfunctions of an operator: an arbitrary wavefunction can be expanded into a superposition of eigenfunctions of an operator. Just as three unit vectors, provided they are not coplanar, span three-dimensional space, it is postulated (number 5, Appendix 1) that all the linearly independent eigenfunctions of an operator span a function space, known as Hilbert space, in the sense that an arbitrary function which satisfies the same boundary conditions can be expanded in terms of them.

Suppose the operator $\hat{Q}$ has a set of eigenfunctions ϕ_n and eigenvalues q_n. Then the postulate states that any wavefunction ψ can be written as

$$\psi=\sum_n a_n\phi_n. \tag{5.30}$$

Using the infinite square-well eigenfunctions as an example:

$$u=a^{-1/2}\begin{Bmatrix}\cos\\ \sin\end{Bmatrix}\frac{n\pi x}{2a} \tag{5.31}$$

we recognize that we are merely restating the Fourier theorem. However, the postulate is quite general and applies, for example, to the Hermite–Gaussian functions as well. Considering the case of the finite square well where there may be very few or only a single bound state, this may be thought an unlikely proposition. However, it should be remembered that in that case there are all the unbounded continuum functions to be included as well.

Taking equation (5.30), multiplying by its conjugate and integrating, we find

$$\int |\psi|^2 \, d\tau = \int \sum_n \sum_m a_n^* a_m \phi_n^* \phi_m \, d\tau$$

$$= \sum_n \sum_m a_n^* a_m \int \phi_n^* \phi_m \, d\tau. \tag{5.32}$$

Because the ϕ are orthonormal this reduces to

$$\int |\psi|^2 \, d\tau = \sum_n \sum_m a_n^* a_m \, \delta_{nm} = \sum_n |a_n|^2 = 1. \tag{5.33}$$

The last equality is true since we assume ψ to be normalized. We interpret this result as follows (postulate 5). If we measure the value of the variable Q (represented by the operator $\hat{Q}$), we must get one of its eigenvalues as the result (postulate 3). Equation (5.33) suggests that if the wavefunction of the system was ψ then the probability of getting the particular result q_n is $|a_n|^2$. This postulate, proposed by Born, is quite crucial to the interpretation of measurements in quantum mechanics.

The calculation of the expansion coefficients in equation (5.30) follows the pattern familiar in Fourier analysis. Suppose we wish to know a_m. We multiply equation (5.30) by ϕ_m^* and integrate:

$$\int \phi_m^* \psi \, d\tau = \int \sum_n a_n \phi_m^* \phi_n \, d\tau. \tag{5.34}$$

Interchanging integral and summation signs, we obtain

$$\int \phi_m^* \psi \, d\tau = \sum_n a_n \int \phi_m^* \phi_n \, d\tau = \sum_n a_n \, \delta_{mn} = a_m \tag{5.35}$$

by virtue of orthonormality. Hence

$$a_m = \int \phi_m^* \psi \, d\tau. \tag{5.36}$$

The type of integral in equation (5.36) is known as an overlap integral. It is a measure of the degree of similarity of the two wavefunctions. The analogy with unit vectors remains valid. If we wished to find the coefficient x of a unit vector $\hat{\boldsymbol{\imath}}$ in the expansion of an arbitrary vector $\boldsymbol{r}$ such that

$$\boldsymbol{r} = x\hat{\boldsymbol{\imath}} + y\hat{\boldsymbol{\jmath}} + z\hat{\boldsymbol{k}} \tag{5.37}$$

we would take the scalar product with $\hat{\imath}$:

$$\hat{\imath} \cdot \boldsymbol{r} = x\hat{\imath} \cdot \hat{\imath} + y\hat{\imath} \cdot \hat{\jmath} + z\hat{\imath} \cdot \hat{k} = x. \tag{5.38}$$

$\hat{\imath} \cdot \boldsymbol{r}$ is analogous to the overlap integral.

〉5.7 Example: the β decay of tritium

The nucleus of the tritium atom ^3H has a single charge, $Z = 1$, and a mass number $A = 3$. It decays, with a half-life of 12.3 d, by emitting a β^- particle together with an antineutrino, with combined energy equal to 18.6 keV, and leaves a singly ionized $^3He^+$ ion with a nuclear charge $Z = 2$. From the point of view of the atomic electron, which was originally in the ground state of the ^3H atom, it now finds itself with a wavefunction which does not fit the new $^3He^+$ potential well since the nuclear charge has increased. More precisely, the old wavefunction is not an eigenfunction of the new Hamiltonian. The decay is a very rapid process and the wavefunction cannot adjust instantly. We can, however, expand the old function in terms of the eigenfunctions of the new well.

The $n = 1$ and $n = 2$ s-state eigenfunctions for hydrogen-like ions are (equations (4.39))

$$u_1 = N_1 \exp(-Zr/a_0) \qquad N_1 = \pi^{-1/2}(Z/a_0)^{3/2} \tag{5.39}$$

$$u_2 = N_2(1 - Zr/2a_0)\exp(-Zr/2a_0) \qquad N_2 = \pi^{-1/2}(Z/2a_0)^{3/2}. \tag{5.40}$$

Using primed symbols to denote $^3He^+$ and unprimed to denote ^3H, we will expand u_1 in terms of u'_n:

$$u_1 = \sum_{n'} a_{n'} u'_{n'}. \tag{5.41}$$

Now using equation (5.36) we have

$$\begin{aligned} a_1 &= N_1 N'_1 \int_0^\infty \exp(-Zr/a_0)\exp(-Z'r/a_0) 4\pi r^2 \, \mathrm{d}r \\ &= 4\pi N_1 N'_2 \int_0^\infty r^2 \exp[-(Z+Z')r/a_0] \, \mathrm{d}r \\ &= 4\pi N_1 N'_2 \, 2! \, a_0^3/(Z+Z')^3 \end{aligned} \tag{5.42}$$

where we have used the standard integral

$$\int_0^\infty x^n \mathrm{e}^{-\alpha x} \, \mathrm{d}x = n!/\alpha^{n+1} \qquad \alpha > 0, \ n \text{ positive}. \tag{5.43}$$

Inserting the various constants we obtain

$$a_1 = 8(ZZ')^{3/2}/(Z+Z')^3. \qquad (5.44)$$

Since $Z = 1$ and $Z' = 2$ we find

$$a_1 = 0.838. \qquad (5.45)$$

The probability that a measurement of the energy of the $^3He^+$ ion would show the system to be in its ground state is therefore

$$P(1) = |a_1|^2 = 0.702. \qquad (5.46)$$

Exactly similar calculations using u_2' and u_3' reveal that $a_2 = -0.5$ and $a_3 = -0.1129$. Hence the probabilities of finding the $^3He^+$ ion in the $n' = 2$ and $n' = 3$ excited states are 0.25 and 0.0127, respectively. Experimentally, this might be verified by the observation of photons emitted a short time after the β particle, as the excited $^3He^+$ ions decay to the ground state. (A selection rule would inhibit an optical transition from the *first* (2s) excited state.)

The excitation process we have described limits the precision of important experiments aimed at determining the rest mass (if any) of the neutrino. This is done by finding the maximum energy of the β^- particles emitted during the decay of tritium. The 18.6 keV of energy available for the decay is shared between a β particle and a neutrino. If a neutrino has a finite rest mass m_ν, then $m_\nu c^2$ sets a lower limit to its share of the energy, whereas if it has zero rest mass, the lower limit is zero and a β^- particle can occasionally acquire the entire 18.6 keV. The excitation of the atomic electron provides an additional channel for a small portion of the energy, and this must be allowed for. Although the probability of this occurring is readily calculated for isolated tritium atoms, for various reasons the actual experiments are performed using tritium bound in rather complex molecules. It is the difficulty of making an accurate estimation of the excitation probability for a tritium atom in its actual experimental environment which is responsible for the current upper limit of ~ 15 eV for the mass of the neutrino.

Computer exercise 5.1: β decay of tritium; overlap integrals

The probabilities of excitation of the $n' = 1, 2, 3$ levels decrease rapidly. The reason lies with the relative sizes of the overlap integrals, which we have found to be in the ratios $0.838 : -0.5 : -0.1129$. Use the program 'Hydrogen-like Ions' to superimpose the function defined by ($Z = 1$, $l = 0$, $n = 1$) with ($Z = 2$, $l = 0$, $n' = 1, 2, 3$) one at a time. Assess visually the relative degree of overlap between the superimposed functions. Obviously places where the functions are of opposite sign contribute negatively to the integral. Note that the volume element in equation (5.42), which is proportional to r^2, weights contributions from larger radii much more heavily. This can be allowed for if $\chi(r)$ is plotted for each state. Since $\chi(r) \propto ru(r)$ the product will then be proportional to r^2.

⟩ 5.8 Expectation values

We have found that if we measure a variable Q for a system whose wavefunction ψ is not an eigenfunction of $\hat{Q}$ we may get a variety of results q_n, these being the eigenvalues of the operator $\hat{Q}$. The probability of obtaining a particular result, q_n say, is $|a_n|^2$, where a_n is the coefficient of ϕ_n in the expansion of ψ. Armed with these facts it is straightforward to deduce that the average value of a series of measurements would be

$$\bar{Q} = \sum_n |a_n|^2 q_n. \tag{5.47}$$

Although equation (5.47) is a simple enough expression it is not very useful, since in order to evaluate it we would need to compute all the a_n, which may be infinite in number. This in fact is the case for the tritium example we have just considered. Fortunately there is a much more convenient method. Consider first of all what we would expect for the mean value of a sequence of measurements of position, x. Since $|\psi(x)|^2$ is the probability density at x we can immediately write

$$\bar{x} = \int_{-\infty}^{\infty} \psi^*(x)\psi(x)x \, \mathrm{d}x. \tag{5.48}$$

Analogously for momentum we might try replacing x by the operator for momentum, $-\mathrm{i}\hbar \, \mathrm{d}/\mathrm{d}x$, in equation (5.48). The problem which then arises is what to allow the differential operator to operate on, ψ or $|\psi|^2$? In fact, the correct answer for any operator is obtained by placing it between the two wavefunctions so that it operates only on the right-hand one, as we will now prove.

Consider the expectation value of Q for the wavefunction $\Psi(r, t)$, defined as

$$\langle \hat{Q} \rangle = \int \Psi(r, t)\hat{Q}\Psi(r, t) \, \mathrm{d}\tau. \tag{5.49}$$

We will examine any time dependence this may have later. For the moment we will consider the situation at a particular instant, $t = 0$ say, replacing $\Psi(r, 0)$ by $\psi(r)$, so that

$$\langle \hat{Q} \rangle = \int \psi^* \hat{Q} \psi \, \mathrm{d}\tau. \tag{5.50}$$

Expanding ψ^* and ψ in terms of ϕ_n, the eigenfunctions of $\hat{Q}$:

$$\begin{aligned}\langle \hat{Q} \rangle &= \int \sum_m a_m^* \phi_m^* \hat{Q} \sum_n a_n \phi_n \, \mathrm{d}\tau \\ &= \sum_m \sum_n a_m^* a_n q_n \int \phi_m^* \phi_n \, \mathrm{d}\tau \\ &= \sum_m \sum_n a_m^* a_n q_n \, \delta_{mn} \\ &= \sum_n | a_n |^2 q_n. \end{aligned} \tag{5.51}$$

We have shown that the expectation value defined in equation (5.49) is identical with the average value we seek, equation (5.47). In quantum mechanics average values are always referred to as expectation values and are given the special symbol we have used on the left of equation (5.49).

If by any chance ψ was in fact an eigenfunction of $\hat{Q}$, ϕ_n say, then measurement of Q would result in q_n with certainty. Equation (5.50) would then reduce to

$$\langle \hat{Q} \rangle = q_n \int \psi^* \psi \, \mathrm{d}\tau = q_n \tag{5.52}$$

as it should.

It is necessary to clarify the meaning of an average value in quantum mechanics. If we take a particular system with wavefunction ψ and make a series of measurements of Q on it, for the first measurement we will obtain the answer q_n with probability $| a_n |^2$, as described above. When, however, we make a second measurement we are no longer dealing with a system with wavefunction ψ but one with eigenfunction ϕ_n. From now on successive measurements will all give the result q_n and this will very soon become the average value! This dilemma is avoided by specifying an average value in quantum mechanics to be an ensemble average. We imagine that we have a large array of identical systems, all of which are described by identical wavefunctions, and proceed to measure Q (once) for each one. The average value will then be $\langle \hat{Q} \rangle$.

As an example of the calculation of expectation values, we will take the harmonic oscillator. If we take a classical oscillator and, without disturbing the system, measure the position x of the particle at a great many random times we expect to find positive values as often as negative values and a zero average. Similarly we would expect to find a zero average for the momentum. The same results are obtained in quantum mechanics, provided we take an ensemble average, since we can no longer hope to make non-perturbing measurements. The expectation value of x for any energy eigenfunction u_n of the harmonic oscillator is

$$\langle x \rangle = \int_{-\infty}^{\infty} u_n^* x u_n \, \mathrm{d}x. \tag{5.53}$$

Now all the eigenfunctions have a definite parity, either even or odd. Examining equation (5.53), we note that the wavefunction appears twice so regardless of parity the combination is even, whereas x is an odd function. Overall then the integrand of equation (5.53) is odd and so the integral is zero.

In the case of the momentum the expectation value is

$$\langle \hat{P}_x \rangle = \int_{-\infty}^{\infty} u_n^* \left(-\mathrm{i}\hbar \frac{\mathrm{d}}{\mathrm{d}x} \right) u_n \, \mathrm{d}x. \tag{5.54}$$

Similar considerations apply, except that the oddness of the differential operator needs some justification. A little thought will reveal that differentiating an even function will turn it into an odd one and vice versa. Again therefore the integrand is odd and the expectation value zero.

The situation is more interesting if we consider the expectation values of x^2 and $\hat{P}^2$. These are even functions and so we expect non-zero results. Take the case of the oscillator in its ground state. Then

$$u_0 = (\alpha/\sqrt{\pi})^{1/2} \exp(-\tfrac{1}{2}\alpha^2 x^2) \qquad \alpha = (m\omega/\hbar)^{1/2}. \tag{5.55}$$

Hence

$$\begin{aligned} \langle x^2 \rangle &= (\alpha/\sqrt{\pi}) \int_{-\infty}^{\infty} \exp(-\tfrac{1}{2}\alpha^2 x^2) x^2 \exp(-\tfrac{1}{2}\alpha^2 x^2) \, \mathrm{d}x \\ &= (\alpha/\sqrt{\pi}) \int_{-\infty}^{\infty} x^2 \exp(-\alpha^2 x^2) \, \mathrm{d}x \\ &= \frac{\alpha}{\sqrt{\pi}} \frac{1}{2} \left(\frac{\pi}{\alpha^6} \right)^{1/2} = \frac{\hbar}{2m\omega}. \end{aligned} \tag{5.56}$$

Similarly

$$\langle \hat{P}_x^2 \rangle = (\alpha/\sqrt{\pi}) \int_{-\infty}^{\infty} \exp(-\tfrac{1}{2}\alpha^2 x^2)\left(-\hbar^2 \frac{d^2}{dx^2}\right)\exp(-\tfrac{1}{2}\alpha^2 x^2)\,dx \quad (5.57)$$

$$= \tfrac{1}{2}\alpha^2\hbar^2 = \tfrac{1}{2}m\hbar\omega. \quad (5.58)$$

To obtain these results we have made use of the standard integral:

$$\int_{-\infty}^{\infty} x^{2n} \exp(-\alpha^2 x^2)\,dx = \frac{1.3.5\ldots(2n-1)\pi^{1/2}}{2^n \alpha^{2n+1}}. \quad (5.59)$$

The results of equations (5.56) and (5.58) are interesting because the Hamiltonian for the oscillator is

$$\hat{H} = \hat{P}^2/2m + \tfrac{1}{2}m\omega^2 x^2 \quad (5.60)$$

and so its expectation value is

$$\begin{aligned}\langle \hat{H} \rangle &= \langle \hat{P}^2/2m + \tfrac{1}{2}m\omega^2 x^2 \rangle \\ &= \frac{1}{2m}\langle \hat{P}^2 \rangle + \frac{m\omega^2}{2}\langle x^2 \rangle \\ &= \tfrac{1}{4}\hbar\omega + \tfrac{1}{4}\hbar\omega = \tfrac{1}{2}\hbar\omega. \end{aligned} \quad (5.61)$$

This shows that, as for a classical oscillator, the mean kinetic and mean potential energies are each just half the total energy, which in this case is $\tfrac{1}{2}\hbar\omega$.

〉5.9 Compatibility of observables: commutators

From our knowledge of the uncertainty principle we are familiar with the impossibility of complete knowledge of, for example, the position and momentum of a particle. The material we have discussed in this chapter throws further light on the matter. Suppose we do in fact measure the position of a particle, and find it to be x_0. After the measurement the wavefunction must be the position eigenfunction $\delta(x - x_0)$. This is a superposition of momentum eigenfunctions $\exp(ip_x x/\hbar)$ which spans all values of momentum. A measurement of the momentum P_x will result in a momentum eigenvalue p_x drawn from this infinite range. In short, the momentum has been rendered completely unpredictable. A similar process operates if the measurements are made in the reverse order. We describe the situation by saying that observations of position x and momentum P_x are incompatible.

On the other hand, after the measurement of the momentum P_x of a free particle, its wavefunction is a momentum eigenfunction, which is also an energy eigenfunction since the Hamiltonian of a free particle is $\hat{P}_x^2/2m$. Consequently, the result of a subsequent measurement of energy is completely predictable. We say that the momentum P_x and the energy E are compatible observables (but only for a free particle). Note that x and P_y provide another example of compatible observables, since the measurement of x only affects P_x.

The preceding discussion shows that two observables will be compatible if their operators have common eigenfunctions. We will now find a formal condition for the compatibility of two observables. Consider operators $\hat{Q}$ and $\hat{R}$ with eigenvalues q_n and r_n, and common eigenfunctions ϕ_n. Then

$$\hat{Q}\phi_n = q_n\phi_n \qquad \text{and} \qquad \hat{R}\phi_n = r_n\phi_n. \tag{5.62}$$

Operating on each equation on the left we find

$$\hat{R}\hat{Q}\hat{\phi}_n = \hat{R}q_n\phi_n \qquad \text{and} \qquad \hat{Q}\hat{R}\phi_n = \hat{Q}r_n\phi_n$$

$$\hat{R}\hat{Q}\hat{\phi}_n = q_nr_n\phi_n \qquad \text{and} \qquad \hat{Q}\hat{R}\phi_n = r_nq_n\phi_n. \tag{5.63}$$

The second pair of equations follows since eigenvalues are simple numbers and unaffected by operators. Subtracting one equation of the pair from the other we have

$$[\hat{Q}\hat{R} - \hat{R}\hat{Q}]\phi_n = 0. \tag{5.64}$$

The expression in square brackets is the commutator of the two operators $\hat{Q}$ and $\hat{R}$ and is usually written as $[\hat{Q}, \hat{R}]$. Now, since any wavefunction ψ can be expanded in terms of the ϕ_n, equation (5.64) can be generalized to

$$[\hat{Q}, \hat{R}]\psi = 0 \tag{5.65}$$

where ψ is quite arbitrary. This being the case we adopt the convention of dropping the wavefunction, the existence of an operand being understood. Hence

$$[\hat{Q}, \hat{R}] = 0. \tag{5.66}$$

What we have shown is that if two observables are compatible and therefore share eigenfunctions the commutator of their operators is zero; the order in which they operate is immaterial. The converse is also true; if two operators commute then they will have common eigenfunctions and so the observables will be compatible. The proof is simpler if the eigenfunctions are non-degenerate, and this will be assumed here. Suppose

$$\hat{Q}\phi_n = q_n\phi_n \tag{5.67}$$

and

$$[\hat{Q}, \hat{R}] = 0. \tag{5.68}$$

Consider

$$\hat{Q}\hat{R}\phi_n = \hat{R}\hat{Q}\phi_n = q_n\hat{R}\phi_n \tag{5.69}$$

where the first equality holds because the operators commute. Writing equation (5.69) as

$$\hat{Q}(\hat{R}\phi_n) = q_n(\hat{R}\phi_n) \tag{5.70}$$

we recognize an eigenvalue equation for $\hat{Q}$ with eigenfunction $(\hat{R}\phi_n)$ and eigenvalue q_n. Since the eigenfunctions of $\hat{Q}$ are not degenerate, by hypothesis, $\hat{R}\phi_n$ can differ from ϕ_n by a constant factor at most. Letting this factor equal r_n we have

$$\hat{R}\phi_n = r_n\phi_n. \tag{5.71}$$

But this implies that ϕ_n is an eigenfunction of $\hat{R}$ with eigenvalue r_n. Hence the ϕ_n are common eigenfunctions of $\hat{Q}$ and $\hat{R}$.

If the ϕ_n are degenerate in $\hat{Q}$ then it cannot be assumed in equation (5.70) that $\hat{R}\phi_n$ is essentially ϕ_n; it could be another degenerate eigenfunction, or a superposition of degenerate eigenfunctions. However, it can be proved that it is always possible to find a set of simultaneous eigenfunctions by constructing appropriate superpositions from the degenerate eigenfunctions.

These theorems will be particularly useful when we come to examine angular momentum in the next chapter.

〉5.10 Non-commuting operators

If two operators commute, then the order in which they operate is irrelevant. Clearly this is not the case for two operators such as x and $\hat{P}$, since $\hat{P} = -i\hbar\, d/dx$. It certainly matters whether we multiply a function by x before or after we differentiate it with respect to x, but not if we differentiate it with respect to y. Interestingly, we can actually quantify the degree to which the order of events matters by evaluating the relevant commutators for an arbitrary wavefunction. Consider

$$[\hat{P}_x, x]\psi = \left(-\mathrm{i}\hbar \frac{\mathrm{d}}{\mathrm{d}x}(x\psi) + x\mathrm{i}\hbar \frac{\mathrm{d}\psi}{\mathrm{d}x}\right)$$

$$= -\mathrm{i}\hbar\left(x\frac{\mathrm{d}\psi}{\mathrm{d}x} + \psi - x\frac{\mathrm{d}\psi}{\mathrm{d}x}\right)$$

$$= -\mathrm{i}\hbar\psi. \tag{5.72}$$

We write this as

$$[\hat{P}, x] = -\mathrm{i}\hbar. \tag{5.73}$$

On the other hand, $[\hat{P}_x, y] = 0$ for the reason given above and also $[\hat{P}_x, \hat{P}_y] = [x, y] = 0$. This can all be summarized as

$$\begin{aligned}[\hat{P}_i, x_j] &= -[\hat{P}_i, x_j] = -\mathrm{i}\hbar\,\delta_{ij}\\ [\hat{P}_i, \hat{P}_j] &= [x_i, x_j] = 0\end{aligned} \tag{5.74}$$

where $i, j = 1, 2, 3$ signify x, y, z components.

Commutators play a crucial role in the development of quantum theory and the non-commutation of observables (through their operators) is a purely quantum mechanical concept. It is found that all commutators are either zero or proportional to $\hbar$ and if this quantity should be reduced to zero classical physics would prevail. In this sense commutators embody the distinction between the quantum and classical world.

〉5.11 Time dependence of expectation values: Ehrenfest's theorem

The expectation value of an operator, $\langle \hat{Q} \rangle$, as defined in equation (5.49), may or may not be time dependent. We can find the conditions which determine this by examining the quantity

$$\frac{\partial}{\partial t}\langle \hat{Q} \rangle = \frac{\partial}{\partial t}\int \Psi^* \hat{Q} \Psi \, \mathrm{d}\tau. \tag{5.75}$$

Differentiating under the integral sign using the chain rule we find

$$\frac{\partial}{\partial t}\langle \hat{Q} \rangle = \int \left(\frac{\partial \Psi^*}{\partial t}\hat{Q}\Psi + \Psi^* \frac{\partial \hat{Q}}{\partial t}\Psi + \Psi^* \hat{Q}\frac{\partial \Psi}{\partial t}\right) \mathrm{d}\tau. \tag{5.76}$$

We can eliminate the time derivatives of $\hat{Q}$ by making use of the TDSE and its complex conjugate,

$$\hat{H}\Psi = \mathrm{i}\hbar \frac{\partial \Psi}{\partial t} \qquad \text{and} \qquad \hat{H}^*\Psi^* = -\mathrm{i}\hbar \frac{\partial \Psi^*}{\partial t}. \tag{5.77}$$

This gives us

$$\frac{\partial}{\partial t}\langle \hat{Q}\rangle = \frac{1}{\mathrm{i}\hbar}\int [-(\hat{H}^*\Psi^*)\hat{Q}\Psi + \Psi^*\hat{Q}(\hat{H}\Psi)]\,\mathrm{d}\tau + \left\langle \frac{\partial \hat{Q}}{\partial t}\right\rangle. \tag{5.78}$$

Since $\hat{H}$ is a Hermitian operator we can rearrange this to give

$$\begin{aligned}\frac{\partial}{\partial t}\langle \hat{Q}\rangle &= \frac{1}{\mathrm{i}\hbar}\int \Psi^*[\hat{Q}\hat{H} - \hat{H}\hat{Q}]\Psi\,\mathrm{d}\tau + \left\langle \frac{\partial \hat{Q}}{\partial t}\right\rangle \\ &= \frac{1}{\mathrm{i}\hbar}\int \Psi^*[\hat{Q}, \hat{H}]\Psi\,\mathrm{d}\tau + \left\langle \frac{\partial \hat{Q}}{\partial t}\right\rangle\end{aligned} \tag{5.79}$$

which is Ehrenfest's theorem. We interpret it as follows. The last term on the right is zero unless the operator $\hat{Q}$ is explicitly time dependent. None of the operators which we encounter in this book is explicitly time dependent, so we will discard this term. The first term on the right is zero if $[\hat{Q}, \hat{H}] = 0$. Hence the equation tells us that the expectation value of any time-independent operator is a constant if it commutes with the Hamiltonian operator. The observable which it represents is then known as a constant of the motion, or a conserved quantity.

The simplest operator we could consider for $\hat{Q}$ is simply the unity operator, in which case the equation reads

$$\frac{\partial}{\partial t}\int \Psi^*\Psi\,\mathrm{d}\tau = \frac{1}{\mathrm{i}\hbar}\int \Psi^*[1, \hat{H}]\Psi\,\mathrm{d}\tau = 0. \tag{5.80}$$

This informs us that once normalized a function remains so for all time, an obviously essential requirement!

If we let $\hat{Q} = \hat{H}$ where $\hat{H}$ is independent of time, we find that

$$\frac{\partial}{\partial t}\langle \hat{H}\rangle = \frac{1}{\mathrm{i}\hbar}\int \Psi^*[\hat{H}, \hat{H}]\Psi\,\mathrm{d}\tau = 0. \tag{5.81}$$

This is the quantum mechanical statement of energy conservation.

If we identify $\hat{Q}$ with $\hat{P}_x$ we must calculate the commutator

$$[\hat{P}_x, \hat{H}] = [\hat{P}_x, (\hat{P}_x^2/2m + V(x))]. \tag{5.82}$$

Since $\hat{P}_x$ and $\hat{P}_x^2$ commute, the commutator equals $[\hat{P}_x, V(x)]$, which we evaluate as follows:

$$[\hat{P}_x, V(x)]\Psi = -i\hbar\left(\frac{\partial}{\partial x}V(x)\Psi - V(x)\frac{\partial}{\partial x}\Psi\right)$$

$$= -i\hbar\left(V(x)\frac{\partial\Psi}{\partial x} + \Psi\frac{\partial V(x)}{\partial x} - V(x)\frac{\partial\Psi}{\partial x}\right).$$

Hence

$$[\hat{P}_x, \hat{H}] = -i\hbar\frac{\partial V(x)}{\partial x}. \tag{5.83}$$

Ehrenfest's theorem thus becomes, for $\hat{P}_x$,

$$\frac{\partial}{\partial t}\langle\hat{P}_x\rangle = -\left\langle\frac{\partial V(x)}{\partial x}\right\rangle. \tag{5.84}$$

If we identify the expectation value $\langle\hat{P}_x\rangle$ with the classical momentum, we find that we have obtained a statement of Newton's second law of motion, equating rate of change of momentum with force, since the term on the right can be identified with the force acting on the particle. If in some particular case $V \neq V(x)$ then $\partial V/\partial x = F_x = 0$. In that case, however, $[\hat{P}_x, \hat{H}] = 0$ and $\hat{P}_x$ is then a constant of the motion, as it should be.

A similar analysis letting $\hat{Q} = x$ leads to the result

$$\frac{\partial}{\partial t}\langle x\rangle = \frac{1}{m}\langle\hat{P}_x\rangle. \tag{5.85}$$

Again, if we equate the expectation value with the classical variable we have a statement of the relation between velocity and momentum which corresponds to the classical one.

The identification of expectation values with classical observables is often but not always valid. The position of an electron in the neighbourhood of the slits of a two-slit diffraction experiment, as in figure 1.1, is badly represented by $\langle x\rangle$ since this position will be just half-way between the two slits, where in fact the electron is never found. On the other hand, if we are considering a typical wavepacket, as in figure 2.5, $\langle x\rangle$ and $\langle\hat{P}_x\rangle$ will certainly be appropriate representations. In this case Ehrenfest's theorem is a statement of the way in which the correspondence principle operates. For a classical particle the wavepacket will shrink to a narrow pulse whose position is highly defined and which moves according to Newton's law.

〉5.12 Abstract operators: the parity operator

Some of the ideas introduced in the preceding paragraphs can be usefully illustrated by introducing an abstract parity operator, $\mathscr{P}(\boldsymbol{r})$. The effect of $\mathscr{P}(\boldsymbol{r})$ is to transform $\boldsymbol{r}$ into $-\boldsymbol{r}$ (i.e $(x, y, z) \Rightarrow (-x, -y, -z)$), so that

$$\hat{\mathscr{P}}(\boldsymbol{r})\Psi(\boldsymbol{r}) = \Psi(-\boldsymbol{r}). \tag{5.86}$$

That $\hat{\mathscr{P}}(\boldsymbol{r})$ is Hermitian can be seen from the following sequence of operations. (For the sake of simplicity from here on the discussion relates to one dimension only, but can be extended in an obvious way to three dimensions.)

$$\int_{-\infty}^{\infty} \psi_1^*(x)\hat{\mathscr{P}}(x)\psi_2(x)\, \mathrm{d}x = \int_{-\infty}^{\infty} \psi_1^*(x)\psi_2(-x)\, \mathrm{d}x$$

$$= \int_{\infty}^{-\infty} \psi_1^*(-x)\psi_2(x)\, \mathrm{d}(-x) = \int_{-\infty}^{\infty} (\mathscr{P}(x)\psi_1(x))^*\psi_2(x)\, \mathrm{d}x. \tag{5.87}$$

The second equality follows on replacing x with $-x$, which cannot affect the value of the integral, which is over all space. The third equality follows on reintroducing $\hat{\mathscr{P}}(x)$ and changing the direction in which the integration is performed.

Now the general form of the Hamiltonian is

$$\hat{H}(x) = \frac{\hbar^2}{2m}\frac{\mathrm{d}^2}{\mathrm{d}x^2} + V(x). \tag{5.88}$$

Hence, provided $V(x) = V(-x)$, the parity operation will leave the Hamiltonian unaltered: $\hat{H}(-x) \equiv \hat{H}(x)$. If this is so then $\hat{\mathscr{P}}(x)$ and $\hat{H}(x)$ commute:

$$\hat{\mathscr{P}}(x)(\hat{H}(x)\psi(x)) = \hat{H}(-x)\psi(-x) = \hat{H}(x)\hat{\mathscr{P}}(x)\psi(x)$$

and

$$[H(x), \mathscr{P}(x)] = 0. \tag{5.89}$$

We now look for the eigenvalues of $\hat{\mathscr{P}}(x)$. The eigenvalue equation is

$$\hat{\mathscr{P}}(x)\phi(x) = \alpha\phi(x) \tag{5.90}$$

but from the definition of the parity operator

$$\hat{\mathscr{P}}(x)\phi(x) = \phi(-x). \tag{5.91}$$

Applying the operator a second time to both equations (5.90) and (5.91) and equating the results we find

$$\hat{\mathscr{P}}^2(x)\phi(x) = \alpha^2\phi(x) = \phi(x). \tag{5.92}$$

We conclude that

$$\alpha = \pm 1. \tag{5.93}$$

These eigenvalues correspond to even- and odd-parity eigenfunctions, respectively. (Strictly speaking, a double application of the parity operator can only be assumed to bring the state back to one which is indistinguishable *physically* from the original, which would allow the appearance of a phase factor of unit magnitude, e.g., $\hat{\mathscr{P}}^2(x)\phi(x) = \alpha^2\phi(x) = \beta\phi(x)$. Then $|\alpha^2| = |\beta| = 1$, in which case we still find $\alpha = \pm 1$ because the eigenvalues of a Hermitian operator must be real.)

If it is the case that $\hat{H}$ and $\hat{\mathscr{P}}$ commute we expect to be able to find simultaneous eigenfunctions of the two operators, and this in fact we have been able to do in all the cases we have studied. Furthermore, from the discussion of the preceding section, the parity of the functions must be constants of the motion, so that no matter how the function evolves in time, its parity will persist.

Introduction of the parity operator has enabled us to confirm and formalize, in an elegant way, our earlier observations on the parity of wavefunctions.

〉5.13 Matrix mechanics

Although our principal concern in this book is with wave mechanics, a formulation of quantum mechanics based upon Schrödinger's wave equation, this is by no means the only formulation and, historically, it was not even the first. It was preceded in 1925 by Heisenberg's matrix mechanics. The two approaches evolved quite independently, and were subsequently shown to be equivalent by Schrödinger, in 1926.

Despite the fact that our wave mechanical approach provides us with an image (i.e. a wavefunction) which we can visualize, we are constantly having to remind ourselves that the wavefunction itself is actually unobservable. Let us review what can be learnt about a quantum mechanical system as a result of observation. The results of all individual measurements of an observable, Q say, are the eigenvalues q_n of the operator $\hat{Q}$ representing the observable. If we make many measurements on an ensemble of systems each with wavefunction $\psi = \Sigma\, a_n\phi_n$ we can also determine the expectation value of the observable, $\langle \hat{Q} \rangle$, and deduce the magnitudes of the expansion coefficients a_n from the frequency of occurrence of each q_n. More sophisticated techniques may allow us to determine the coefficients themselves, to within an overall phase factor. Beyond this there is not much more we can learn about the system. Matrix mechanics can be regarded as a sort of book-keeping system which keeps track of all the measurable quantities we have just mentioned, but which pays almost no attention at all to the unobservable wavefunction. This is actually its great strength since it enables us to deal with physical systems which, while perfectly capable of being observed and measured, are inherently incapable of being described by an explicit wavefunction. This occurs whenever the variables (e.g. coordinates) describing the system are not known. An example of such a case is that of electron spin, which will be described in Chapter 6.

We will deduce the elements of matrix mechanics from our knowledge of wave mechanics, starting from the expansion postulate. If ϕ_j form a complete orthonormal set of functions such that

$$\int \phi_j \phi_k \, \mathrm{d}\tau = \delta_{jk} \tag{5.94}$$

then an arbitrary wavefunction ψ can be expressed as

$$\psi = \sum_j a_j \phi_j \tag{5.95}$$

where

$$a_j = \int \phi_j^* \psi \, \mathrm{d}\tau. \tag{5.96}$$

We note that, once a basis set of functions ϕ_j has been defined, it suffices to know the a_j in order to specify the wavefunction completely. This is quite analogous to the case of a vector $\boldsymbol{r}$. Provided the unit vectors $\hat{\boldsymbol{x}}$, $\hat{\boldsymbol{y}}$, $\hat{\boldsymbol{z}}$ have been defined, $\boldsymbol{r}$ is specified completely by its cartesian components (x, y, z). Because of the strength of this analogy in matrix mechanics we refer to wavefunctions such as ψ as *state vectors*, and the basis functions ϕ_j as *basis vectors*.

We will now show that all the operations to which we have become accustomed in this chapter can be performed in matrix language without the need for specific functional forms for the basis vectors or the wavefunction. First, consider the fundamental operator equation

$$\hat{Q}\psi = \psi'. \tag{5.97}$$

The most general result of allowing an operator to operate on a wavefunction is that the function is changed into a different one. If $\psi' = \alpha\psi$, with α a constant, then we have the special case in which ψ is an eigenfunction of $\hat{Q}$, but this is *not* the present intention. We now expand both sides of the equation in terms of a basis set of functions ϕ_j, using equation (5.95):

$$\hat{Q} \sum_j a_j\phi_j = \sum_j a_j' \phi_j. \tag{5.98}$$

Note that, again for generality, we will *not* assume the ϕ_j to be eigenfunctions of $\hat{Q}$. We now multiply both sides of the equation on the left by ϕ_k^* and integrate:

$$\int \phi_k^* \hat{Q} \sum_j a_j\phi_j \, \mathrm{d}\tau = \int \phi_k^* \sum_j a_j' \phi_j \, \mathrm{d}\tau. \tag{5.99}$$

Interchanging the order of integration and summation, and using the orthonormality of the basis functions, we obtain

$$\sum_j \int \phi_k^* \hat{Q} \phi_j \, \mathrm{d}\tau \; a_j = \sum_j a_j' \delta_{kj} \tag{5.100}$$

or

$$\sum_j Q_{kj} a_j = a_k' \tag{5.101}$$

where

$$Q_{kj} = \int \phi_k^* \hat{Q} \phi_j \, \mathrm{d}\tau. \tag{5.102}$$

Equation (5.101) can be recognized as the rule for matrix multiplication, for the matrix equation

$$\begin{bmatrix} Q_{11} & Q_{12} & \cdots \\ Q_{21} & Q_{22} & \cdots \\ \vdots & & \ddots \end{bmatrix} \begin{bmatrix} a_1 \\ a_2 \\ \vdots \end{bmatrix} = \begin{bmatrix} a_1' \\ a_2' \\ \vdots \end{bmatrix}. \tag{5.103}$$

The Q_{kj}, defined in equation (5.102), are referred to as the matrix elements of the operator $\hat{Q}$. Clearly their value will depend upon our choice of basis vectors ϕ_j. The two column-matrix state vectors have replaced the wavefunctions ψ and ψ'. The entries in these columns are simply the coefficients of the basis vectors in the expansions of the wavefunctions.

Now let us examine the type of integral expression which plays many roles in quantum mechanics:

$$\int \psi_a^* \psi_b \, \mathrm{d}\tau = \int \sum_j a_j^* \phi_j^* \sum_k b_k \phi_k \, \mathrm{d}\tau = \sum_{j,k} a_j^* b_k \, \delta_{jk} = \sum_j a_j^* b_j$$

$$= [a_1^* a_2^* \quad \ldots] \begin{bmatrix} b_1 \\ b_2 \\ \vdots \end{bmatrix} \qquad = \mathbf{a}^\dagger \mathbf{b}. \tag{5.104}$$

From this we learn that the matrix equivalent of multiplying on the left with ψ^* and integrating is to premultiply by the transposed conjugate of the matrix $\mathbf{a}$ representing ψ. This matrix is referred to as the Hermitian adjoint of $\mathbf{a}$, and denoted by $\mathbf{a}^\dagger$. Matrices representing operators also have Hermitian adjoints. Thus the elements of $\mathbf{Q}^\dagger$ are

$$Q_{kj}^\dagger = Q_{jk}^*. \tag{5.105}$$

All the operators we will meet in this book are Hermitian and so satisfy the condition

$$\int \phi_k^* \hat{Q} \phi_j \, \mathrm{d}\tau = \int \phi_j \hat{Q}^* \phi_k^* \, \mathrm{d}\tau$$

or

$$Q_{kj} = Q_{jk}^* \tag{5.106}$$

and we find that Hermitian operators are represented by matrices which are equal to their own Hermitian adjoints, or 'self-Hermitian-adjoint'.

The diagonal elements of the matrix of $\hat{Q}$ have a special significance. Since

$$Q_{jj} = \int \phi_j^* \hat{Q} \phi_j \, \mathrm{d}\tau = \langle \hat{Q} \rangle \tag{5.107}$$

we see that Q_{jj} is the expectation value of $\hat{Q}$ when the system is in the state ϕ_j. When that is the case, the state vector $\mathbf{a}$ becomes a column vector in which all of the entries are zero except for $a_j = 1$.

If it so happens that the basis set of functions which we choose are actually the eigenfunctions of $\hat{Q}$ then the diagonal elements become the eigenvalues of $\hat{Q}$, and all the other elements are equal to zero:

$$Q_{kj} = \int \phi_k^* \hat{Q} \phi_j \, \mathrm{d}\tau = q_j \, \delta_{kj}. \tag{5.108}$$

In matrix mechanics, therefore, an eigenvalue problem consists of finding those basis functions in which the matrix of the operator is diagonal. The eigenvalues are then the elements of that diagonal. Conversely, if the eigenvalues of an operator are known it is a trivial matter to write down its matrix for the basis consisting of its eigenfunctions whether or not we have explicit expressions for those eigenfunctions. Electron spin operators will provide an example of this process in the next chapter.

A difficulty which arises with matrix mechanics is that many of the operators with which we deal have an infinite number of eigenvalues, so that the matrices are infinite in dimensions. Worse, in many cases the eigenvalues are continuous! Continuous, infinite matrices are not the most convenient entities to work with. Fortunately the only example for which we will need matrices in this book, namely angular momentum, can be dealt with using finite matrices.

〉 Problems

5.1 Given that the operator $\hat{Q}$ is Hermitian, show that $\hat{Q}^n$ is also Hermitian. (Hint: consider first $\hat{Q}^2 = \hat{Q}\hat{Q}$ and perform the Hermitian operation sequentially on the two $\hat{Q}$s.)

5.2 Show that the expectation value of an Hermitian operator is real.

5.3 Given the commutator $[x, \hat{P}_x] = \mathrm{i}\hbar$, show that $[x^n, \hat{P}_x] = \mathrm{i}n\hbar x^{n-1}$. Similarly show that $[x, \hat{P}_x^n] = \mathrm{i}n\hbar \hat{P}_x^{n-1}$.

Hence show that if $f(x)$ can be expressed as a polynomial in x, then

$$[f(x), \hat{P}_x] = \mathrm{i}\hbar \frac{\mathrm{d}}{\mathrm{d}x} f(x)$$

and if $F(\hat{P}_x)$ can be expressed as a polynomial in $\hat{P}_x$, then

$$[x, F(\hat{P}_x)] = \mathrm{i}\hbar \frac{\mathrm{d}}{\mathrm{d}\hat{P}_x} F(\hat{P}_x).$$

5.4 If Q is an Hermitian operator and u_i is an eigenfunction of $\hat{H}$ with energy E_i, show that

$$\int u_n^* [\hat{H}, \hat{Q}] u_m \,\mathrm{d}\tau = (E_n - E_m) \int u_n^* \hat{Q} u_m \,\mathrm{d}\tau.$$

Hence show that for a particle with the Hamiltonian

$\hat{P}_x^2/2m + V(x)$

$$\int u_n^* \hat{P}_x u_m \, d\tau = \frac{im}{\hbar} (E_n - E_m) \int u_n^* \hat{x} u_m \, d\tau.$$

(Hint: one of the results of problem 3 may be useful here.)

5.5 Show that, even though the operators $\hat{Q}$ and $\hat{R}$ are Hermitian, the combined operator $\hat{Q}\hat{R}$ is only Hermitian provided $[\hat{Q}, \hat{R}] = 0$.
Show that $\frac{1}{2}(\hat{Q}\hat{R} + \hat{R}\hat{Q})$ is Hermitian even if $[\hat{Q}, \hat{R}] \neq 0$. (This device is used when it is desired to calculate quantities such as $\overline{xp_x}$. In general, $\langle x\hat{P}_x \rangle \neq \langle \hat{P}_x x \rangle$, and neither is real. However, $\frac{1}{2}\langle xP_x + P_x x \rangle$ *is* real, and also unambiguous.)

5.6 Demonstrate that the energy eigenfunctions u_1, u_2 and u_3 for the infinite square well, given in equation (3.20), are orthogonal. (Hint: show that functions of opposite parity are automatically orthogonal and exploit this result where possible. Explicit integration should be used only as a last resort!)

5.7 A particle is confined in an infinite square well of length L. Its wavefunction is

$$\psi = Ax^2(x - L) \qquad 0 \leqslant x \leqslant L$$
$$\psi = 0 \qquad x > L, x < 0.$$

What are the probabilities that a measurement of its energy would yield the $n = 1, 2$ eigenvalues? What is the expectation value of the energy? Compare this with the $n = 1, 2$ eigenvalues of the infinite square well. Sketch the given function and the $n = 1, 2$ eigenfunctions and try to rationalize your results.

5.8 For any inverse square law classical orbit $\bar{T} = -\frac{1}{2}\bar{V}$, where $\bar{T}$ and $\bar{V}$ are the mean values of the kinetic and potential energies of the particle. The equivalent quantum mechanical result is $\langle \hat{P}^2/2\mu \rangle = -\frac{1}{2}\langle V(r) \rangle$. Show that this is true for the ground state of the hydrogen atom.

5.9 The electron of a hydrogen atom has the wavefunction $\psi = A \exp(-2r/3a_0)$. What is the probability that a measurement of its energy would yield the ground-state value? What is the expectation value of the energy?

5.10 A particle of mass m moves on the x axis in a potential well $V(x)=\frac{1}{2}m\omega^2x^2$. Initially the particle is in the lowest quantum state of the well. As a result of an external influence the potential abruptly changes to $\frac{1}{2}m\omega'^2x^2$.

Find expressions (in terms of $\alpha=\sqrt{m\omega/\hbar}$ and $\alpha'=\sqrt{m\omega'/\hbar}$) for the following quantities:

(a) The probability that after the potential change the particle will be found in the lowest quantum state of the new well.
(b) The probability that after the potential change the particle will be found in the first excited state of the new well.
(c) The expectation value of the energy of the particle after the change.

In each case evaluate your expressions for the case where $\omega'^2=\frac{1}{2}\omega^2$.

5.11 A particle is known to be in the ground state of a potential well with infinite barriers at $x=\pm a$. The barriers are abruptly moved to $x=\pm(x+\delta)$. Show that a measurement of the particle's energy would show it to be in the ground state of the modified well with a probability given by

$$P=\frac{4a(a+\delta)}{\pi^2}\left(\frac{1}{\delta}\sin\frac{\pi\delta}{2(a+\delta)}+\frac{1}{(2a+\delta)}\sin\frac{\pi(2a+\delta)}{2(a+\delta)}\right)^2.$$

What is the probability of finding the particle in the first excited state of the modified well?

5.12 A particle is known to have the wavefunction

$$\psi(x)=a^{-1/2}\cos n\pi x/2a \qquad |x|\leqslant a$$
$$\psi(x)=0 \qquad |x|>a$$

where $n=1,3,5,\ldots$.

Show that the probability that a measurement of momentum will result in a value lying between $\hbar k$ and $\hbar(k+\mathrm{d}k)$ is given by

$$P(k)\,\mathrm{d}k=\frac{\pi}{2a^3}\frac{\cos^2 ka\,\mathrm{d}k}{(n^2\pi^2/4a^2-k^2)^2}.$$

Sketch this distribution, showing how it depends upon the value of n. (Hint/discussion: these wavefunctions are of course the even-parity energy eigenfunctions of the infinite square well. Inasmuch as we can express them as

$$\cos\frac{n\pi x}{2a} = \tfrac{1}{2}[\exp(\mathrm{i}n\pi x/2a) + \exp(-\mathrm{i}n\pi x/2a)]$$

which seems to be a sum of two momentum eigenfunctions, at first sight it appears that a measurement of momentum must yield $p = \hbar k = \pm n\pi\hbar/a$ with equal probabilities. However, they are *not* momentum eigenfunctions because they are truncated, being zero outside the range $\pm a$. Calculation of the momentum distribution requires Fourier transformation (see sections 5.3 and 5.6). Examination of the result should reveal that as n increases the distribution approaches the classical one of two δ functions, located at $\hbar k = \pm n\pi\hbar/a$, because the waves become better approximations of the untruncated momentum eigenfunctions.)

〉Chapter 6

〉Angular Momentum

〉6.1 Introduction

Angular momentum plays a crucial role in quantum mechanics, just as it does in classical mechanics, and for the same reason. Like energy and momentum it is a conserved quantity for an isolated system. The total energy plays an important role in the analysis of any problem in dynamics. Linear momentum, whilst important in scattering problems, does not usually play much part in the analysis of bound systems, such as atoms and molecules, since we nearly always employ a reference frame in which the structure is at rest, and so the momenta of its component parts are zero on average. This is not the case for angular momentum, and knowledge of the angular momentum components of an atom or molecule is vital in characterizing its quantum state.

If the position of a particle, such as an atomic electron, is described by $\boldsymbol{r}$, and its momentum by $\boldsymbol{P}$ (figure 6.1) then, according to classical mechanics, its angular momentum is $\boldsymbol{L}$, where

$$\boldsymbol{L} = \boldsymbol{r} \times \boldsymbol{P}. \tag{6.1}$$

$\boldsymbol{L}$ satisfies the angular form of Newton's second law of motion

$$\mathrm{d}\boldsymbol{L}/\mathrm{d}t = \boldsymbol{r} \times \boldsymbol{F} = \boldsymbol{\Gamma} \tag{6.2}$$

where $\boldsymbol{\Gamma}$ is the torque, or moment of force, acting on the particle. Both the torque and the angular momentum contain $\boldsymbol{r}$, which requires an origin. In the case of an atom this may conveniently be taken as the position of the nucleus or, more precisely, the position of the centre of mass of the system.

Equation (6.2) makes it obvious that, in the absence of any torque, $\boldsymbol{L}$ will be a constant of the motion. In many cases of interest there *is* no torque, the most important example being that of a particle moving under the action of a central force. An electron moving in the Coulomb field of a nucleus is just such a case. From equation (6.2) we can see that if $\boldsymbol{F}$ lies along $\boldsymbol{r}$ then $\boldsymbol{\Gamma} = 0$, and the orbital angular momentum of the electron about the centre of mass of the system will be a 'constant of the motion'.

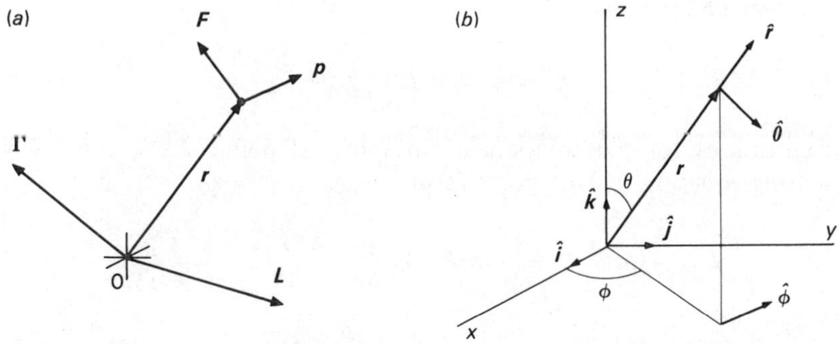

Figure 6.1 (*a*) The angular momentum of a particle about a point, and associated vectors. (*b*) Relations between cartesian and spherical polar unit vectors.

〉6.2 Angular momentum operators

According to postulate number 4 (Appendix 1), the operator for $\boldsymbol{L}$ will be

$$\hat{\boldsymbol{L}} = -\mathrm{i}\hbar \boldsymbol{r} \times \nabla \tag{6.3}$$

which is obtained from equation (6.1) by replacing $\boldsymbol{r}$ and $\boldsymbol{P}$ by their own operators. In practice one seldom works with $\hat{\boldsymbol{L}}$ itself, but rather with the cartesian components $\hat{L}_x$, $\hat{L}_y$ and $\hat{L}_z$. We shall require expressions for these in terms of both cartesian and spherical polar coordinates. Writing equation (6.1) in determinant form we have

$$\boldsymbol{L} = \begin{bmatrix} \hat{\imath} & \hat{\jmath} & \hat{k} \\ x & y & z \\ \hat{P}_x & \hat{P}_y & \hat{P}_z \end{bmatrix}$$

where $\hat{\boldsymbol{i}}$, $\hat{\boldsymbol{j}}$ and $\hat{\boldsymbol{k}}$ are cartesian unit vectors. Hence

$$\begin{aligned}
\hat{L}_x &= (y\hat{P}_z - z\hat{P}_y) = -\mathrm{i}\hbar\left(y\frac{\partial}{\partial z} - z\frac{\partial}{\partial y}\right)\\
\hat{L}_y &= (z\hat{P}_x - x\hat{P}_z) = -\mathrm{i}\hbar\left(z\frac{\partial}{\partial x} - x\frac{\partial}{\partial z}\right)\\
\hat{L}_z &= (x\hat{P}_y - y\hat{P}_x) = -\mathrm{i}\hbar\left(x\frac{\partial}{\partial y} - y\frac{\partial}{\partial x}\right).
\end{aligned}\tag{6.4}$$

Note in addition that

$$\hat{L}^2 = \hat{\boldsymbol{L}}\cdot\hat{\boldsymbol{L}} = \hat{L}_x^2 + \hat{L}_y^2 + \hat{L}_z^2. \tag{6.5}$$

Expressions for the components in spherical polars are obtained by writing equation (6.3) in polar form:

$$\begin{aligned}
\hat{\boldsymbol{L}} = -\mathrm{i}\hbar\boldsymbol{r}\times\boldsymbol{\nabla} &= -\mathrm{i}\hbar r\hat{\boldsymbol{r}}\times\left(\hat{\boldsymbol{r}}\frac{\partial}{\partial r} + \frac{\hat{\boldsymbol{\theta}}}{r}\frac{\partial}{\partial\theta} + \frac{\hat{\boldsymbol{\phi}}}{r\sin\theta}\frac{\partial}{\partial\phi}\right)\\
&= -\mathrm{i}\hbar\left(\hat{\boldsymbol{\phi}}\frac{\partial}{\partial\theta} - \frac{\hat{\boldsymbol{\theta}}}{\sin\theta}\frac{\partial}{\partial\phi}\right).
\end{aligned}\tag{6.6}$$

The symbols $\hat{\boldsymbol{r}}$, $\hat{\boldsymbol{\theta}}$ and $\hat{\boldsymbol{\phi}}$ denote unit vectors in the spherical polar coordinate system. (NB $\hat{\boldsymbol{r}}$ does *not* represent the operator for r, which we write simply as $\boldsymbol{r}$.) We will need the following relations between the unit vectors of the two coordinate systems (see figure 6.1(*b*)):

$$\begin{aligned}
\hat{\boldsymbol{i}} &= \hat{\boldsymbol{r}}\sin\theta\cos\phi + \hat{\boldsymbol{\theta}}\cos\theta\cos\phi - \hat{\boldsymbol{\phi}}\sin\phi\\
\hat{\boldsymbol{j}} &= \hat{\boldsymbol{r}}\sin\theta\sin\phi + \hat{\boldsymbol{\theta}}\cos\theta\sin\phi + \hat{\boldsymbol{\phi}}\cos\phi\\
\hat{\boldsymbol{k}} &= \hat{\boldsymbol{r}}\cos\theta - \hat{\boldsymbol{\theta}}\sin\theta
\end{aligned}\tag{6.7}$$

and the inverse relations

$$\begin{aligned}
\hat{\boldsymbol{r}} &= \hat{\boldsymbol{i}}\cos\phi\sin\theta + \hat{\boldsymbol{j}}\sin\phi\sin\theta + \hat{\boldsymbol{k}}\cos\theta\\
\hat{\boldsymbol{\theta}} &= \hat{\boldsymbol{i}}\cos\phi\cos\theta + \hat{\boldsymbol{j}}\sin\phi\cos\theta - \hat{\boldsymbol{k}}\sin\theta\\
\hat{\boldsymbol{\phi}} &= -\hat{\boldsymbol{i}}\sin\phi + \hat{\boldsymbol{j}}\cos\phi.
\end{aligned}\tag{6.8}$$

It follows immediately from equations (6.6) and (6.7) that

$$\hat{L}_x = \hat{\boldsymbol{i}} \cdot \hat{\boldsymbol{L}} = \mathrm{i}\hbar\left(\sin\phi \frac{\partial}{\partial\theta} + \cot\theta\cos\phi\frac{\partial}{\partial\phi}\right)$$

$$\hat{L}_y = \hat{\boldsymbol{j}} \cdot \hat{\boldsymbol{L}} = -\mathrm{i}\hbar\left(\cos\phi \frac{\partial}{\partial\theta} - \cot\theta\sin\phi\frac{\partial}{\partial\phi}\right) \tag{6.9}$$

$$\hat{L}_z = \hat{\boldsymbol{k}} \cdot \hat{\boldsymbol{L}} = -\mathrm{i}\hbar\frac{\partial}{\partial\phi}.$$

An expression for $\hat{L}^2 = \hat{\boldsymbol{L}} \cdot \hat{\boldsymbol{L}}$ can be obtained from equation (6.6), but this will require the differentiation of the unit vectors $\boldsymbol{\theta}$ and $\hat{\boldsymbol{\phi}}$ with respect to θ and ϕ. The expressions we need can be obtained by differentiating equations (6.8), bearing in mind that $\hat{\boldsymbol{i}}$, $\hat{\boldsymbol{j}}$ and $\hat{\boldsymbol{k}}$ are constant vectors. We find that

$$\frac{\partial\hat{\boldsymbol{\theta}}}{\partial\theta} = -\hat{\boldsymbol{r}} \qquad \frac{\partial\hat{\boldsymbol{\theta}}}{\partial\phi} = \hat{\boldsymbol{\phi}}\cos\theta$$

$$\frac{\partial\hat{\boldsymbol{\phi}}}{\partial\theta} = 0 \qquad \frac{\partial\hat{\boldsymbol{\phi}}}{\partial\phi} = -\hat{\boldsymbol{r}}\sin\theta - \hat{\boldsymbol{\theta}}\cos\theta. \tag{6.10}$$

Hence

$$\hat{L}^2 = -\hbar^2\left(\hat{\boldsymbol{\phi}}\frac{\partial}{\partial\theta} - \frac{\hat{\boldsymbol{\theta}}}{\sin\theta}\frac{\partial}{\partial\phi}\right) \cdot \left(\hat{\boldsymbol{\phi}}\frac{\partial}{\partial\theta} - \frac{\hat{\boldsymbol{\theta}}}{\sin\theta}\frac{\partial}{\partial\phi}\right)$$

$$= -\hbar^2\left[\frac{1}{\sin\theta}\frac{\partial}{\partial\theta}\left(\sin\theta\frac{\partial}{\partial\theta}\right) + \frac{1}{\sin^2\theta}\frac{\partial^2}{\partial\phi^2}\right]. \tag{6.11}$$

〉6.3 Angular momentum commutators

It will emerge that the components of angular momentum are not compatible observables. In order to show this we must calculate the commutators of the components of $\hat{\boldsymbol{L}}$ and $\hat{L}^2$. These are most easily found using the cartesian forms of these operators expressed in equation (6.4). Consider the commutator

$$[\hat{L}_x, \hat{L}_y] = \hat{L}_x\hat{L}_y - \hat{L}_y\hat{L}_x$$

$$= (y\hat{P}_z - z\hat{P}_y)(z\hat{P}_x - x\hat{P}_z) - (z\hat{P}_x - x\hat{P}_z)(y\hat{P}_z - z\hat{P}_y). \tag{6.12}$$

In expanding the products in equation (6.12) it is essential to preserve the order of non-commuting factors, such as x and $\hat{P}_x$. Some writing can be saved by noting that in the first product the term $z\hat{P}_y z\hat{P}_x$ is cancelled by $z\hat{P}_x z\hat{P}_y$ from the second product, since in this case all factors commute. A similar cancellation occurs for the terms $y\hat{P}_z x\hat{P}_z$. The remaining four terms factorize to give

$$\begin{aligned}[\hat{L}_x, \hat{L}_y] &= (\hat{P}_z z - z\hat{P}_z)(y\hat{P}_x - x\hat{P}_y) \\ &= [\hat{P}_z, z](-\hat{L}_z) \\ &= \mathrm{i}\hbar\hat{L}_z \end{aligned} \tag{6.13}$$

using equation (5.74) in the last step. The commutators of the other components are obtained from this by cyclic permutation so that

$$\begin{aligned}[\hat{L}_x, \hat{L}_y] &= \mathrm{i}\hbar\hat{L}_z \\ [\hat{L}_y, \hat{L}_z] &= \mathrm{i}\hbar\hat{L}_x \\ [\hat{L}_z, \hat{L}_x] &= \mathrm{i}\hbar\hat{L}_y. \end{aligned} \tag{6.14}$$

We will also require the commutators $[\hat{L}^2, \hat{L}_z]$, etc. Now

$$[\hat{L}^2, \hat{L}_z] = [\hat{L}_x^2, \hat{L}_z] + [\hat{L}_y^2, \hat{L}_z] + [\hat{L}_z^2, \hat{L}_z] \tag{6.15}$$

in which the last term is zero. This is most easily evaluated by noting that

$$[\hat{L}_x^2, \hat{L}_z] = \hat{L}_x[\hat{L}_x, \hat{L}_z] + [\hat{L}_x, \hat{L}_z]\hat{L}_x \qquad \text{etc} \tag{6.16}$$

as can be verified by expanding the commutators. Hence, inserting the commutator values from equation (6.14), we find

$$[\hat{L}^2, \hat{L}_z] = \mathrm{i}\hbar[-\hat{L}_x\hat{L}_y - \hat{L}_y\hat{L}_x + \hat{L}_y\hat{L}_x + \hat{L}_x\hat{L}_y] = 0. \tag{6.17}$$

Similarly $[\hat{L}^2, \hat{L}_x] = [\hat{L}^2, \hat{L}_y] = 0$. The commutators (6.14) and (6.17) and the theorems proved in section 5.9 tell us that $\hat{L}^2$ and any one component $\hat{L}_x$, $\hat{L}_y$, or $\hat{L}_z$ are compatible observables, and equivalently that it must be possible to find simultaneous eigenfunctions of $\hat{L}^2$ and $\hat{L}_x$, or $\hat{L}^2$ and $\hat{L}_y$, or $\hat{L}^2$ and $\hat{L}_z$, but not of $\hat{L}_x$ and $\hat{L}_y$ or $\hat{L}_y$ and $\hat{L}_z$ or $\hat{L}_z$ and $\hat{L}_x$. We will see later that this is another manifestation of the uncertainty principle.

〉6.4 Angular momentum eigenvalues and eigenfunctions

As we have seen, the angular momentum of a particle will be constant if it moves under the action of a central force. A central force can be represented by a potential function $V(r)$ which depends only upon r, the distance from the centre of force. The TISE for such a system is then

$$\hat{H}u(r,\theta,\phi) = Eu(r,\theta,\phi) \tag{6.18}$$

where the Hamiltonian operator is

$$\begin{aligned}\hat{H} &= -\frac{\hbar^2}{2m}\nabla^2 + V(r)\\ &= -\frac{\hbar^2}{2m}\left[\frac{1}{r^2}\frac{\partial}{\partial r}r^2\frac{\partial}{\partial r} + \frac{1}{r^2\sin\theta}\frac{\partial}{\partial\theta}\left(\sin\theta\frac{\partial}{\partial\theta}\right) + \frac{1}{r^2\sin^2\theta}\frac{\partial^2}{\partial\phi^2}\right] + V(r).\end{aligned} \tag{6.19}$$

Examining this expression we note that the angular part of it is proportional to the operator $\hat{L}^2$, equation (6.11). We can therefore write the Hamiltonian (6.19) as

$$\hat{H} = -\frac{\hbar^2}{2mr^2}\frac{\partial}{\partial r}r^2\frac{\partial}{\partial r} + \frac{\hat{L}^2}{2mr^2} + V(r). \tag{6.20}$$

From this it is apparent that the Hamiltonian will commute with $\hat{L}^2$, and $\hat{L}_x$, $\hat{L}_y$ and $\hat{L}_z$, since the angular momentum operators are functions of θ and ϕ only, and the Hamiltonian consists of terms in r only apart from the term in $\hat{L}^2$, which will of course commute with any of the angular momentum operators. Ehrenfest's theorem (equation (5.78)) now tells us that the components of angular momentum and its magnitude will be constants of the motion. We also know that we can expect to find simultaneous eigenfunctions of, say, $\hat{H}$, $\hat{L}^2$ and $\hat{L}_z$. (There is no special physical significance attached to the z axis, but the polar coordinate system endows $\hat{L}_z$ with the simplest operator, so it is convenient to choose this set of commuting operators.) We therefore seek a set of simultaneous eigenfunctions which satisfy the following eigenvalue equations. In writing them we anticipate that we will be able to express the eigenfunctions as products of radial and angular factors:

$$u(r,\theta,\phi) = R(r)\,Y(\theta,\phi). \tag{6.21}$$

The eigenvalue equations are as follows. For energy,

$$\hat{H}u(r,\theta,\phi) = Eu(r,\theta,\phi). \tag{6.22}$$

For squared magnitude of angular momentum,

$$\hat{L}^2\,Y(\theta,\phi) = \lambda\hbar^2\,Y(\theta,\phi). \tag{6.23}$$

For the z component of angular momentum

$$\hat{L}_z Y(\theta, \phi) = m\hbar Y(\theta, \phi). \tag{6.24}$$

The factor $R(r)$ has cancelled in equations (6.23) and (6.24) since the operators are purely angular. The eigenvalues of $\hat{L}^2$ and $\hat{L}_z$ have been written as $\lambda\hbar^2$ and $m\hbar$, respectively.

Since the operator $\hat{L}_z$ depends solely upon ϕ, equation (6.24) will be satisfied by a function Y of the form

$$Y(\theta, \phi) = \Theta(\theta)\Phi(\phi) \tag{6.25}$$

provided that

$$\hat{L}_z\Phi = m\hbar\Phi \tag{6.26}$$

or

$$-\mathrm{i}\hbar \frac{\mathrm{d}\Phi}{\mathrm{d}\phi} = m\hbar\Phi. \tag{6.27}$$

This equation can be solved immediately and we find

$$\Phi = N_\phi \exp(\mathrm{i}m\phi) \tag{6.28}$$

where N_ϕ is a normalization constant to be calculated shortly. The eigenvalues $m\hbar$ are determined by the requirement that the wavefunction be single valued. Recalling the nature of the polar coordinate we realize that ϕ and $(\phi + 2\pi)$ represent the same angle and so we require that

$$N_\phi \exp(\mathrm{i}m\phi) = N_\phi \exp[\mathrm{i}m(\phi + 2\pi)]$$

or

$$\exp(2\mathrm{i}m\pi) = 1. \tag{6.29}$$

This in turn requires that m is integral, so that the eigenvalues of the operator $\hat{L}_z$ are

$$L_z = m\hbar = 0, \pm\hbar, \pm 2\hbar, \pm 3\hbar, \ldots. \tag{6.30}$$

We note in passing that this argument is flawed, since in fact it would suffice to require only measurable quantities such as the probability density and expectation values in general to be single valued, the wavefunction itself not being observable. Since both the wavefunction and its conjugate appear in these quantities measurable values will be unchanged provided that an increment of 2π in ϕ leaves the wavefunction unchanged apart from a phase factor. For example, it would be acceptable if

$$\exp(im\phi) = \pm\exp[im(\phi + 2\pi)] \tag{6.31}$$

in which case it is easy to see that half-integral values of m would also be permissible, and the wavefunction would be double valued. Nevertheless, it seems that nature does not avail herself of this opportunity and orbital angular momentum, which we are considering here, invariably exhibits eigenvalues which are integral multiples of $\hbar$. Half-integral values do occur when particle spin is considered, and then the wavefunction, although not expressible in terms of ordinary angular variables, is indeed double valued. Spin will be considered in section 6.9. A more sophisticated treatment of angular momentum based upon the properties of the angular momentum commutators reveals that integral and half-integral values exhaust the possible values for this quantity. Analytic proof that orbital angular momentum is restricted to integral values, though possible, is beyond the scope of this book. See for example Merzbacher (1962).

We have factored the wavefunction into a product of functions of r, θ and ϕ, and it is conventional to normalize each factor independently. Accordingly we require

$$\int_0^{2\pi} \Phi^*\Phi \, \mathrm{d}\phi = N_\phi^2 \int_0^{2\pi} \mathrm{d}\phi = 2\pi N_\phi^2 = 1. \tag{6.32}$$

Hence

$$\Phi = \frac{1}{\sqrt{2\pi}} \mathrm{e}^{im\phi}. \tag{6.33}$$

Equations (6.30) and (6.33) are respectively the eigenvalues and normalized eigenfunctions of the operator $\hat{L}_z$ representing the z component of angular momentum for a particle moving in a central field. Before we can usefully discuss their interpretation we must also solve equation (6.23), the eigenvalue equation for $\hat{L}^2$, the operator representing the squared magnitude of the angular momentum. Inserting first the factored wavefunction (6.25) and then the explicit form of the operator given in equation (6.11), equation (6.23) becomes

$$\hat{L}^2\Theta(\theta)\frac{1}{\sqrt{2\pi}}\mathrm{e}^{im\phi} = \lambda\hbar^2\Theta(\phi)\frac{1}{\sqrt{2\pi}}\mathrm{e}^{im\phi} \tag{6.34}$$

$$-\hbar^2\left[\frac{1}{\sin\theta}\frac{\partial}{\partial\theta}\left(\sin\theta\frac{\partial}{\partial\theta}\right) + \frac{1}{\sin^2\theta}\frac{\partial^2}{\partial\phi^2}\right]\Theta(\theta)\mathrm{e}^{im\phi} = \lambda\hbar^2\Theta(\theta)\mathrm{e}^{im\phi}. \tag{6.35}$$

Performing the indicated differentiation with respect to ϕ leads to

$$\left[\frac{1}{\sin\theta}\frac{\partial}{\partial\theta}\left(\sin\theta\frac{\partial}{\partial\theta}\right)-\frac{m^2}{\sin^2\theta}\right]\Theta(\theta)=-\lambda\Theta(\theta). \tag{6.36}$$

This equation is simplified by a change of variable, so that

$$\Theta(\theta)\Rightarrow P(v) \tag{6.37}$$

where

$$v=\cos\theta. \tag{6.38}$$

The resulting equation is

$$\frac{\mathrm{d}}{\mathrm{d}v}\left((1-v^2)\frac{\mathrm{d}P(v)}{\mathrm{d}v}\right)+\left(\lambda-\frac{m^2}{1-v^2}\right)P(v)=0 \tag{6.39}$$

which may be recognized as the associated Legendre equation. (Setting $m=0$ reduces it to Legendre's equation itself.) Solution follows the now familiar technique used in solving Hermite's and Laguerre's equations in earlier chapters. On examining equation (6.39) we can see immediately that difficulties may be anticipated in the regions $v\to\pm1$, $(\theta\to0,\ \pi)$, since the coefficients in the equation exhibit singularities there. On studying the behaviour of the equation near these points a suitable series solution is found to be

$$P(v)=(1-v^2)^{|m|/2}\sum_0^\infty a_r v^r \tag{6.40}$$

and this leads to the recurrence relation

$$\frac{a_{r+2}}{a_r}=\left(\frac{(r+|m|)(r+|m|+1)-\lambda}{(r+1)(r+2)}\right). \tag{6.41}$$

However this leads to a solution which persists in diverging at $v=\pm1$ unless the series terminates. From equation (6.41) it can be seen that truncation will occur after the rth term if

$$\lambda=(r+|m|)(r+|m|+1)=l(l+1) \tag{6.42}$$

where

$$l=r+|m|. \tag{6.43}$$

Hence

$$l\geqslant|m| \tag{6.44}$$

and

$$l - |m| = 0, 1, 2, 3, \ldots. \qquad (6.45)$$

Equation (6.45) shows that l must be integral since m is integral, as discussed above. We have shown therefore that the eigenvalues of the operator $\hat{L}^2$ are

$$\lambda\hbar^2 = l(l+1)\hbar^2 \qquad l = 0, 1, 2, 3, \ldots. \qquad (6.46)$$

Eigenstates for which $l = 0, 1, 2, 3, 4, 5, \ldots$ are referred to as s, p, d, f, g, h, ... states, following long-standing spectroscopic practice. The functions which satisfy the associated Legendre equation and remain finite throughout the range $-1 \leqslant v \leqslant 1$, $(0 \leqslant \theta \leqslant \pi)$, are the associated Legendre functions, $P_l^{|m|}(v)$. They are independent of the sign of m since only m^2 appears in the differential equation. They can be found from equations (6.40) and (6.41), or alternatively from the Rodrigues formula,

$$P_l^{|m|}(v) = \frac{(-1)^{|m|}}{2^l l!} \frac{(l+|m|)!}{(l-|m|)!} (1-v^2)^{-|m|} \frac{d^{l-|m|}}{dv^{l-|m|}} (v^2-1)^l. \qquad (6.47)$$

The function $\Theta(\theta)$ (equation (6.25)) differs from $P_l^{|m|}(\cos\theta)$ only in normalization constant and is given by

$$\Theta_l^{|m|}(\theta) = \left(\frac{2l+1}{2} \frac{(l-|m|)!}{(l+|m|)!}\right)^{1/2} P_l^{|m|}(\cos\theta). \qquad (6.48)$$

These functions are normalized so that

$$\int_0^\infty |\Theta_l^{|m|}(\theta)|^2 \sin\theta \, d\theta = 1. \qquad (6.49)$$

The complete angular part of the wavefunction, $Y_l^{|m|}(\theta, \phi)$ (equation (6.25)), for the first few allowed values of l and m is provided in table 6.1. These functions, which appear in this context as the eigenfunctions of orbital angular momentum, crop up in many other branches of physics, and are more generally known as the spherical harmonics. Combining equations (6.25), (6.33) and (6.49) we have

$$Y_l^m(\theta, \phi) = \left(\frac{2l+1}{4\pi} \frac{(l+|m|)!}{(l-|m|)!}\right)^{1/2} \left(\frac{-m}{|m|}\right)^{|m|} e^{im\phi} P_l^{|m|}(\cos\theta) \qquad (6.50)$$

where we have introduced the customary sign convention. The spherical harmonics are mutually orthogonal, and normalized so that

$$\int_0^\pi \int_0^{2\pi} Y_{l'}^{m'*}(\theta, \phi) Y_l^m(\theta, \phi) \sin\theta \, d\theta \, d\phi = \delta_{l'l}\delta_{m'm}. \qquad (6.51)$$

Table 6.1 Normalized spherical harmonics for $l = 0, 1, 2$.

$$Y_0^0 = \left(\frac{1}{4\pi}\right)^{1/2}$$

$$Y_1^1 = -\left(\frac{3}{8\pi}\right)^{1/2} \sin\theta\, e^{i\phi}$$

$$Y_1^0 = \left(\frac{3}{4\pi}\right)^{1/2} \cos\theta$$

$$Y_1^{-1} = \left(\frac{3}{8\pi}\right)^{1/2} \sin\theta\, e^{-i\phi}$$

$$Y_2^2 = \left(\frac{15}{32\pi}\right)^{1/2} \sin^2\theta\, e^{2i\phi}$$

$$Y_2^1 = -\left(\frac{15}{8\pi}\right)^{1/2} \cos\theta \sin\theta\, e^{i\phi}$$

$$Y_2^0 = \left(\frac{5}{16\pi}\right)^{1/2} (3\cos^2\theta - 1)$$

$$Y_2^{-1} = \left(\frac{15}{8\pi}\right)^{1/2} \cos\theta \sin\theta\, e^{-i\phi}$$

$$Y_2^{-2} = \left(\frac{15}{32\pi}\right)^{1/2} \sin^2\theta\, e^{-2i\phi}$$

It is important for many purposes to know the parity of the spherical harmonics, and we can establish this as follows. The parity operation of reflection through the origin ($\boldsymbol{r} \to -\boldsymbol{r}$) is described in spherical polars by the operations

$$r \to r \qquad \theta \to \pi - \theta \qquad \phi \to \phi + \pi. \tag{6.52}$$

Hence

$$v = \cos\theta \to -v \tag{6.53}$$

and equation (6.47) reveals that

$$P_l^{|m|}(v) \to (-1)^{l-|m|} P_l^{|m|}(v). \tag{6.54}$$

However,

$$e^{im\phi} \to e^{im\pi} e^{im\phi} = (-1)^{|m|} e^{im\phi} \tag{6.55}$$

and so, combining the last two results (and recalling equations (6.25) and (6.37)) we find, finally

$$Y_l^{|m|} \to (-1)^l Y_l^{|m|} \tag{6.56}$$

and the parity of the spherical harmonics is found to depend entirely upon the value of l.

〉6.5 Interpretation of angular momentum eigenvalues

We have shown that the magnitude and z component of the orbital angular momentum vector are each quantized and have the following eigenvalues:

$$|\boldsymbol{L}| = \sqrt{l(l+1)}\hbar \qquad l = 1, 2, 3, \ldots \tag{6.57}$$

$$L_z = m\hbar \qquad m = \pm l, \pm(l-1), \ldots, 0. \tag{6.58}$$

We can interpret this as follows. Knowledge only of the magnitude $|\boldsymbol{L}|$ of a vector tells us nothing at all about its direction in space. Knowledge of its magnitude and one cartesian component, L_z say, restricts the vector to those which are the generators of a cone, as shown in figure 6.2(a). Clearly all vectors which start at the origin and whose tips lie on the circular rim of the cone will have a common magnitude and a common z component. But the commutation relations discussed earlier tell us that this amount of information is the most that we are permitted to know about the angular momentum, in the sense that $\hat{L}^2$ and $\hat{L}_z$ are compatible observables but $\hat{L}_x$, $\hat{L}_y$ and $\hat{L}_z$ are not. When $\hat{L}^2$ and $\hat{L}_z$ have been measured, the results of subsequent measurements of $\hat{L}_x$ or $\hat{L}_y$ are unpredictable, just as they would be if we knew only that the vector lay somewhere on the cone, but had no idea of its azimuthal angle. The fact that, for a given l, m can take integral values between l and $-l$ tells us that the cone may take any of $2l+1$ different vertical angles, as shown in figure 6.2(b), which is drawn for the case $l = 3$. This curious state of affairs is referred to as space quantization since it seems to imply that only certain directions in space (as well as only certain values) are permitted for the angular momentum vector $\hat{\boldsymbol{L}}$. This startling result becomes more so when we remember that our choice of axes was arbitrary, and we are considering a system of spherical symmetry. However, if we recall the role of eigenvalues as the only possible result of measurements (postulate 3) and realize that any experiment to measure a

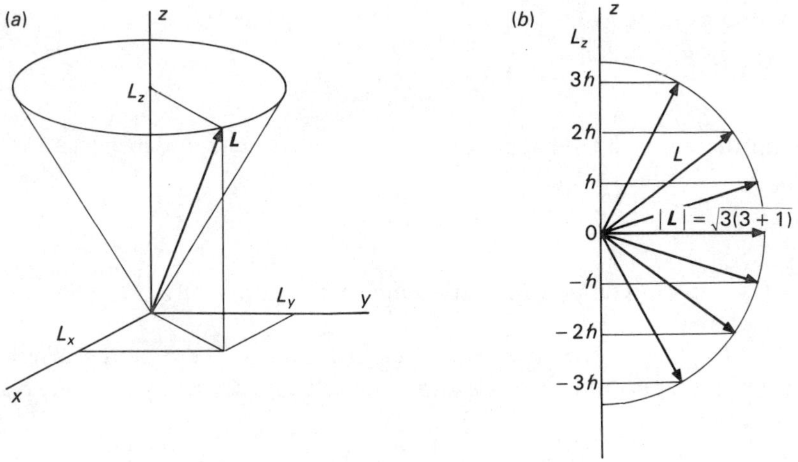

Figure 6.2 (a) $|L|$ and L_z determine the cone upon which L must lie. (b) Space quantization. The orientations of L for each of the permitted values of L_z. Drawn for $l = 3$. (For each polar angle shown all azimuthal angles are equally likely, as in (a).)

component of angular momentum must, by its very nature, single out a direction in space, then the physical picture becomes less paradoxical. We can think of the analysis in the previous section as a calculation to provide predictions for the possible results of experiments to measure $|\hat{L}|$ and $\hat{L}_z$, where z is a direction we have predetermined for the orientation of our angular momentum measuring instrument. (The instrument most commonly used is a Stern–Gerlach apparatus, in which the orientation is defined by a magnetic field. Its use will be discussed shortly.) The point is that, whatever orientation we choose, the possible results will always be $m\hbar$, where $m = \pm l, \pm(l-1), \ldots, 0$, and for convenience of analysis we may as well define that direction as the z direction. If, after measuring the value of $\hat{L}_z$, we choose to make a further measurement, of say $\hat{L}_x$ or $\hat{L}_y$ (or indeed a component of quite arbitrary orientation), the result will be unpredictable, except that we can be sure it will again be one of the values $m\hbar$, as above. The frequency with which these values appear in an ensemble measurement is not random, however, since we know from equation (6.5) that

$$\hat{L}_x^2 + \hat{L}_y^2 = \hat{L}^2 - \hat{L}_z^2 = l(l+1)\hbar^2 - m^2\hbar^2.$$

Consequently the expectation value $\langle \hat{L}_x^2 + \hat{L}_y^2 \rangle$ must be given by

$$\langle \hat{L}_x^2 + \hat{L}_y^2 \rangle = l(l+1)\hbar^2 - m^2\hbar^2 \qquad (6.59)$$

and by symmetry

$$\langle \hat{L}_x^2 \rangle = \langle \hat{L}_y^2 \rangle = \tfrac{1}{2}\,[l(l+1)\hbar^2 - m^2\hbar^2]. \tag{6.60}$$

On the other hand, if we measure $\hat{L}_z$ and then immediately repeat the measurement, we will get the same answer, since after the first measurement the value is known with certainty and the wavefunction becomes the corresponding eigenfunction.

Since the maximum value of L_z is $l\hbar$, and this is less than $\sqrt{l(l+1)}\hbar$, we can see that the angle of the cone in figure 6.2(a) can never be zero. If a measurement of $\hat{L}_z$ were to result in the value $|\boldsymbol{L}|$, then

$$\boldsymbol{L} = \hat{k}\,|\boldsymbol{L}| \tag{6.61}$$

and we would know that $\hat{L}_x$ and $\hat{L}_y$ were each precisely zero. The wavefunction would then have to be a simultaneous eigenfunction of $\hat{L}_x$, $\hat{L}_y$ and $\hat{L}_z$, contrary to the commutator argument of section 6.3. This behaviour can be understood as a manifestation of the uncertainty principle. Suppose that both the magnitude and direction of $\boldsymbol{L}$ were known precisely, as in equation (6.61). This would imply that the motion of the particle was entirely confined to the xy plane, since any motion perpendicular to the plane would give rise to transverse components of angular momentum. It follows that both z and P_z must be zero precisely, in violation of the uncertainty principle. It also follows that the planar orbital motions predicted classically for particles in central potentials are not permitted in quantum mechanics. However, we can see that as the value of l increases the discrepancy between $l\hbar$ and $\sqrt{l(l+1)}\hbar$ decreases, so motion more closely confined to a plane becomes possible, and in this way the classical limit is approached in accordance with the correspondence principle.

〉6.6 Interpretation of angular momentum eigenfunctions

The spherical harmonic function $Y_l^m(\theta, \phi)$ forms the angular part of the wavefunction for a particle in an angular momentum eigenstate with quantum numbers l and m. Hence the probability of finding the particle in the element of solid angle $\mathrm{d}\Omega = \sin\theta\,\mathrm{d}\theta\,\mathrm{d}\phi$, lying in the direction (θ, ϕ), is given by

$$Y_l^{m*}(\theta,\phi)\,Y_l^m(\theta,\phi)\sin\theta\,\mathrm{d}\theta\,\mathrm{d}\phi = \frac{1}{2\pi}\,|\Theta(\theta)|^2 \sin\theta\,\mathrm{e}^{-\mathrm{i}m\phi}\,\mathrm{e}^{\mathrm{i}m\phi}\,\mathrm{d}\theta\,\mathrm{d}\phi. \tag{6.62}$$

The first thing to notice is that, since the ϕ dependence is given by $e^{im\phi}$, in all such eigenstates the probability density is independent of ϕ and the system has axial symmetry. The angular dependence of the probability density is therefore determined by $\Theta_l^{|m|}(\theta)$. The functions $Y_l^m(\theta, \phi)$ are plotted for the first few values of l and m in figure 6.3. Many more can be plotted using the computer program provided.

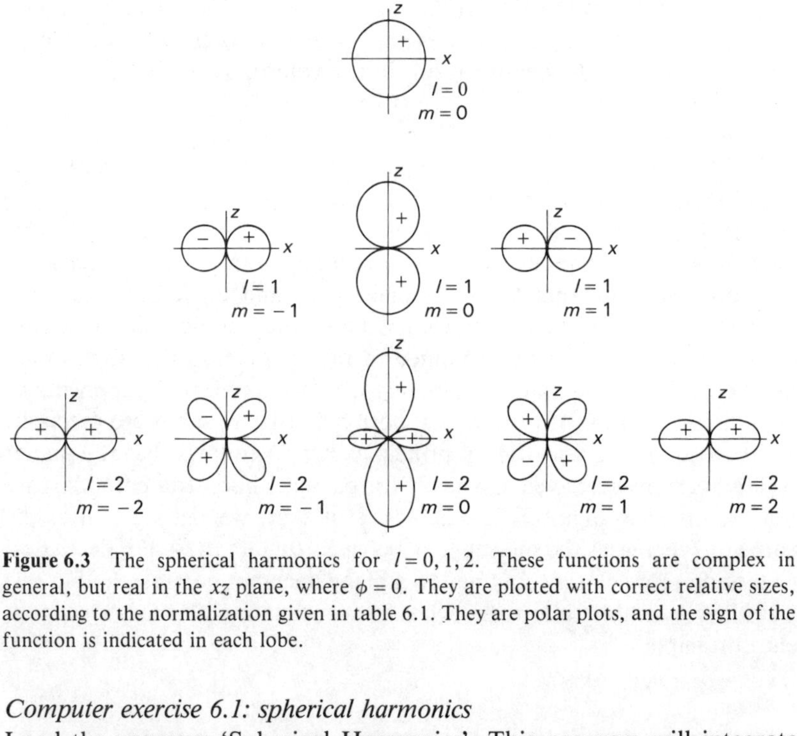

Figure 6.3 The spherical harmonics for $l = 0, 1, 2$. These functions are complex in general, but real in the xz plane, where $\phi = 0$. They are plotted with correct relative sizes, according to the normalization given in table 6.1. They are polar plots, and the sign of the function is indicated in each lobe.

Computer exercise 6.1: spherical harmonics

Load the program 'Spherical Harmonics'. This program will integrate and plot the solution of the associated Legendre equation for any values of l and m which you specify, but preferably with $l, |m| < 10$. The plotting, which is polar, is done in the xz plane because the spherical harmonics are real in this plane ($\phi = 0$) and identical, apart from a factor of $(2\pi)^{-1/2}$, to the functions $\Theta_l^{|m|}(\theta)$. The equation is more awkward to deal with than the one-dimensional Schrödinger equation and even with allowed values of l and m residual errors usually produce an unaesthetic divergence as $\theta \to 0, \pi$, unless very small step sizes are used, in which case the plotting is very slow. An option to suppress this divergence is provided but can be switched off in order to examine the genuine divergences which occur when forbidden combinations of l and m are entered. The program decides the appropriate initial values for integration for allowed values of l and m but these can be over-ridden for the purposes of investigation. Once you have familiarized yourself with the program, starting with the default values, and trying out the various options, try out the following, always bearing in mind the requirements of equations (6.56) and (6.57).

(*a*) Plot an s state, for which $l = m = 0$. Note that it is spherically symmetric. The existence of such states is excluded in Bohr's theory of the hydrogen atom, and also in Sommerfeld's elliptical orbit theory, since motion with zero angular momentum requires passage through the origin. Such states are quite permissible in quantum mechanics however.

(*b*) Make a sequence of plots with $l = m$: $(l, m) = (1, 1), (2, 2)$, etc. Note that the general shape is similar in each case ($\propto \sin^l\theta$) but the particle becomes confined more and more closely to the xy plane as $l = m$ increases. This illustrates the approach to a classical orbit, for which $\Theta(\theta)$ would be a δ function in the $\theta = \frac{1}{2}\pi$ plane.

(*c*) Make some plots for which $m = 0$, $l \neq 0$. Note that only for $m = 0$ is there a probability of the particle being on the z axis. This can only occur if there is no z component of angular momentum, as can be seen from the relation for L_z in equation (6.5). This result holds classically also, and corresponds to an orbit which is in a vertical plane (which includes the z axis) so that $\boldsymbol{L}$ lies in the xy plane.

(*d*) Make a variety of plots using allowed values of l and m. Verify that, discounting $\theta = 0, \pi$, there are $l - |m|$ values of θ for which the probability drops to zero. Bearing in mind the axial symmetry, these represent conical surfaces, and are referred to as nodal cones.

(*e*) Try a few illegal values of l and m. $(l, m) = (3, 4)$, for example, leads to obviously divergent behaviour, whatever initial values are used.

(*f*) Now try $(l, m) = (\frac{1}{2}, \frac{1}{2})$, $(\frac{3}{2}, \frac{3}{2})$, $(\frac{3}{2}, \frac{1}{2})$, or even $(2.1, 1.1)$. These values satisfy equation (6.45) and hence represent bounded solutions of the associated Legendre equation. Nevertheless, they do not represent valid wavefunctions, as discussed in section 6.5.

〉6.7 The Stern–Gerlach apparatus and electron spin

Anyone who has ever played with a gyroscope or watched a precessing top is aware that, even classically, angular momentum behaves in a fashion quite contrary to our intuition. We now find that quantum mechanically the situation is even more bizarre. At least some of the mystery is dispelled once it is understood how a quantum mechanical measurement of angular momentum is carried out. Such measurements were first made in 1921 by W Stern and O Gerlach. The experimental method hinges upon the fact that any charged body such as an electron which has angular momentum will also possess a magnetic dipole moment. The experiment actually measures this dipole moment by the application of a magnetic field, and the angular momentum is deduced from this.

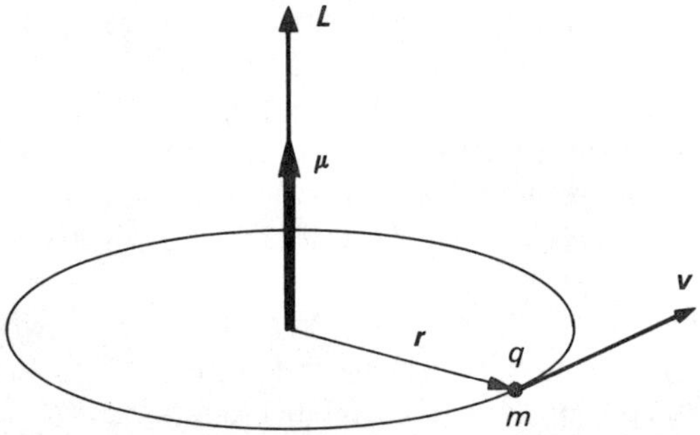

Figure 6.4 An orbiting charge has a magnetic dipole moment associated with its angular momentum.

The first thing to do is to calculate the relationship between orbital angular momentum and magnetic dipole moment. This is very easy to do for the simple case of a charge executing a circular orbit, as in figure 6.4. The basic result we require from electromagnetic theory is that, for a current loop with current I and area A, the magnetic dipole moment μ is

$$\mu = IA. \tag{6.63}$$

The current can be thought of as the quantity of charge passing a given point in unit time. For a circular orbit of radius r the charge q has an orbital period of $2\pi r/v$, where v is its speed. The current is therefore

$$I = \frac{qv}{2\pi r} \tag{6.64}$$

and the magnetic dipole moment is

$$\mu = \pi r^2 \frac{qv}{2\pi r} = \frac{qvr}{2}. \tag{6.65}$$

We can eliminate the vr product by introducing the angular momentum

$$L = mvr \tag{6.66}$$

with the result

$$\boldsymbol{\mu} = \frac{q}{2m}\boldsymbol{L} = \gamma\boldsymbol{L}. \tag{6.67}$$

The vector notation has been introduced to indicate that the magnetic dipole will have the same direction as the angular momentum for a positive charge. The quantity γ is the *magneto-gyric ratio*. This elementary derivation disguises the fact that this is a very fundamental result of wide applicability. It is valid for any angular momentum conserving orbit, and any charge distribution. It is also valid in quantum mechanics as long as we confine the discussion to orbital angular momentum.

A magnetic dipole $\boldsymbol{\mu}$, placed in a magnetic field $\boldsymbol{B}$, experiences a torque $\boldsymbol{\Gamma}$, such that

$$\boldsymbol{\Gamma} = \boldsymbol{\mu} \times \boldsymbol{B}. \tag{6.68}$$

We might expect the result of this to be that $\boldsymbol{\mu}$ would oscillate about $\boldsymbol{B}$ and, provided there was some dissipation mechanism, eventually come to rest pointing in its direction. This fails to take account of the angular momentum associated with the dipole, which must obey equation (6.2). The effect of this is to cause $\boldsymbol{\mu}$ and $\boldsymbol{L}$ to precess about $\boldsymbol{B}$, as shown in figure 6.5. An interesting feature of this behaviour is that the component of $\boldsymbol{L}$ along $\boldsymbol{B}$ is a constant of the motion.

One other result from electromagnetism is required in order to understand the Stern–Gerlach (SG) experiment. If the magnetic field in which the magnetic dipole is situated is inhomogeneous, then as well as experiencing a torque the dipole will feel a force tending to displace it equal to

$$\boldsymbol{F} = (\boldsymbol{\mu} \cdot \nabla)\boldsymbol{B}. \tag{6.69}$$

In the SG apparatus (figure 6.6) a collimated beam of atoms travelling in the y direction passes between the pole pieces of a powerful electromagnet. The pole pieces of this magnet are shaped so as to provide a large field gradient in the same direction as the field, which we will define to be the z direction. With this arrangement the z component of the force on the dipole is

$$F_z = \mu_z \frac{\partial B_z}{\partial z}. \tag{6.70}$$

The other eight terms in the force equation (6.69) all vanish. The beam is confined to the yz plane where B_x and B_y are zero, so $\partial B_x/\partial z$ and $\partial B_y/\partial z$ vanish, and the magnetic dipole precesses about z so that μ_x and μ_y average to zero.

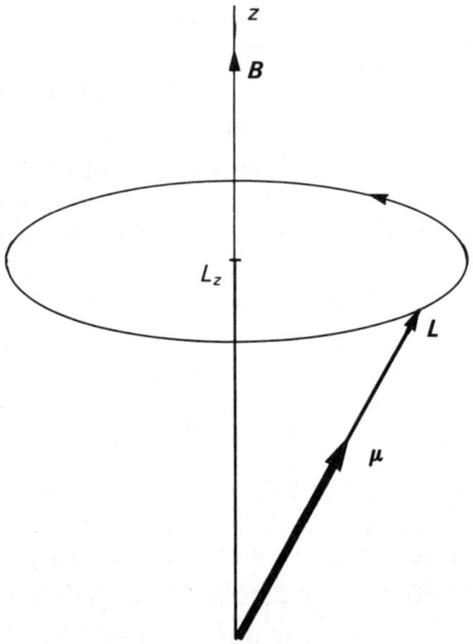

Figure 6.5 The effect of a magnetic field is to cause the angular momentum and associated magnetic dipole to precess about the direction of the field, leaving $|L|$ and L_z unchanged.

Taking the z component of equation (6.67) we have

$$\mu_z = \gamma L_z \tag{6.71}$$

and so the deflecting force is

$$F_z = \gamma L_z \frac{\partial B_z}{\partial z}. \tag{6.72}$$

As we note above, L_z remains constant throughout the trajectory. We now incorporate the quantum mechanical result that $L_z = m_l \hbar$, so that

$$F_z = \gamma \hbar \frac{\partial B_z}{\partial z} m_l. \tag{6.73}$$

(The need for the subscript in m_l will shortly become apparent.) This tells us that we can expect the beam to split into $2l + 1$ subsidiary beams, one for each value of m_l. In this way we can determine both l and m_l, the former from the number of beams, and the latter from the particular beam in which an atom emerges from the apparatus. In addition, from the angular deflection of the beams, and knowing $\partial B_z/\partial z$, we can deduce the magnitude of the magnetic dipole moment, and confirm equation (6.67).

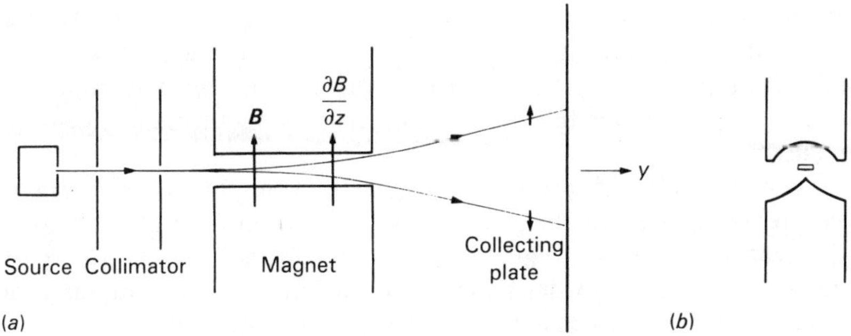

Figure 6.6 (*a*) A schematic of the Stern–Gerlach apparatus. (*b*) A cross section through the electromagnet pole pieces. These were designed to produce the greatest possible magnetic field gradient.

When first performed, the experiment worked as expected except for the startling result that only two beams emerged. We now know that the reason for this was that the chosen atoms, which were silver, happen to have zero overall orbital angular momentum in their ground state. The magnetic dipole moment responsible for the deflection arose from the spin angular momentum $\boldsymbol{S}$ of the single unpaired atomic electron. However, if this type of angular momentum is to satisfy rules of quantization similar to those for $\boldsymbol{L}$ (and more advanced quantum mechanical treatments of angular momentum tell us it must), then the spin quantum number s must equal $\frac{1}{2}$, so that $2s + 1 = 2$, giving two beams, which correspond to $m_s = \pm\frac{1}{2}$. Spin angular momentum is anomalous in other ways also. Unlike l, s is fixed in value. It is impossible to impart more or less spin to an electron. Spin also carries with it twice the normal amount of magnetic dipole moment! These results can be summarized as follows:

$$|\boldsymbol{S}| = [s(s+1)]^{1/2}\hbar = \frac{\sqrt{3}}{2}\hbar \qquad (s = \tfrac{1}{2}) \tag{6.74}$$

$$S_z = m_s\hbar \qquad m_s = \pm\tfrac{1}{2} \tag{6.75}$$

$$\boldsymbol{\mu}_s = 2\gamma\boldsymbol{S}. \tag{6.76}$$

In no sense do these properties emerge from the Schrödinger wave equation, which has nothing to say about electron spin. If we wish to employ the Schrödinger formalism to describe atomic structure, the effects of spin must be added in an empirical fashion. By contrast, the Dirac equation, which it may be recalled is a wave equation which takes relativistic effects into account, predicts these results precisely, so to that extent we can regard electron spin as necessitated by the requirements of relativity.

Despite the surprising results relating to electron spin, in other respects the Stern–Gerlach experiment triumphantly confirmed the predictions of the quantum theory of angular momentum, especially as regards space quantization, and provides an excellent illustration of the way angular momentum measurements can be made. We will develop the theory of spin further later in the chapter, but first we will examine another effect which depends upon the magnetic properties of atoms.

〉6.8 The Zeeman effect

The magnetic dipole moments associated with the orbital and spin components of atomic angular momenta manifest themselves in a number of ways. One of these, the Zeeman effect, was very important in unravelling the mysteries of atomic structure. It concerns the changes which occur in atomic spectral lines when the emitting atoms are situated in a magnetic field. In what follows we will ignore the existence of electron spin and consider an atom with orbital angular momentum only, which undergoes a transition from a state for which $l = l_1$ to $l = l_2$, as in figure 6.7(a). Because of the orbital angular momentum the atom has a magnetic dipole moment $\boldsymbol{\mu}$ given by equation (6.67). If the atom is placed in a magnetic field $\boldsymbol{B}$, $\boldsymbol{\mu}$ and hence $\boldsymbol{L}$ precess about $\boldsymbol{B}$, as discussed in connection with equation (6.68) and illustrated in figure 6.5. Now any dipole so situated possesses potential energy given by

$$E_{\text{mag}} = -\boldsymbol{\mu}\cdot\boldsymbol{B}. \tag{6.77}$$

This represents the work which could be done if the dipole were to respond to the torque by aligning itself with the magnetic field.

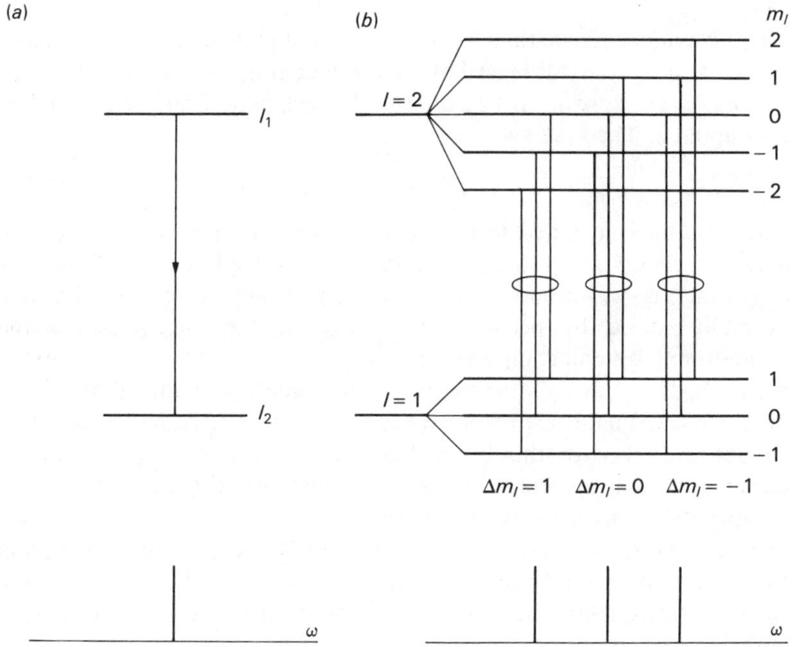

Figure 6.7 The normal Zeeman effect. (*a*) A transition between states of angular momentum l_1 and l_2. (*b*) The effects of applying a magnetic field to the system shown in (*a*). Drawn for $l_1 = 2$ and $l_2 = 1$. Although nine transitions occur, only three spectral lines are seen because the splittings of the energy levels are the same in the upper and lower states.

We now insert the value $\gamma \boldsymbol{L}$ for $\boldsymbol{\mu}$ and assume $\boldsymbol{B}$ to be directed along the z axis. We find

$$E_{\text{mag}} = -\gamma \boldsymbol{L} \cdot \boldsymbol{B} = -\gamma L_z B. \tag{6.78}$$

The second equality follows because the scalar product requires us to take the component of $\boldsymbol{L}$ along the z axis. This energy is additional to the electrostatic and kinetic energies of the atom and in order to treat the effect quantum mechanically we must introduce a term into the Hamiltonian to represent it. This is readily done following the prescription of postulate 4; we simply replace L_z by its operator $\hat{L}_z$. The magnetic field in this case is simply a constant.

Let us suppose that the Hamiltonian for the atom in the absence of the magnetic field is $\hat{H}_0$, so that the TISE is

$$\hat{H}_0 u_{n,l,m_l} = E_{n,l,m_l} u_{n,l,m_l}. \tag{6.79}$$

The subscripts indicate that the u_{n,l,m_l} are simultaneous eigenfunctions of $\hat{H}_0$, $\hat{\boldsymbol{L}}^2$ and $\hat{L}_z$. We will regard this as a solved problem, so that the u_{n,l,m_l} and E_{n,l,m_l} are known in the absence of a magnetic field. The field $\boldsymbol{B}$ is now applied. The TISE becomes

$$(\hat{H}_0 - \gamma \hat{L}_z B) u'_{n,l,m_l} = E'_{n,l,m_l} u'_{n,l,m_l} \tag{6.80}$$

where we have included in the Hamiltonian the operator form of the magnetic energy given in equation (6.78). Now we have to find the new eigenfunctions u'_{n,l,m_l} and eigenvalues E'_{n,l,m_l}. Problems of this nature are normally solved by perturbation theory. If the extra term in the Hamiltonian is small compared with the original one we may assume that the changes in the eigenfunctions are also small, and this allows us to obtain approximate expressions for these changes. Perturbation theory is beyond the scope of this book, but it is not needed in this instance, because in fact the eigenfunctions, though not the eigenvalues, are completely unchanged by the perturbation: the u'_{n,l,m_l} and u_{n,l,m_l} are identical. This is because the u_{n,l,m_l} are already eigenfunctions of $\hat{L}_z$ and the extra term in the Hamiltonian is simply $\hat{L}_z$ multiplied by a constant factor. Consequently the new eigenvalues are easily found by operating with $\hat{H}_0 - \gamma \hat{L}_z B$ on u_{n,l,m_l}:

$$(\hat{H}_0 - \gamma \hat{L}_z B) u_{n,l,m_l} = (E_{n,l,m_l} - \gamma m_l \hbar B) u_{n,l,m_l}. \tag{6.81}$$

The effect of the magnetic field is to increment the original energy by an amount $E_{\text{mag}} = -\gamma m_l \hbar B$. For an electron of charge $-e$ and mass m we have

$$E_{\text{mag}} = \frac{e\hbar}{2m} B m_l. \tag{6.82}$$

The quantity $e\hbar/2m$ is a natural unit of magnetic moment and is referred to as the *Bohr magneton*, μ_{B}.

Examination of equations (6.81) and (6.82) reveals that any energy level characterized by the quantum numbers n, l and m_l is shifted by an amount which depends only on m_l and is proportional to it. In the absence of a magnetic (or any other) field all directions in space are equivalent and so it is not possible for the energy of an atom to depend upon m_l since this quantum number has to do only with the orientation of the atom in space. States with the same value of m_l are therefore degenerate. Application of the field lifts this degeneracy and states with different values of m_l acquire different energies. It should now be clear why m_l is known as the magnetic quantum number.

In the presence of the magnetic field the two energy levels of figure 6.7(a) are split into $2l_1 + 1$ and $2l_2 + 1$ levels respectively. One might expect therefore that the original spectral line would split into $(2l_1 + 1)(2l_2 + 1)$ lines corresponding to all possible combinations of upper and lower levels. In fact only three lines are seen. This comes about for two reasons. In the first place the splitting of the energy levels is independent of n and l, and so is the same in both the upper and lower levels. The separation between adjacent levels, for which m_l differs by 1, is

$$\Delta E_{\text{mag}} = \mu_{\text{B}} B. \tag{6.83}$$

Secondly, two selection rules operate which restrict the transitions which can occur during the emission of electric dipole radiation. (Other processes can occur but they are far weaker.) The rules can be expressed as

$$\Delta l = \pm 1 \qquad \Delta m_l = \pm 1, 0. \tag{6.84}$$

These selection rules relate to the fact that in this process a photon can carry off only an amount $\hbar$ of angular momentum. This then restricts the extent of the changes which can occur in both the magnitude and orientation of the angular momentum of the atom. Figure 6.7(b) is drawn for $l_1 = 3$ and $l_2 = 2$ which satisfies the first rule and only those transitions which satisfy the second rule are shown. It is easy to see that only nine transitions can occur, and that these fall into three energy groups, corresponding to $\Delta m_l = 1$, $\Delta m_l = 0$ and $\Delta m_l = -1$. Hence only three spectral lines are expected.

Examples of the Zeeman effect which exhibit precisely this behaviour, which is known as the *normal* Zeeman effect, can be observed, but frequently more complicated patterns are seen, containing more than three spectral lines. The effect is then referred to as the *anomalous* Zeeman effect. It arises because of the complicating effects of electron spin, which we have ignored in the above discussion. When the normal Zeeman effect does occur it is because the effects of several spins have compensated each other in a multi-electron atom. When this compensation does not occur, the effect of the anomalously large magnetic moment of the electron is to cause the splitting in the upper and lower levels to be unequal, leading to a multiplicity of components in the Zeeman pattern. Studies of the number and position of the lines observed, together with their intensities and polarization properties, have enabled many of the details of the structure of multi-electron atoms to be discovered.

〉6.9 Spin operators: the Pauli spin matrices

No detailed semiclassical model of the inner workings of a spinning electron exists, nor is one thought possible. An example of the sort of difficulty which arises if such a model is attempted is the following. It is possible to define a 'classical radius' for an electron in which the entire electron mass arises from the electrostatic potential self-energy of a spherical cloud of charge. Assuming a uniformly distributed charge, this radius is equal to

$$r_c = \frac{1}{4\pi\varepsilon_0}\frac{3}{5}\frac{e^2}{m_e c^2} = 1.7 \times 10^{-15}\ \mathrm{m}. \tag{6.85}$$

Let us now attempt to extend this model by ascribing the electron spin angular momentum to the rotation of this 'mass' distribution. The moment of inertia of a uniform sphere of mass m and radius r_c is $I = 2mr_c^2/5$, and if it has angular velocity $\omega = v_s/r_c$ where v_s is the surface velocity, the angular momentum is

$$|S| = \tfrac{2}{5} m_e r_c^2 \frac{v_s}{r_c} = \frac{\sqrt{3}}{2}\hbar. \tag{6.86}$$

Solving this for v_s,

$$v_s = \frac{5\hbar}{2m_e r_c} = 2.9 \times 10^{11}\ \mathrm{m\,s^{-1}} \tag{6.87}$$

we find that the rotation rate required implies surface velocities v_s far in excess of the speed of light. The situation is made even worse if we take into account the results of modern electron scattering experiments which indicate that to all intents and purposes electrons are point-like particles!

The lack of a model for the electron spin places us in some difficulty when we try to develop quantum mechanical operators to represent this property. To proceed in the manner we used in the case of orbital angular momentum we would require a set of coordinates to describe the rotation of the electron, and these we do not have. Fortunately the matrix approach provides us with a solution, which we will now examine.

Since we find that for electron spin a component such as S_z can take only two values, namely $\pm\frac{1}{2}\hbar$, the operators we seek can be represented by 2×2 matrices, and they will be Hermitian. We will choose for our basis vectors the simultaneous eigenvectors of $\hat{S}^2$ and S_z. We therefore expect these two operators to be diagonal, and the diagonal elements to be the eigenvalues of the operators (equations (6.74) and (6.75)). This is sufficient information to specify the operators as

$$\hat{S}_z = \begin{bmatrix} \frac{1}{2}\hbar & 0 \\ 0 & -\frac{1}{2}\hbar \end{bmatrix} = \frac{1}{2}\hbar \begin{bmatrix} 1 & 0 \\ 0 & -1 \end{bmatrix} \tag{6.88}$$

and

$$\hat{S}^2 = \begin{bmatrix} \frac{3}{4}\hbar^2 & 0 \\ 0 & \frac{3}{4}\hbar^2 \end{bmatrix} = \frac{3}{4}\hbar^2 \begin{bmatrix} 1 & 0 \\ 0 & 1 \end{bmatrix}. \tag{6.89}$$

We now require operators for $\hat{S}_x$ and $\hat{S}_y$ which have the following properties in common with the orbital angular momentum operators:

$$\hat{S}^2 = \hat{S}_x^2 + \hat{S}_y^2 + \hat{S}_z^2 \tag{6.90}$$

$$[\hat{S}_x, \hat{S}_y] = \mathrm{i}\hbar\hat{S}_z \qquad \text{and cyclic permutations} \tag{6.91}$$

$$\langle\hat{S}_x\rangle = \langle\hat{S}_y\rangle = 0. \tag{6.92}$$

The expectation values in the final equation above are zero because they are evaluated for states which are eigenvectors of $\hat{S}_z$. Recalling that for any matrix operator the diagonal elements will be the expectation values of the operator with respect to the basis vectors, this implies that the matrices we seek must have diagonal elements equal to zero. This, together with the Hermitian condition, requires the matrices to have the forms

$$\hat{S}_x = \frac{1}{2}\hbar \begin{bmatrix} 0 & a \\ a^* & 0 \end{bmatrix} \qquad \hat{S}_y = \frac{1}{2}\hbar \begin{bmatrix} 0 & b \\ b^* & 0 \end{bmatrix}. \tag{6.93}$$

Inserting equations (6.88) and (6.93) into equation (6.91) and performing the matrix multiplications, we find that the commutation relations require that

$$a = \mathrm{i}b. \tag{6.94}$$

Making the (customary) arbitrary choice $a = 1$, the operators $\hat{S}_x$ and $\hat{S}_y$ become

$$\hat{S}_x = \frac{1}{2}\hbar \begin{bmatrix} 0 & 1 \\ 1 & 0 \end{bmatrix} \qquad \hat{S}_y = \frac{1}{2}\hbar \begin{bmatrix} 0 & -\mathrm{i} \\ \mathrm{i} & 0 \end{bmatrix}. \tag{6.95}$$

It may be verified that these operators satisfy all requirements including equation (6.90).

It is instructive to investigate the eigenvalue equations satisfied by the operators we have found. Taking $\hat{S}_x$ as an example, denoting its eigenvalues by $\frac{1}{2}\lambda\hbar$, and its eigenvectors by $\left[\begin{smallmatrix} a_1 \\ a_2 \end{smallmatrix}\right]$, the eigenvalue equation is

$$\tfrac{1}{2}\hbar\begin{bmatrix} 0 & 1 \\ 1 & 0 \end{bmatrix}\begin{bmatrix} a_1 \\ a_2 \end{bmatrix} = \tfrac{1}{2}\lambda\hbar\begin{bmatrix} a_1 \\ a_2 \end{bmatrix}. \tag{6.96}$$

Hence

$$\tfrac{1}{2}\hbar\begin{bmatrix} -\lambda & 1 \\ 1 & -\lambda \end{bmatrix}\begin{bmatrix} a_1 \\ a_2 \end{bmatrix} = 0. \tag{6.97}$$

This equation can only be satisfied provided

$$\begin{vmatrix} -\lambda & 1 \\ 1 & -\lambda \end{vmatrix} = 0$$

or

$$\lambda^2 = 1 \qquad \lambda = \pm 1. \tag{6.98}$$

We find therefore that the eigenvalues of $\hat{S}_x$ are $\pm\frac{1}{2}\hbar$. Had we started with $\hat{S}_y$, or indeed *any* component of $\hat{S}$, we would have obtained the same result. This is just what we should expect since all directions in space are equivalent.

We can obtain the eigenvectors of $\hat{S}_x$ from equation (6.97). For the case $\lambda = 1$ we find $a_1 = a_2$, while for $\lambda = -1$, $a_1 = -a_2$. The eigenvectors can therefore be written

$$\alpha_x = \frac{1}{\sqrt{2}}\begin{bmatrix} 1 \\ 1 \end{bmatrix} \qquad \beta_x = \frac{1}{\sqrt{2}}\begin{bmatrix} -1 \\ 1 \end{bmatrix}. \tag{6.99}$$

These have been normalized so that, for example,

$$\alpha_x^\dagger \alpha_x = \tfrac{1}{2}\,[1 \quad 1]\begin{bmatrix} 1 \\ 1 \end{bmatrix} = 1. \tag{6.100}$$

The eigenvectors α_x and β_x can be thought of as representing states which have 'spin up' and 'spin down' in the x direction.

The eigenvalues and normalized eigenvectors of the three spin component operators are listed in table 6.2. All the eigenvectors in the table are also eigenvectors of $\hat{S}^2$, with eigenvalue $\frac{3}{4}\hbar^2$. Any of the pairs (α_x, β_x), (α_y, β_y) and (α_z, β_z) form a complete orthonormal set of basis functions, in terms of which any general spin function can be written. For example, using α_z and β_z as basis vectors a general spin function can be written as

$$\psi = \begin{bmatrix} a_1 \\ a_2 \end{bmatrix} = a_1 \begin{bmatrix} 1 \\ 0 \end{bmatrix} + a_2 \begin{bmatrix} 0 \\ 1 \end{bmatrix}. \tag{6.101}$$

Table 6.2 Spin-$\frac{1}{2}$ eigenvalues and eigenvectors.

Operator	Eigenvalue	Eigenvector
$\hat{S}_x$	$\frac{1}{2}\hbar$	$\alpha_x = \frac{1}{\sqrt{2}} \begin{bmatrix} 1 \\ 1 \end{bmatrix}$
$\hat{S}_x$	$-\frac{1}{2}\hbar$	$\beta_x = \frac{1}{\sqrt{2}} \begin{bmatrix} -1 \\ 1 \end{bmatrix}$
$\hat{S}_y$	$\frac{1}{2}\hbar$	$\alpha_y = \frac{1}{\sqrt{2}} \begin{bmatrix} 1 \\ \mathrm{i} \end{bmatrix}$
$\hat{S}_y$	$-\frac{1}{2}\hbar$	$\beta_y = \frac{1}{\sqrt{2}} \begin{bmatrix} 1 \\ -\mathrm{i} \end{bmatrix}$
$\hat{S}_z$	$\frac{1}{2}\hbar$	$\alpha_z = \begin{bmatrix} 1 \\ 0 \end{bmatrix}$
$\hat{S}_z$	$-\frac{1}{2}\hbar$	$\beta_z = \begin{bmatrix} 0 \\ 1 \end{bmatrix}$

Normalization requires that

$$[a_1^* \quad a_2^*] \begin{bmatrix} a_1 \\ a_2 \end{bmatrix} = 1$$

or

$$|a_1|^2 + |a_2|^2 = 1. \tag{6.102}$$

It is a simple matter, using postulate 4, to find operators representing any arbitrary component of spin. Let us suppose we require the operator $\hat{S}_n$ for the component lying in the direction of the unit vector $\hat{\boldsymbol{n}}$. Then, classically, $S_n = \boldsymbol{S} \cdot \hat{\boldsymbol{n}}$ and so the required operator is

$$\hat{S}_n = \hat{\boldsymbol{S}} \cdot \hat{\boldsymbol{n}}. \tag{6.103}$$

Suppose now that $\hat{n}$ lies in the xz plane, and makes an angle θ with the z axis (figure 6.8). This would be the case if, for example, we were considering the beam of a Stern–Gerlach apparatus to be propagating in the y direction, and the measurement axis of the equipment to be tilted at an angle θ with respect to the z axis. The operator we require is then

$$\begin{aligned}\hat{S}_n &= \hat{S}_z \cos\theta + \hat{S}_x \sin\theta \\ &= \tfrac{1}{2}\hbar\begin{bmatrix}\cos\theta & 0 \\ 0 & -\cos\theta\end{bmatrix} + \tfrac{1}{2}\hbar\begin{bmatrix}0 & \sin\theta \\ \sin\theta & 0\end{bmatrix} \\ &= \tfrac{1}{2}\hbar\begin{bmatrix}\cos\theta & \sin\theta \\ \sin\theta & -\cos\theta\end{bmatrix}.\end{aligned} \tag{6.104}$$

The eigenvalues $\frac{1}{2}\lambda\hbar$ of this operator can be determined in the usual way; they must satisfy

$$\begin{vmatrix}\cos\theta - \lambda & \sin\theta \\ \sin\theta & -\cos\theta - \lambda\end{vmatrix} = 0$$

or

$$\cos^2\theta + \sin^2\theta = \lambda^2 = 1. \tag{6.105}$$

Hence $\lambda = \pm 1$ and again the eigenvalues are $\pm\frac{1}{2}\hbar$.

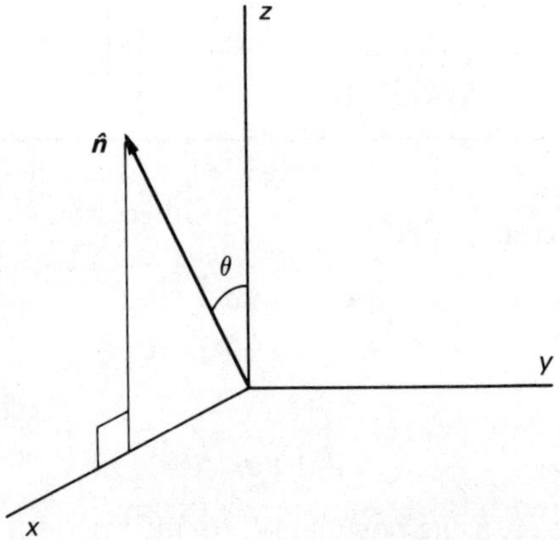

Figure 6.8 Polarization unit vector for a particle travelling along the y axis.

The eigenvector α_n is found from the eigenvalue equation:

$$\tfrac{1}{2}\hbar\begin{bmatrix}\cos\theta & \sin\theta \\ \sin\theta & -\cos\theta\end{bmatrix}\begin{bmatrix}a_1 \\ a_2\end{bmatrix} = \tfrac{1}{2}\hbar\begin{bmatrix}a_1 \\ a_2\end{bmatrix}. \tag{6.106}$$

Hence

$$\begin{bmatrix} a_1 \cos\theta + a_2 \sin\theta \\ a_1 \sin\theta - a_2 \cos\theta \end{bmatrix} = \begin{bmatrix} a_1 \\ a_2 \end{bmatrix}$$

and

$$\frac{a_1}{a_2} = \frac{\sin\theta}{1-\cos\theta} = \frac{2\sin\frac{1}{2}\theta\cos\frac{1}{2}\theta}{2\sin^2\frac{1}{2}\theta} = \frac{\cos\frac{1}{2}\theta}{\sin\frac{1}{2}\theta} \tag{6.107}$$

with a similar analysis for β_n. The normalized eigenvectors can therefore be written

$$\alpha_n = \begin{bmatrix} \cos\frac{1}{2}\theta \\ \sin\frac{1}{2}\theta \end{bmatrix} \qquad \beta_n = \begin{bmatrix} -\sin\frac{1}{2}\theta \\ \cos\frac{1}{2}\theta \end{bmatrix}. \tag{6.108}$$

It is easy to see that in the limits $\theta \to 0$ and $\theta \to \frac{1}{2}\pi$ these eigenvectors become α_z, β_z and α_x, β_x respectively, as they should.

These eigenvectors have the surprising property of being double valued! If we imagine θ increasing from θ to $\theta + 2\pi$, our reference axes have undergone one complete revolution about the y axis, and the final state should be indistinguishable from the initial one. However, the eigenvectors depend only upon $\frac{1}{2}\theta$, and the rotation has the effect of reversing their signs:

$$\alpha_{\theta+2\pi} = -\alpha_\theta \qquad \beta_{\theta+2\pi} = -\beta_\theta. \tag{6.109}$$

Although for many purposes the absolute sign of the wavefunction has no significance, the *relative* sign is significant in superposition states. Interference experiments have been designed in which particles of spin $\frac{1}{2}$ (neutrons) traverse the particle equivalent of a Michelson interferometer (Greenberger 1983). In one arm of the interferometer a magnetic field induces one complete rotation of the particle. The resulting fringe shift when the beams are recombined confirms the remarkable prediction above, namely that a rotation of 2π reverses the sign of the wavefunction of a spin-$\frac{1}{2}$ particle!

〉6.10 Measurements in quantum mechanics

A spin-$\frac{1}{2}$ system provides an ideal model with which to demonstrate the principles of quantum mechanical measurement, particularly the content of postulate 5, the expansion postulate. To do so we will make extensive use of Stern–Gerlach instruments, which we will regard simply as measuring devices (not black boxes though; we do understand their internal workings) and whose function is to measure specific components of the angular momentum of spin-$\frac{1}{2}$ particles. These devices each have an input channel, two output channels and a specified orientation (see figure 6.9(a)). Spin-$\frac{1}{2}$ particles travelling along the y axis enter the input channel and emerge from one or other of the output channels, which are labelled α and β. If the orientation of the box is 'z' then particles coming out of the channel labelled α are 'spin up' in the z direction, have $L_z = \frac{1}{2}\hbar$ and state vector α_z, and particles emerging from the β channel are 'spin down' in the z direction, have $L_z = -\frac{1}{2}\hbar$ and state vector β_z. If one or other of the output channels is closed, then the device can be regarded as a spin filter, accepting any particle, but transmitting only, for example, α_z particles.

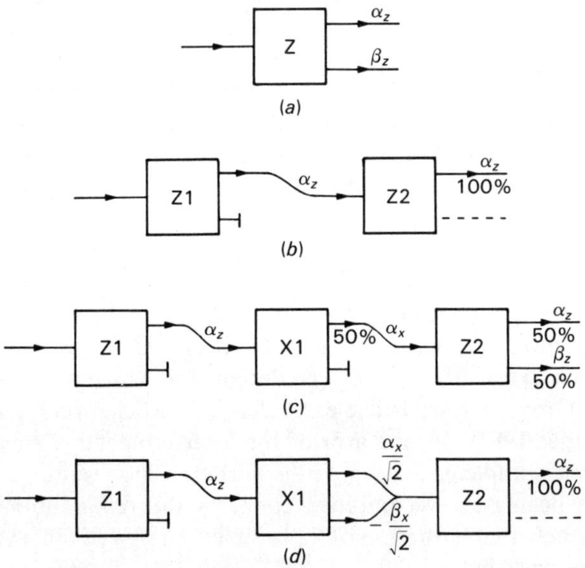

Figure 6.9 Illustrating the quantum theory of measurement using Stern–Gerlach measuring instruments.

In this description we have already illustrated some important aspects of quantum mechanical measurement. For example *measurement of an arbitrary spin-$\frac{1}{2}$ particle with a z device will always result in one of two values for the z component of angular momentum, namely the eigenvalues $\pm\frac{1}{2}\hbar$*. Nothing gets lost in the device, and nothing emerges except in one of the two channels.

We now close off the β channel of the z device we have been discussing, Z1, and direct the output of the α channel into a second z device, Z2 (figure 6.9(*b*)). We observe that all the input particles emerge from the α channel. This illustrates that *following a measurement the state of the system is the eigenstate corresponding to the eigenvalue result, and that a subsequent measurement will yield the same eigenvalue with certainty.*

We now insert a third instrument, X1, between Z1 and Z2, and arrange that only the α_z beam enters X1 and only the α_x beam enters Z2 (figure 6.9(*c*)). We find that equal numbers of particles emerge in the final α_z and β_z channels. We can explain this as follows. The particles entering Z2 have state vector α_x. But an examination of table 6.2 reveals that α_x can be expanded in terms of the basis vectors α_z and β_z as

$$\alpha_x = \frac{1}{\sqrt{2}}\,\alpha_z + \frac{1}{\sqrt{2}}\,\beta_z. \tag{6.110}$$

This therefore illustrates that *in the expansion of a wavefunction or state vector in terms of a basis set the squared magnitude of a coefficient of one member of the basis set represents the probability that a measurement would yield the eigenvalue corresponding to that member.* In this case the squared coefficients are each equal to $\frac{1}{2}$, and so any given particle has an equal probability of emerging in either channel. A similar output would result if the β_x beam was directed into Z2, since this can be expanded as

$$\beta_x = -\frac{1}{\sqrt{2}}\,\alpha_z + \frac{1}{\sqrt{2}}\,\beta_z. \tag{6.111}$$

It is also noteworthy that the second (X1) measurement alters the state of the system. Prior to the X1 measurement it was in the state α_z, and would have traversed the Z2 α_z channel with certainty. After the X1 measurement it is in the state α_x and has only a 50% chance of traversing the Z2 α_z channel. We cannot maintain that the particles entering X1 are of two kinds, α_x and β_x, *prior* to the x measurement, since then only 50% of them would have traversed the α_z channel *without* the x device in place! This follows from the expansions (6.110) and (6.111).

Finally we become extremely devious, and arrange to direct both the α_x and β_x beams emerging from X1 into Z2 (figure 6.9(*d*)). We do this without making any observation as to which channel any particular particle emerges from. The remarkable result is that all the particles now emerge in the α_z channel! *No measurement of L_x has been made, and so the state vector remains α_z, just as it was on emerging from Z1.*

In order to account for this observation formally we need to consider carefully the state vectors at various stages in the sequence. After the first measurement, the state vector is α_z, which we can expand as

$$\alpha_z = \frac{1}{\sqrt{2}}\,\alpha_x - \frac{1}{\sqrt{2}}\,\beta_x. \tag{6.112}$$

The state vectors applicable to the particles emerging from the two channels of the x device will therefore be $(1/\sqrt{2})\alpha_x$ and $-(1/\sqrt{2})\beta_x$ respectively. We could then simply combine them again prior to Z2, regaining α_z. However, it is more interesting to regard them as two beams in their own right and follow them through the second z device. We expand them as follows:

$$\frac{1}{\sqrt{2}}\,\alpha_x = \frac{1}{2}\alpha_z + \frac{1}{2}\beta_z \qquad -\frac{1}{\sqrt{2}}\,\beta_x = \frac{1}{2}\alpha_z - \frac{1}{2}\beta_z. \tag{6.113}$$

We conclude that coming out of the final α_z channel we have $\frac{1}{2}\alpha_z + \frac{1}{2}\alpha_z = \alpha_z$, whereas coming out of the β_z channel we have $\frac{1}{2}\beta_z - \frac{1}{2}\beta_z = 0$.

This throws a little light upon what constitutes a measurement. It will be observed that for the recomposition of the state vectors into α_z the amplitudes and phases of the component state vectors are crucial. If they were to be altered, they would recombine into something different. But any attempt to observe the particles emerging from the x device would do precisely that; it would alter the amplitudes and relative phases in an uncontrollable way. It is not the *presence* of X1 which affects the outcome of the Z2 measurement; it is the act of *observing the outcome* of the X1 experiment which does it!

It will be realized at this point that we are describing experiments closely analogous to the two-slit interference experiment. There we found that it was impossible to observe interference fringes (which of course depend upon the amplitudes and phases of the component beams), whilst also determining which slit the particles pass through.

There is one further illuminating discovery we can make with the aid of the spin-$\frac{1}{2}$ formalism. Consider a Stern–Gerlach device N, which measures the component of angular momentum in the direction $\hat{\boldsymbol{n}}$, defined in equation (6.103) and figure 6.8. What output should we expect from N if it takes its input from the α_z channel of Z (figure 6.10)? The eigenvectors α_n and β_n of the operator $\hat{S}_n$ are given in equation (6.108).

We can therefore expand α_z in terms of them:

$$\alpha_z = \begin{bmatrix} 1 \\ 0 \end{bmatrix} = c_\alpha \begin{bmatrix} \cos \frac{1}{2}\theta \\ \sin \frac{1}{2}\theta \end{bmatrix} + c_\beta \begin{bmatrix} -\sin \frac{1}{2}\theta \\ \cos \frac{1}{2}\theta \end{bmatrix}. \tag{6.114}$$

The expansion coefficients can be found by inspection; they are

$$c_\alpha = \cos \tfrac{1}{2}\theta \qquad c_\beta = -\sin \tfrac{1}{2}\theta. \tag{6.115}$$

Evidently the intensity in the output channels α_n and β_n will be proportional to P_α and P_β, where

$$P_\alpha = \cos^2 \tfrac{1}{2}\theta \qquad P_\beta = \sin^2 \tfrac{1}{2}\theta. \tag{6.116}$$

This is precisely similar to the way in which the output beams from a polarizing beam splitter vary with orientation angle in the optical case. There the behaviour is explicable entirely in terms of wave optics. It is remarkable how a system as essentially quantum in its nature as a spin-$\frac{1}{2}$ particle contrives to exhibit such analogous behaviour!

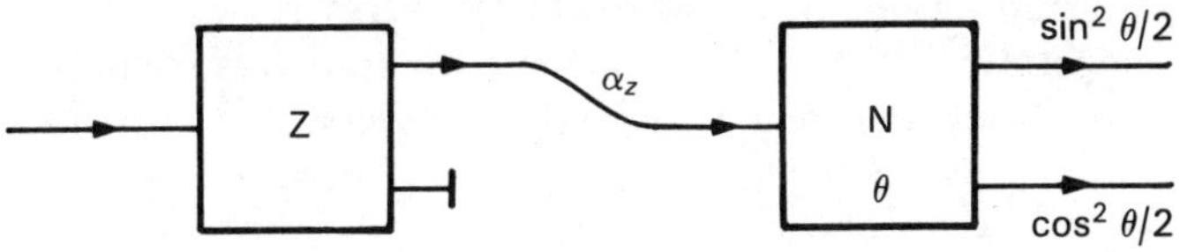

Figure 6.10 The output of a rotatable Stern–Gerlach spin analyser varies in the same way as the output of a polarizing beam splitter in optics.

〉 Problems

6.1 (a) Verify that the spherical harmonics $Y_l^m = Y_2^2, Y_2^1, Y_1^1$ are eigenfunctions of $\hat{L}^2$ and $\hat{L}_z$ with the appropriate eigenvalues.
(b) Verify that Y_2^2, Y_2^1, Y_1^1 are mutually orthogonal.

(Hint: in (b) it is sufficient to prove *either* $\Phi(\phi)$ *or* $\Theta(\theta)$ orthogonal. Try the easiest integral first!)

6.2 Verify that $(2l+1)^{-1} \Sigma_m \mid Y_l^m \mid^2 = 1/4\pi$ for the cases $l = 1, 2$. (One consequence of this is that, if an electron subshell of an atom is full, the overall charge density of the subshell is spherically symmetric.)

6.3 A system is known to be in the angular momentum eigenstate Y_1^1.
(a) Show explicitly that the expectation value of $\hat{L}_x$ is zero.

(b) What is the expectation value for a measurement of the angular momentum component in the direction $(\theta, \phi) = (\alpha, 0)$?

(Hint: the relevant operator is $\hat{L}_\alpha = \hat{L}_x \sin\alpha + \hat{L}_z \cos\alpha$.)

6.4 Show that $[\hat{L}_x, \hat{P}_y] = i\hbar\hat{P}_z$ and $[\hat{L}_x, y] = i\hbar z$.

6.5 In a version of the Stern–Gerlach experiment a collimated beam of silver atoms emerges from a furnace at temperature T and passes between the pole pieces of an electromagnet, of length a. The magnetic moment of a silver atom is entirely due to the spin of a single unpaired electron, and so equal to 1 Bohr magneton. If the magnetic field gradient is $\partial B_z/\partial z$ show that the angle through which the beam is deflected is given by

$$\Delta\theta = \frac{\partial B_z}{\partial z}\frac{\mu_B a}{4kT}.$$

Evaluate this for $a = 0.1$ m, $\partial B_z/\partial z = 10^4\ \mathrm{T\,m^{-1}}$ and $T = 1500$ K. (Hint: in a *beam* of atoms the mean energy is $2kT$, where T is the temperature of the oven. Calculate the transverse impulse applied to the beam and equate this to the transverse momentum acquired. Assume that the deflection angle will be small.)

6.6 An attempt is made to perform the Stern–Gerlach experiment with electrons. The first difficulty this presents is that the beam is deflected by the Lorentz force, because the particles are charged. This is dealt with by curving the magnet poles to match the trajectory (which will be in the horizontal plane if the magnetic field is in the vertical z direction). Prove that despite this precaution it is impossible to obtain results from the experiment for fundamental reasons connected with the uncertainty principle.
(Hint: one approach is as follows:

(a) Find the vertical deflection for a particle of magnetic moment μ in a field gradient $\partial B_z/\partial z$.
(b) The existence of $\partial B_z/\partial z$ implies the existence of a horizontal field component everywhere except on the beam axis ($\nabla \cdot \boldsymbol{B} = 0$). Calculate the vertical (Lorentz) spreading of the beam this produces for a beam of width Δx.
(c) Show that effect (b) will swamp effect (a) unless the uncertainty principle is violated.)

〉 Chapter 7

〉 Spherically Symmetric Potential Wells

〉 7.1 Introduction

In this chapter we will study the bound states of particles confined in spherically symmetric potential wells. In Chapter 6 we developed the theory of angular momentum. We found that whenever the potential energy function could be expressed as $V(r)$, a function of r only, then it is possible to find eigenfunctions of Schrödinger's equation of the form

$$u(r,\theta,\phi) = R(r)\,Y_l^m(\theta,\phi) \tag{7.1}$$

where the spherical harmonics $Y_l^m(\theta,\phi)$ are eigenfunctions of the angular momentum operators $\hat{L}^2$ and $\hat{L}_z$. Our task now is to find the form of the radial factor $R(r)$ in equation (7.1), for a variety of specific forms for the potential $V(r)$.

〉 7.2 The rigid rotator

Imagine an isolated system consisting of a particle of mass m constrained to move upon the surface of a sphere of radius a. The constraint may take the form of a massless rigid rod attached at one end to a fixed universal pivot, with the mass mounted on its other end. It is apparent that for this system the radial factor in equation (7.1) is simply $\delta(r-a)$, which we will suppress and take it as understood that the particle moves so that $r = a$. Since radial motion of the mass is excluded and the system has no way of storing potential energy, the total energy of the system arises entirely from the rotational motion of the mass and can be written

$$E = L^2/2I = L^2/2ma^2 \tag{7.2}$$

where I is the moment of inertia of the mass about the pivot. The time-independent Schrödinger equation

$$\hat{H}u(r, \theta, \phi) = Eu(r, \theta, \phi) \tag{7.3}$$

becomes for this system

$$\frac{\hat{L}^2}{2ma^2}\, u(\theta, \phi) = Eu(\theta, \phi) \tag{7.4}$$

which upon a trivial rearrangement can be recognized as the eigenvalue equation for $\hat{L}^2$:

$$\hat{L}^2 u(\theta, \phi) = 2ma^2 Eu(\theta, \phi). \tag{7.5}$$

The eigenvalues and eigenfunctions for this equation were discussed in Chapter 6 and are

$$L^2 = 2ma^2 E = l(l+1)\hbar^2 \qquad l = 0, 1, 2, \ldots \tag{7.6}$$

$$u(\theta, \phi) = Y_l^m(\theta, \phi) \tag{7.7}$$

respectively.

These functions are simultaneous eigenfunctions of the Hamiltonian and angular momentum operators, $\hat{H}$, $\hat{L}^2$ and $\hat{L}_z$. Equation (7.6) shows that the energy of the system is entirely determined by its angular momentum, and is

$$E = \frac{l(l+1)\hbar^2}{2ma^2} \qquad l = 0, 1, 2, \ldots. \tag{7.8}$$

The rigid rotator described above is completely unrealistic, but nevertheless the rotational motion of molecules can be modelled quite precisely by this analysis. Consider, for example, the diatomic molecule shown in figure 7.1(a), consisting of atoms of masses m_1 and m_2. If we assume that the separation of the two atoms is a and cannot change, and that we are using a frame of reference in which the centre of mass of the molecule is at rest, then again the energy of the system arises entirely from its rotational motion. As usual (see Appendix 3) we can describe the two-body system rotating about its centre of mass in terms of a fictitious single particle of reduced mass $\mu = m_1 m_2/(m_1 + m_2)$, moving about a fixed centre, in this case at the fixed distance a, as in figure 7.1(b). The above analysis then applies to the molecule provided we replace m by μ.

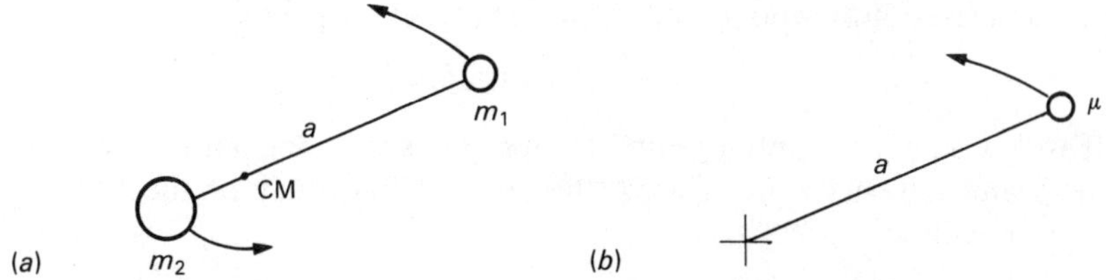

Figure 7.1 (*a*) A rotating diatomic molecule. (*b*) An equivalent reduced mass system.

The energy level diagram for such a molecule is shown in figure 7.2. The emission spectrum is found to consist of a number of equally spaced lines, also shown in figure 7.2. These lines correspond to transitions in which the angular momentum quantum number l changes by 1. The absence of transitions between any but adjacent levels can be understood as a manifestation of the correspondence principle. According to this principle, the radiation emitted by the system when it is in a state of large l should correspond to the radiation which would be emitted classically.

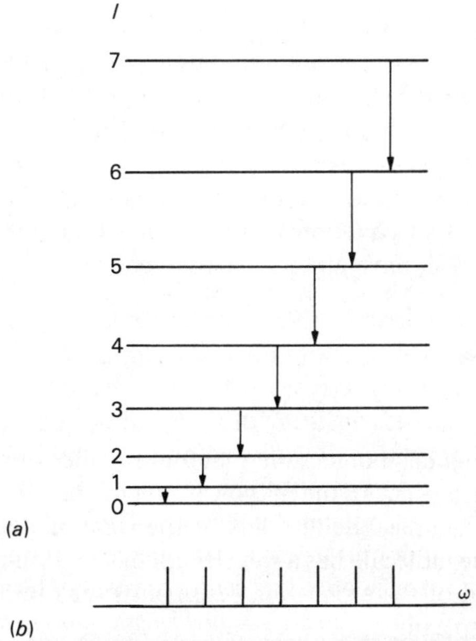

Figure 7.2 (*a*) The energy levels of a rotator, and (*b*) the resulting spectrum.

A classical rotator would emit at the frequency of rotation,

$$\omega = L/I = L/\mu a^2. \tag{7.9}$$

If we consider a quantum jump between states with angular momentum quantum numbers l and $l - \Delta l$, the angular frequency of the emitted photon will be

$$\omega = (E_l - E_{l-\Delta l})/\hbar. \tag{7.10}$$

Substituting for the energies using (7.8) we find

$$\begin{aligned}\omega &= [l(l+1) - (l-\Delta l)(l-\Delta l+1)]\hbar/2\mu a^2 \\ &= [2l\Delta l - \Delta l(\Delta l - 1)]\hbar/2\mu a^2.\end{aligned} \tag{7.11}$$

Two conditions are required to make this frequency correspond to that of equation (7.9). They are that

$$\Delta l = 1 \tag{7.12}$$

and that the quantum number l is large, so that

$$L \approx l\hbar. \tag{7.13}$$

Equation (7.12) is an example of a quantum mechanical selection rule, which observations confirm. It is interesting that the requirement that the system's behaviour should approach an appropriate classical limit allows us to deduce characteristics which remain true even for very small quantum numbers. Rigorous proofs of such rules require time-dependent perturbation theory and are beyond the scope of this book.

Inserting $\Delta l = 1$ in equation (7.11) we find that quite generally the emitted photons will have an angular frequency

$$\omega = l\hbar/\mu a^2 \tag{7.14}$$

and a wavelength

$$\lambda = 2\pi c\mu a^2/l\hbar \tag{7.15}$$

where l is the angular momentum quantum number of the upper level. Rotational spectra occur in the microwave region of the electromagnetic spectrum. For example, the first line of the rotational spectrum of the carbon monoxide molecule has a wavelength of 2.602 mm. Knowledge of this quantity and of the atomic masses permits the direct calculation of the interatomic spacing a, using equation (7.15). For carbon monoxide a is 0.1131 nm.

It should be noted that no radiation at all is to be expected classically unless the molecule has an electric dipole moment. Emission requires an oscillating dipole component, and this is provided by the rotation of the molecular dipole. This restriction applies in quantum mechanics also and one consequence is that homonuclear molecules such as H_2, O_2, etc, have no rotational spectrum, since symmetry dictates that there can be no dipole moment. In a molecule such as CO, on the other hand, the centres of mass and of charge will not coincide, and rotation will result in the required oscillating dipole moment.

Analysis of a molecule on the basis that it is a rigid rotator is an approximation. In reality molecules are elastic and the interatomic separation can vary as a result of molecular vibrations, studied later in this chapter. Even in the absence of vibration the spacing will vary as a result of centrifugal force. The molecule will stretch as the angular momentum increases. This removes the strict equality of spacing between the spectral lines.

〉7.3 The radial equation

Analysis of systems less restricted than the rigid rotator require the solution of the radial equation. This is obtained as follows. As a result of the analysis of section 6.4 we know that the TISE for a particle moving under the influence of a central force can be written

$$\hat{H}R(r)\,Y_l^m(\theta,\phi) = ER(r)\,Y_l^m(\theta,\phi) \tag{7.16}$$

where the Hamiltonian has the form

$$\hat{H} = -\frac{\hbar^2}{2m}\frac{1}{r^2}\frac{\partial}{\partial r}r^2\frac{\partial}{\partial r} + \frac{\hat{L}^2}{2mr^2} + V(r). \tag{7.17}$$

Allowing the operator $\hat{L}^2$ to operate on the function Y_l^m in equation (7.16) we arrive at the radial wave equation:

$$\left[-\frac{\hbar^2}{2mr^2}\frac{\mathrm{d}}{\mathrm{d}r}\left(r^2\frac{\mathrm{d}}{\mathrm{d}r}\right) + \frac{l(l+1)\hbar^2}{2mr^2} + V(r)\right]R(r) = ER(r). \tag{7.18}$$

This equation can be simplified by expressing it in terms of $\chi(r)$, where

$$\chi(r) = rR(r). \tag{7.19}$$

Making this substitution and performing the indicated differentiations leads to

$$\left(-\frac{\hbar^2}{2m}\frac{\mathrm{d}^2}{\mathrm{d}r^2} + \frac{l(l+1)\hbar^2}{2mr^2} + V(r)\right)\chi(r) = E\chi(r). \tag{7.20}$$

Let us first consider the term $l(l+1)\hbar^2/2mr^2$. Although this term occurs in the Hamiltonian representing the kinetic energy associated with the angular motion of the particle, now that the operator aspect of the term has disappeared and it is simply a function of r it has all the appearance of a potential energy term. It is only present if the angular momentum is non-zero and is usually referred to as the centrifugal potential, or centrifugal barrier. The reason is that it represents an effective repulsive force, which becomes very large near the origin, and acts to reduce the likelihood of finding the electron in this region. Since the term actually becomes infinite for $r=0$ we expect zero probability for the electron to be found there, unless the angular momentum is zero. This parallels the classical situation. Since the angular momentum $\boldsymbol{L}=\boldsymbol{r}\times\boldsymbol{p}$ is constant, $\boldsymbol{r}$ could not be zero unless $\boldsymbol{p}$ was infinite.

We will invariably be dealing with potentials which are no more singular at the origin than $1/r$ and so for small r the centrifugal potential will be the dominant term in equation (7.20) (unless of course $l=0$). We can examine the behaviour of the wavefunction in this region by considering the approximation to equation (7.20) obtained by neglecting the terms in $V(r)$ and E:

$$\frac{d^2}{dr^2}\chi(r)=\frac{l(l+1)}{r^2}\chi(r). \tag{7.21}$$

This equation is satisfied by

$$\chi(r)=br^s \qquad b=\text{constant} \tag{7.22}$$

provided

$$s=-l \qquad \text{or} \qquad s=l+1. \tag{7.23}$$

Since we cannot allow the wavefunction to be infinite anywhere, only the value $s=l+1$ is acceptable, and we find the general result

$$\chi(r)\propto r^{l+1} \qquad \text{and} \qquad R(r)\propto r^l \qquad \text{as } r\to 0. \tag{7.24}$$

This result indicates that the probability density is zero at the origin and small over a distance which increases with the value of l, just as we would expect from the angular momentum argument above. If $l=0$ the analysis does not hold and the form of the wavefunction near the origin is determined by the form of $V(r)$.

〉7.4 The hydrogen atom and hydrogen-like ions

We are now in a position to complete our discussion of the eigenfunctions and eigenvalues of the hydrogen atom and hydrogen-like ions begun in section 4.4. In that section we considered only spherically symmetric solutions, thereby excluding all but zero angular momentum states.

Inserting the appropriate Coulomb potential energy term into equation (7.20) and introducing the reduced mass we have

$$\left(-\frac{\hbar^2}{2\mu}\frac{\mathrm{d}^2}{\mathrm{d}r^2}+\frac{l(l+1)\hbar^2}{2\mu r^2}-\frac{Ze^2}{4\pi\varepsilon_0 r}\right)\chi(r)=E\chi(r). \tag{7.25}$$

Making the substitutions

$$\rho=(8\mu\,|\,E\,|\,)^{1/2}r/\hbar \tag{7.26}$$

and

$$\beta=(\mu/2\,|\,E\,|)^{1/2}Ze^2/4\pi\varepsilon_0\hbar \tag{7.27}$$

we arrive at

$$\left(\frac{\mathrm{d}^2}{\mathrm{d}\rho^2}-\frac{l(l+1)}{\rho^2}+\frac{\beta}{\rho}-\frac{1}{4}\right)\chi(\rho)=0 \tag{7.28}$$

which is identical to equation (4.26) apart from the term in $l(l+1)$. Following previous practice we first examine the behaviour as $\rho\to\infty$. In that case the middle two terms in (7.28) may be neglected and the equation is satisfied by

$$\chi(\rho)=\mathrm{e}^{-\rho/2} \tag{7.29}$$

as was the case for the s-state solutions.

We have now found the behaviour for $\rho\to 0$ (7.24) and $\rho\to\infty$ (7.29). Accordingly we attempt to find an exact solution of the form

$$\chi(\rho)=\rho^{l+1}L(\rho)\mathrm{e}^{-\rho/2} \tag{7.30}$$

where

$$L(\rho)=\sum_{p=0}^{\infty}a_p\rho^p. \tag{7.31}$$

Substitution of equation (7.30) into (7.28) leads to the following equation for $L(\rho)$:

$$\rho \frac{\mathrm{d}^2}{\mathrm{d}\rho^2} L(\rho) + (2l + 2 - \rho) \frac{\mathrm{d}}{\mathrm{d}\rho} L(\rho) + (\beta - l - 1)L(\rho) = 0. \quad (7.32)$$

Inserting equation (7.31) into (7.32) we arrive at the recurrence relation

$$\frac{a_{p+1}}{a_p} = \frac{p + l + 1 - \beta}{(p+1)(p + 2l + 2)}. \quad (7.33)$$

As was the case for the s states, this relation indicates that the series for $L(\rho)$ behaves as $\exp(\rho)$ for large p, so that the wavefunction (7.30) diverges unless $L(\rho)$ can be truncated. Examination of equation (7.33) shows that the series will terminate after a_p if

$$\beta = p + l + 1 = n \quad (7.34)$$

where we have introduced the quantum number n, known as the principal quantum number. Since p is an integer $\geqslant 0$ it is apparent that n must be an integer and satisfy

$$n \geqslant l + 1. \quad (7.35)$$

Substituting $\beta = n$ into equation (7.27) we obtain an expression for the energy eigenvalues

$$E_n = \frac{-Z^2 e^4 \mu}{2(4\pi\varepsilon_0)^2 \hbar^2} \frac{1}{n^2}. \quad (7.36)$$

This is identical to the values obtained previously for the s states, as it should be since this more general treatment is equally valid for $l = 0$. More surprisingly, the energies are independent of l. This result is unique to the Coulomb potential. What we have discovered is that, for each value of n, there are n values of l, namely $n - 1$, $n - 2, \ldots, 0$. Although these correspond to distinct states they have the same energy and are thus degenerate. In fact the degeneracy is even higher than this since for each value of l there are $2l + 1$ values of the quantum number m. The energy is independent of m since this quantum number does not even appear in the radial equation. This is to be expected because m is related to the orientation of the angular momentum and since the system has spherical symmetry the energy could not depend upon orientation (see section 6.8 on the Zeeman effect). We therefore find that the overall degeneracy for a particular value of n is given by the arithmetical progression

$$\sum_{l=0}^{n-1} (2l + 1) = \tfrac{1}{2} n[2(n-1) + 1 + 1] = n^2. \quad (7.37)$$

Later we will find that inclusion of electron spin raises this to $2n^2$.

An energy level diagram for hydrogen is shown in figure 7.3. Following common practice, states of the same n but different l are shown separated horizontally, although they have the same energy. The exact degeneracy we have been discussing holds only for the pure Coulomb potential. Inclusion of the effects of electron spin and relativistic effects leads to small differences in energy, referred to as fine structure, which however will not be discussed further in this book.

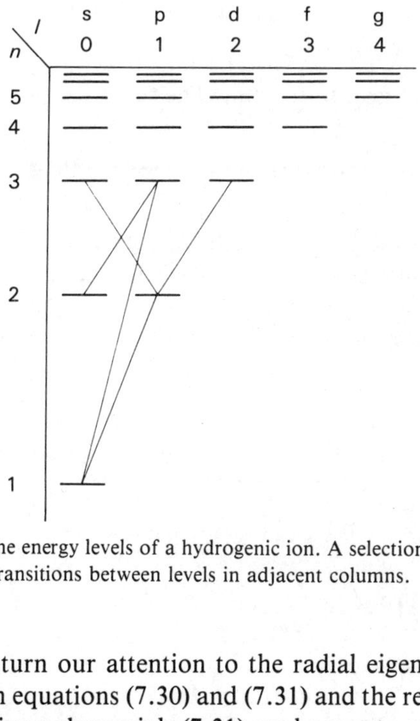

Figure 7.3 The energy levels of a hydrogenic ion. A selection rule $\Delta l = \pm 1$ permits only transitions between levels in adjacent columns.

We must now turn our attention to the radial eigenfunctions. These are obtained from equations (7.30) and (7.31) and the recurrence relation (7.33). The resulting polynomials (7.31) are known to mathematicians as the associated Laguerre polynomials

$$L(\rho) = L_{n+l}^{2l+1}(\rho). \tag{7.38}$$

They can be expressed explicitly as

$$L_{n+l}^{2l+1}(\rho) = \sum_{k=0}^{n-l-1} (-1)^{k+1} \frac{[(n+l)!]^2}{(n-l-1-k)!(2l+1+k)!k!} \rho^k. \tag{7.39}$$

(Pure mathematicians use a slightly different nomenclature, $L_{n+1}^{(2l+1)}(\rho)$, which is equivalent to our $L_{n+l}^{2l+1}(\rho)$.)

The radial wavefunction now takes the form

$$R_{n,l}(\rho) = -N_{n,l}\rho^l L_{n+l}^{2l+1}(\rho)\,\mathrm{e}^{-\rho/2} \tag{7.40}$$

where

$$\rho = 2Zr/na_0 \tag{7.41}$$

and

$$N_{n,l} = \left[\left(\frac{2Z}{na_0}\right)^3 \frac{(n-l-1)!}{2n[(n+l)!]^3}\right]^{1/2}. \tag{7.42}$$

(The minus sign in equation (7.40) is introduced by physicists who prefer $R(\rho)$ to be positive near the origin.) The functions are normalized so that

$$\int_0^\infty r^2 R_{n,l}^2(r)\,\mathrm{d}r = \int_0^\infty D_{n,l}(r)\,\mathrm{d}r = 1. \tag{7.43}$$

It will be recalled that the radial distribution function $D(r)$ represents the probability per unit radial distance of the electron being found at a distance r from the origin, integrated over all angles. Table 7.1 lists the first few radial wavefunctions $R_{n,l}(r)$, and they are illustrated in figure 7.4.

Table 7.1 Normalized radial wavefunctions for hydrogen-like ions.

$$R_{1,0}(r) = \left(\frac{Z}{a_0}\right)^{3/2} 2\exp\left(-\frac{Zr}{a_0}\right)$$

$$R_{2,0}(r) = \left(\frac{Z}{2a_0}\right)^{3/2} 2\left(1-\frac{Zr}{2a_0}\right)\exp\left(-\frac{Zr}{2a_0}\right)$$

$$R_{2,1}(r) = \left(\frac{Z}{2a_0}\right)^{3/2} \frac{2}{\sqrt{3}}\frac{Zr}{2a_0}\exp\left(-\frac{Zr}{2a_0}\right)$$

$$R_{3,0}(r) = \left(\frac{Z}{3a_0}\right)^{3/2} 2\left[1-2\frac{Zr}{3a_0}+\frac{2}{3}\left(\frac{Zr}{3a_0}\right)^2\right]\exp\left(-\frac{Zr}{3a_0}\right)$$

$$R_{3,1}(r) = \left(\frac{Z}{3a_0}\right)^{3/2} 4\sqrt{2}\,\frac{Zr}{3a_0}\left(1-\frac{1}{2}\frac{Zr}{3a_0}\right)\exp\left(-\frac{Zr}{3a_0}\right)$$

$$R_{3,2}(r) = \left(\frac{Z}{3a_0}\right)^{3/2} \frac{2\sqrt{2}}{3\sqrt{5}}\left(\frac{Zr}{3a_0}\right)^2\exp\left(-\frac{Zr}{3a_0}\right)$$

In the case of the s states we noted that $D_{n,0}(r)$ showed a maximum at $r = n^2 a_0/Z$, which is the radius of the corresponding Bohr orbit. This property is common to all states for which $l = n - 1$. Such states have the maximum allowed angular momentum for a given n, and correspond most closely to circular orbits. To show this we notice first that, for such states, $L_{n+l}^{2l+1}(\rho)$ is a constant (equation (7.39)), so that

$$D_{n,n-1}(\rho) = N_{n,n-1}^2 \rho^{2n} \, e^{-\rho}. \tag{7.44}$$

Hence

$$\frac{dD}{d\rho} = N_{n,n-1}^2 \rho^{2n-1} \, e^{-\rho}(2n - \rho) \tag{7.45}$$

showing that D has a maximum at $\rho = 2n$, or $r = n^2 a_0/Z$, as before.

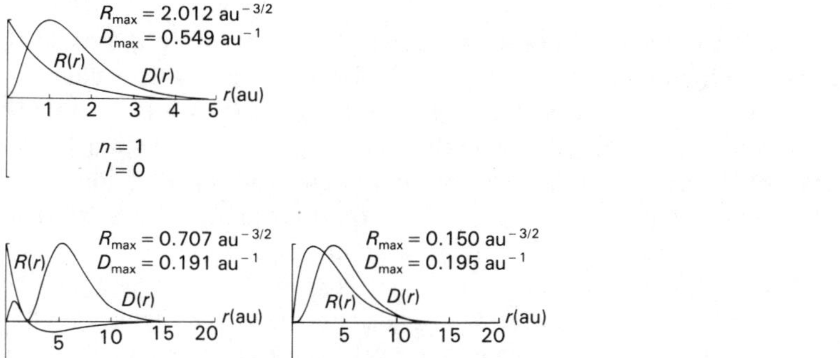

Figure 7.4 The radial-wave wavefunctions and probability densities for hydrogenic ions.

Computer exercise 7.1: hydrogen-like ions

Run the program 'Hydrogen-like Ions' again. You have used it already to examine the s-state wavefunctions. The program plots the potential energy function before it plots the wavefunction. In the case $l \neq 0$ what is plotted is the effective potential, which includes the centrifugal potential.

(*a*) Plot in turn the radial wavefunctions for $n = 3$ and $l = 0, 1, 2$. Some of the things to observe are:

(i) The centrifugal barrier, absent for $l = 0$, appears for $l = 1$ and becomes more pronounced for $l = 2$.

(ii) Only the function for $l = 0$ is finite at the origin. The other functions are zero at the origin and become progressively flatter near the origin as l increases, in accordance with the analysis leading to equation (7.24).

(iii) The wavefunction exhibits $n - l - 1$ nodes. This is to be expected since the functions $L_{n+l}^{2l+1}(\rho)$ are polynomials of degree $n - l - 1$. (See equation (7.34).)

(iv) In all cases the wavefunction decays at large distances in an exponential way, in accordance with equation (7.40).

(*b*) Use the same quantum numbers as before but have a look at the radial distribution function, $D_{n,l}(r)$. Notice how the r^2 factor causes all functions to fall to zero at the origin, even for $l = 0$. It also weights the function heavily for large r. Note that $D_{3,2}(r)$ peaks at $r = 9$, which is the radius of the third Bohr orbit (measured in units of a_0). Plot some more radial density functions for $l = n - 1$ states (maximum l for a given n). They always peak at $r = n^2 a_0$, as shown above.

(*c*) While plotting $D_{n,l}(r)$ notice the relationship between it and the effective potential. A marker is provided to show the energy level. A horizontal line at this level would intersect the effective potential curve at the classical limits of motion. The bulk of the distribution $D_{n,l}(r)$ lies between these points. There is in fact a point of inflection in the curves at the classical turning points.

(*d*) Try some illegal values of n and l, i.e. $l \geqslant n$ or l, n non-integral. You should find that the wavefunction always diverges in such cases.

We have studied the angular and radial parts of the hydrogen wavefunctions separately; it remains to assemble the complete wavefunction and attempt to interpret it. The complete normalized hydrogenic wavefunctions can be expressed as

$$u_{n,l,m}(r, \theta, \phi) = N_{n,l}\left(\frac{2Z}{na_0} r\right)^l L_{n+l}^{2l+1}\left(\frac{2Z}{na_0}\right) \exp(-Zr/na_0) Y_l^m(\theta, \phi). \tag{7.46}$$

The overall parity of the wavefunctions can be found by noting that the parity operation leaves r unchanged. Consequently the parity is that of $Y_l^m(\theta,\phi)$, which we have already shown to be $(-1)^l$. Table 7.2 lists all the eigenfunctions for $n \leqslant 3$. Visualizing the resulting probability distribution is not easy and here the computer can help us.

Table 7.2 Normalized wavefunctions for hydrogen-like ions.

$$u_{1,0,0}(r,\theta,\phi) = \frac{1}{\sqrt{\pi}}\left(\frac{Z}{a_0}\right)^{3/2} \exp(-Zr/a_0)$$

$$u_{2,0,0}(r,\theta,\phi) = \frac{1}{\sqrt{\pi}}\left(\frac{Z}{2a_0}\right)^{3/2}\left(1-\frac{Zr}{2a_0}\right)\exp(-Zr/2a_0)$$

$$u_{2,1,0}(r,\theta,\phi) = \frac{1}{\sqrt{\pi}}\left(\frac{Z}{2a_0}\right)^{3/2}\left(\frac{Zr}{2a_0}\right)\exp(-Zr/2a_0)\cos\theta$$

$$u_{2,1,\pm1}(r,\theta,\phi) = \frac{1}{\sqrt{2\pi}}\left(\frac{Z}{2a_0}\right)^{3/2}\left(\frac{Zr}{2a_0}\right)\exp(-Zr/2a_0)\sin\theta\, e^{\pm i\phi}$$

$$u_{3,0,0}(r,\theta,\phi) = \frac{1}{3\sqrt{\pi}}\left(\frac{Z}{3a_0}\right)^{3/2}\left[3-6\left(\frac{Zr}{3a_0}\right)+2\left(\frac{Zr}{3a_0}\right)^2\right]\exp(-Zr/3a_0)$$

$$u_{3,1,0}(r,\theta,\phi) = \frac{\sqrt{2}}{\sqrt{3\pi}}\left(\frac{Z}{3a_0}\right)^{3/2}\left(2-\frac{Zr}{3a_0}\right)\left(\frac{Zr}{3a_0}\right)\exp(-Zr/3a_0)\cos\theta$$

$$u_{3,1,\pm1}(r,\theta,\phi) = \frac{1}{\sqrt{3\pi}}\left(\frac{Z}{3a_0}\right)^{3/2}\left(2-\frac{Zr}{3a_0}\right)\left(\frac{Zr}{3a_0}\right)\exp(-Zr/3a_0)\sin\theta\, e^{\pm i\phi}$$

$$u_{3,2,0}(r,\theta,\phi) = \frac{1}{3\sqrt{2\pi}}\left(\frac{Z}{3a_0}\right)^{3/2}\left(\frac{Zr}{3a_0}\right)^2 \exp(-Zr/3a_0)(3\cos^2\theta-1)$$

$$u_{3,2,\pm1}(r,\theta,\phi) = \frac{1}{\sqrt{3\pi}}\left(\frac{Z}{3a_0}\right)^{3/2}\left(\frac{Zr}{3a_0}\right)^2 \exp(-Zr/3a_0)\sin\theta\cos\theta\, e^{\pm i\phi}$$

$$u_{3,2,\pm2}(r,\theta,\phi) = \frac{1}{2\sqrt{3\pi}}\left(\frac{Z}{3a_0}\right)^{3/2}\left(\frac{Zr}{3a_0}\right)^2 \exp(-Zr/3a_0)\sin^2\theta\, e^{\pm 2i\phi}$$

Computer exercise 7.2: electron densities in hydrogen

The program 'Electron Densities in Hydrogen' simulates and plots the results of a series of imaginary experiments in which the positions of the electrons in an ensemble of hydrogen atoms, each known to be in the state $u_{n,l,m}$, are measured. The result of each measurement is a precise position, but this changes from measurement to measurement since the electrons are not in position eigenstates. In this way we build up a picture of the probability distribution. Since we must present the results in two dimensions what we see is actually a projection of the distribution onto a plane, and we lose depth perception.

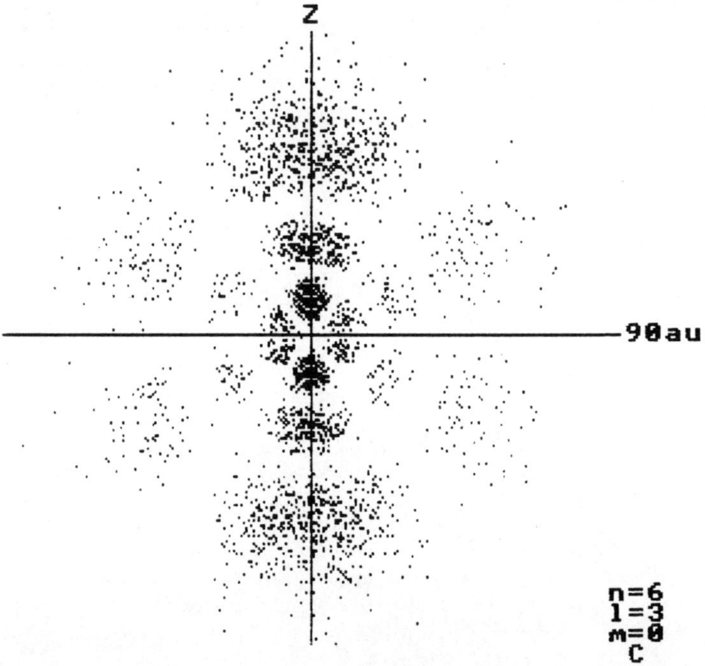

Figure 7.5 A computer simulation of the electron density distribution in the $n = 6$, $l = 3$, $m_l = 0$ eigenstate of the hydrogen atom.

The preceding paragraph is carefully worded so as to be in accord with the accepted quantum mechanical interpretation of the measurement process. Nevertheless it is hard to resist the idea that the picture we build up represents what we would 'see' if we could somehow observe an atom in a non-perturbing fashion. Naturally in that case we would find the outer layers obstructing our view of the inner layers, which is frustrating. Fortunately the computer can deal with this difficulty and present us with a cross-sectional view of the atom, much as computerized x-ray tomography can give us a cross section of the human body. An example is shown in figure 7.5.

The program stores all the wavefunctions for $n \leqslant 6$, and any distribution can be plotted as a projection or as a cross section. Much more detail is visible in the cross-sectional plots. It should be remembered that in all cases the distributions exhibit axial symmetry.

(*a*) Examine the distributions for some s states ($l = m = 0$). $|Y_0^0|^2 = 1/4\pi$, a constant, for these states, which therefore exhibit spherical symmetry. The distribution is a series of n concentric shells. Note the rapid increase in the scale of the atom with n ($\propto n^2$).

(*b*) Compare the distribution for $n = 2$, $l = 1$, $m = 1$ with that for $n = 6$, $l = 5$, $m = 5$. States with $l = n - 1$ and $l = m$ correspond most closely to a classical orbit in the xy plane. Note how much more compact the $n = 6$ distribution is compared with the $n = 2$, in accordance with the correspondence principle. The distribution cross section for a classical orbit would appear as single spots at $r = n^2 a_0$, $\theta = \frac{1}{2}\pi$.

(*c*) Plot some distributions with $m = 0$. Only in such cases is the probability on the z axis finite.

(*d*) Sketch in advance the general features of the distribution for $n = 6$, $l = 4$, $m = 0$. Do this on the basis that there should be $n - l - 1$ nodal spheres, and $l - m$ nodal cones. For this purpose $r = 0, \infty$ are not counted as nodes. Nor do $\theta = 0, \pi$ count even though the density may or may not be zero there depending upon the value of m. Now plot the distribution and compare it with your prediction.

It should be realized that, although the $u_{n,l,m}(r, \theta, \phi)$ which we have discovered for the hydrogen atom form a complete set of eigenfunctions, there is no requirement for the wavefunction of given atom to be one of them. On the other hand, by the expansion postulate, an arbitrary hydrogenic wavefunction can be expressed in terms of these functions. For specific situations it may be appropriate to take particular linear combinations. This is the case for example when electron spin and its effects are introduced. As a simpler example, consider the case of a hydrogen atom in an environment which exhibits symmetry with respect to the x, y and z axes, as might occur in a crystal or a molecule. It would then be desirable to adopt hydrogenic wavefunctions which exhibit the same symmetry. If we further suppose that the energy and angular momentum of the atom are specified to be those of the 2p states, we can construct suitable wavefunctions as linear combinations of $u_{2,1,1}$, $u_{2,1,0}$ and $u_{2,1,-1}$ as follows.

The basis functions are

$$\begin{aligned} u_{2,1,1} &= Ar\,\mathrm{e}^{-r/2a_0}\sin\theta\,\mathrm{e}^{\mathrm{i}\phi} \\ u_{2,1,0} &= \sqrt{2}Ar\,\mathrm{e}^{-r/2a_0}\cos\theta \\ u_{2,1,-1} &= Ar\,\mathrm{e}^{-r/2a_0}\sin\theta\,\mathrm{e}^{-\mathrm{i}\phi}. \end{aligned} \tag{7.47}$$

Consider the normalised linear superposition:

$$\begin{aligned} u_{2,1,x} &= \frac{1}{\sqrt{2}}(u_{2,1,1}+u_{2,1,-1}) \\ &= \frac{A}{\sqrt{2}}r\sin\theta(\mathrm{e}^{\mathrm{i}\phi}+\mathrm{e}^{-\mathrm{i}\phi})\,\mathrm{e}^{-r/2a_0} \\ &= \sqrt{2}Ax\,\mathrm{e}^{-r/2a_0}. \end{aligned} \tag{7.48}$$

Similarly

$$\begin{aligned} u_{2,1,y} &= -\frac{\mathrm{i}}{\sqrt{2}}(u_{2,1,1}-u_{2,1,-1}) \\ &= \sqrt{2}Ay\,\mathrm{e}^{-r/2a_0} \end{aligned} \tag{7.49}$$

and

$$\begin{aligned} u_{2,1,z} &= u_{2,1,0} \\ &= \sqrt{2}Az\,\mathrm{e}^{-r/2a_0}. \end{aligned} \tag{7.50}$$

These three functions are identical apart from the x, y and z factors and represent distributions oriented along these three axes. The distribution $|u_{2,1,z}|^2 \equiv |u_{2,1,0}|^2$ can be examined with the program 'Electron Densities in Hydrogen', and the other two are identical apart from orientation.

The new functions remain eigenfunctions of $\hat{H}$ and $\hat{L}^2$, but only $u_{2,1,z}$ is an eigenfunction of $\hat{L}_z$. It may be verified that $u_{2,1,x}$ and $u_{2,1,y}$ are eigenfunctions of $\hat{L}_x$ and $\hat{L}_y$ respectively. The operators for these observables can be found in equations (6.5) and (6.9).

〉7.5 The isotropic harmonic oscillator

A particle free to oscillate harmonically in three dimensions has a potential energy which can be written as

$$V(x,y,z) = \tfrac{1}{2}m\omega_x^2x^2 + \tfrac{1}{2}m\omega_y^2y^2 + \tfrac{1}{2}m\omega_z^2z^2. \tag{7.51}$$

A classical example would be a particle constrained by three mutually orthogonal springs of unequal strength. (For small oscillations we assume that motion in one dimension leaves the other two spring lengths unaffected.) The TISE for the system is

$$\left(-\frac{\hbar^2}{2m}\nabla^2 + V(x,y,z)\right)u(x,y,z) = Eu(x,y,z) \qquad (7.52)$$

and this is readily separable in cartesian coordinates leading to the three one-dimensional harmonic oscillator equations:

$$\left(-\frac{\hbar^2}{2m}\frac{\mathrm{d}^2}{\mathrm{d}x^2} + \tfrac{1}{2}m\omega_x^2x^2\right)X(x) = E_xX(x)$$

$$\left(-\frac{\hbar^2}{2m}\frac{\mathrm{d}^2}{\mathrm{d}y^2} + \tfrac{1}{2}m\omega_y^2y^2\right)Y(y) = E_yY(y) \qquad (7.53)$$

$$\left(-\frac{\hbar^2}{2m}\frac{\mathrm{d}^2}{\mathrm{d}z^2} + \tfrac{1}{2}m\omega_z^2z^2\right)Z(z) = E_zZ(z).$$

The solutions to these equations were shown in section 3.15 to be the Hermite–Gaussian functions. It follows (section 4.2) that the eigenvalues and eigenfunctions for the three-dimensional case are

$$E = E_x + E_y + E_z$$

$$= (n_x + \tfrac{1}{2})\hbar\omega_x + (n_y + \tfrac{1}{2})\hbar\omega_y + (n_z + \tfrac{1}{2})\hbar\omega_z \qquad n_{x,y,z} = 0, 1, 2, \ldots \quad (7.54)$$

and

$$u_{n_xn_yn_z} = X_{n_x}(x)Y_{n_y}(y)Z_{n_z}(z)$$

$$= N_xN_yN_zH_{n_x}(x)H_{n_y}(y)H_{n_z}(z)\exp(-\tfrac{1}{2}\alpha_x^2x^2 - \tfrac{1}{2}\alpha_y^2y^2 - \tfrac{1}{2}\alpha_z^2z^2) \quad (7.55)$$

where

$$\alpha_x = \sqrt{m\omega_x/\hbar} \qquad \text{etc.} \qquad (7.56)$$

If we now consider the special case where $\omega_x = \omega_y = \omega_z = \omega$, the oscillator becomes isotropic since the potential can be written

$$V(r) = \tfrac{1}{2}m\omega^2(x^2 + y^2 + z^2) = \tfrac{1}{2}m\omega^2r^2. \qquad (7.57)$$

The eigenvalues and eigenfunctions can then be written

$$E = (n_x + n_y + n_z + \tfrac{3}{2})\hbar\omega = (n + \tfrac{3}{2})\hbar\omega \qquad n = 0, 1, 2, \ldots \quad (7.58)$$

and

$$u_{n_xn_yn_z} = N_xN_yN_zH_{n_x}(x)H_{n_y}(y)H_{n_z}(z)\exp(-\tfrac{1}{2}\alpha^2r^2). \qquad (7.59)$$

From equation (7.58) it can be seen that the energy levels are degenerate, since n can be made up in a number of different ways by taking different combinations of n_x, n_y and n_z. The degeneracy g_n is thus equal to the number of ways of partitioning n into three components, which is given by

$$g_n = \tfrac{1}{2}(n+1)(n+2). \tag{7.60}$$

Since the potential is now spherically symmetric, according to the analysis of section 6.3 it must be possible to find solutions of the form

$$u_{n,l,m}(r,\theta,\phi) = R_{n,l}(r)\,Y_l^m(\theta,\phi) = r^{-1}\chi_{n,l}(r)\,Y_l^m(\theta,\phi) \tag{7.61}$$

where $\chi(r)$ satisfies the radial equation (7.20), with the potential (7.57):

$$\left(-\frac{\hbar^2}{2m}\frac{\mathrm{d}^2}{\mathrm{d}r^2} + \frac{l(l+1)\hbar^2}{2mr^2} + \tfrac{1}{2}m\omega^2 r^2\right)\chi(r) = E\chi(r). \tag{7.62}$$

The new solutions must be expressible in terms of the old ones, and simply represent an alternative set of basis functions. Solution of equation (7.62) follows the same pattern as before. First we reduce the equation to dimensionless form using the substitutions we used for the one-dimensional oscillator:

$$\rho = \alpha r \qquad \alpha = \sqrt{m\omega/\hbar} \tag{7.63}$$

and

$$\lambda = 2E/\hbar\omega. \tag{7.64}$$

This results in

$$\left(\frac{\mathrm{d}^2}{\mathrm{d}\rho^2} - \frac{l(l+1)}{\rho^2} + \lambda - \rho^2\right)\chi(\rho) = 0 \tag{7.65}$$

which is the same as equation (3.104) apart from the centrifugal potential term, although the function we are considering is not the wavefunction, but $\chi(r)$. The asymptotic behaviour can be found by neglecting the middle two terms in equation (7.65) which are insignificant compared with the others at large ρ. Hence

$$\left(\frac{\mathrm{d}^2}{\mathrm{d}\rho^2} - \rho^2\right)\chi(\rho) = 0 \qquad \rho \to \infty \tag{7.66}$$

and

$$\chi(\rho) = \mathrm{e}^{-\rho^2/2}. \tag{7.67}$$

This prompts us to look for a solution of the form

$$\chi(\rho) = F(\rho)\,\mathrm{e}^{-\rho^2/2} \tag{7.68}$$

and substitution of this into equation (7.65) leads to the equation which must be satisfied by $F(\rho)$:

$$F''(\rho) - 2\rho F'(\rho) + F(\rho)(\lambda - 1) - F(\rho)l(l+1)/\rho^2 = 0. \tag{7.69}$$

For $\rho \to 0$, $\chi(\rho)$ must behave like ρ^{l+1} (equation (7.24)), so we attempt to satisfy equation (7.69) with a series solution:

$$F(\rho) = \rho^{l+1} \sum_{p=0}^{\infty} a_p \rho^p = \sum_{p=0}^{\infty} a_p \rho^{p+l+1}. \tag{7.70}$$

Substitution into equation (7.69) yields

$$\begin{aligned}[(p+l+1)(p+l) - l(l+1)] &\sum_{p=0}^{\infty} a_p \rho^{p+l-1} \\ &= [2(p+l+1) - (\lambda - 1)] \sum_{p=0}^{\infty} a_p \rho^{p+l+1}.\end{aligned} \tag{7.71}$$

Setting the coefficient of ρ^l equal to zero tells us that

$$a_1 = 0. \tag{7.72}$$

In general we find that

$$\frac{a_{p+2}}{a_p} = \frac{2(p+l) + 3 - \lambda}{(p+l+3)(p+l+2) - l(l+1)}. \tag{7.73}$$

For large values of p this series behaves like e^{ρ^2} and so $\chi(\rho)$, equation (7.68), diverges at large ρ. The divergence can only be avoided if the series terminates, which it will do after a_p, provided

$$\lambda = 2(p+l) + 3. \tag{7.74}$$

Now the recurrence formula relates alternate terms of the series and, since $a_1 = 0$, only terms with p even occur. We therefore write $n_r = \frac{1}{2}p$, with n_r a positive integer or zero, whence the $F(\rho)$ are polynomials of degree n_r in ρ^2. Equation (7.74) becomes

$$\lambda = 2(2n_r + l) + 3. \tag{7.75}$$

Because n and l are both integral equation (7.75) has the form

$$\lambda = 2n + 3 \tag{7.76}$$

so that, recalling equation (7.64)

$$E = (n + \tfrac{3}{2})\hbar\omega \tag{7.77}$$

where

$$n = 2n_r + l. \tag{7.78}$$

Now equations (7.77) and (7.58) agree, as they should, since the energies cannot depend upon the way in which we choose to analyse the system. We also require the degeneracies of the levels, given by equation (7.60), to be the same in the two schemes, since the eigenfunctions $u_{n,l,m}$ must be superpositions of u_{n_x,n_y,n_z}. Equation (7.78) tells us that $n - l$ is even, so that n and l are either both odd or both even, and $n \geqslant l$. This information is sufficient to determine the permitted values of l for any n. Table 7.3 gives the allowed combinations of n and l, together with the level degeneracy and the radial quantum number n_r. Because the polynomials are of order n_r in ρ^2, this quantity equals the number of nodes in the radial wavefunction.

Table 7.3 Quantum numbers and degeneracies for the isotropic harmonic oscillator.

n	l	n_r	$2l+1$	g
0	0	0	1	1
1	1	0	3	3
2	0	1	1	6
	2	0	5	
3	1	1	3	10
	3	0	7	
4	0	2	1	15
	2	1	5	
	4	0	9	

The polynomials $F_{n,l}(\rho)$ are in fact confluent hypergeometric functions, and can be generated by means of the recurrence relation (7.73) with $a_0 = 1$. The complete radial wavefunctions take the form

$$R_{n,l}(\rho) = N_{n,l}F(\rho)\,\mathrm{e}^{-\rho^2/2} \tag{7.79}$$

where the $N_{n,l}$ are normalization factors. The functions are listed for $n \leqslant 4$ in table 7.4.

Table 7.4 Radial wavefunctions for the isotropic harmonic oscillator in terms of variable $\rho = \alpha r = (m\omega/\hbar)^{1/2} r$.

$$R_{0,0}(\rho) = N_{0,0} \exp(-\tfrac{1}{2}\rho^2)$$
$$R_{1,1}(\rho) = N_{1,1}\rho \exp(-\tfrac{1}{2}\rho^2)$$
$$R_{2,0}(\rho) = N_{2,0} (1 - \tfrac{2}{3}\rho^2) \exp(-\tfrac{1}{2}\rho^2)$$
$$R_{2,2}(\rho) = N_{2,2}\rho^2 \exp(-\tfrac{1}{2}\rho^2)$$
$$R_{3,1}(\rho) = N_{3,1}\rho(1 - \tfrac{2}{5}\rho^2) \exp(-\tfrac{1}{2}\rho^2)$$
$$R_{3,3}(\rho) = N_{3,3}\rho^3 \exp(-\tfrac{1}{2}\rho^2)$$
$$R_{4,0}(\rho) = N_{4,0}(1 - \tfrac{4}{3}\rho^2 + \tfrac{4}{15}\rho^4) \exp(-\tfrac{1}{2}\rho^2)$$
$$R_{4,2}(\rho) = N_{4,2}\rho^2(1 - \tfrac{2}{7}\rho^2) \exp(-\tfrac{1}{2}\rho^2)$$
$$R_{4,4}(\rho) = N_{4,4}\rho^4 \exp(-\tfrac{1}{2}\rho^2)$$

Computer exercise 7.3: isotropic harmonic oscillator

Load the program 'General 3-D Potential (Num.)'. Num. refers to the integration technique used in the program, which is the Numerov method. The program will accept any potential function which must be entered in BASIC. (The function must not be more singular at the origin than $1/r$.) For the isotropic harmonic oscillator enter R*R/2 for the potential. (Use upper case R.) This function is valid for all R so both break points should be set to some large value such as 100. (These are in fact the default values.) The atomic units option should be taken, so that the allowed energies are simply $\frac{3}{2}, \frac{5}{2}, \frac{7}{2}, \ldots$, corresponding to $n = 0, 1, 2, \ldots$.

(*a*) Set $l = 0$ and verify that only the values $\frac{3}{2}, \frac{7}{2}, \frac{11}{2}, \ldots$ (corresponding to even values of n) give acceptable wavefunctions. Observe that the wavefunction is finite at the origin in these cases. Compare the form of the function near the origin with the shape for the s states of hydrogen. The latter shows a cusp at the origin, but the oscillator function is smooth, since it has zero gradient at $r = 0$. The difference in behaviour arises because of the singular nature of the hydrogen potential for $r = 0$. Note the number of nodes in the wavefunction; $\frac{1}{2}n$.

(*b*) Verify generally that, for a given value of n, only values of l such that n and l are both odd or both even with $l \leqslant n$ lead to acceptable behaviour. Observe the number of nodes in the wavefunction. There should be $n_r = \frac{1}{2}(n - l)$.

〉7.6 Molecular vibrations: the Morse potential

In section 3.14 it was observed that diatomic molecular vibrations could be described in terms of a quantum oscillator, in which a particle of reduced mass μ oscillated in one dimension about a mean position x_m. More generally the molecule will be able to vibrate in three dimensions and in addition it may rotate, so that the molecular angular momentum must be taken into account. This problem can be treated within the formalism which we have developed here provided we can find a suitable form of spherically symmetric potential to describe the interatomic force. It will be recalled that we require a potential which has the general shape of figure 3.13, but now we regard this as having spherical symmetry, figure 7.6. The central peak is required to prevent the atoms from approaching too closely, and the asymptote at $V(r) = D$ represents the empirical possibility of dissociating the molecule on providing the dissociation energy, D.

There have been many empirical attempts to model this potential by means of an analytic function. One of the most popular is due to Morse, who introduced the following form:

$$V(r) = D\{1 - \exp[-a(r - r_e)]\}^2. \tag{7.80}$$

It can be shown that with this potential and for zero angular momentum Schrödinger's equation (7.20) is exactly soluble, and has energy eigenvalues

$$E_n = (n + \tfrac{1}{2})\hbar\omega_e - (n + \tfrac{1}{2})^2 b\hbar\omega_e \tag{7.81}$$

where

$$\omega_e = a\sqrt{2D/\mu} \tag{7.82}$$

and

$$b = (\hbar\omega_e/4D). \tag{7.83}$$

b is the anharmonicity constant, and ω_e is the angular frequency of oscillation which a classical particle would exhibit if it executed oscillations of infinitesimal amplitude about r_e. We must specify the amplitude because the frequency of oscillation is amplitude dependent for an anharmonic oscillator.

The second term in equation (7.81) describes the departure of the vibrational levels from equal spacing, resulting from the anharmonicity. The higher levels become more crowded. Unlike the harmonic oscillator there is now a finite number of levels, since the well is of finite depth. By differentiating equation (7.81) we find that n has a maximum value

$$n_{\mathrm{max}} = \mathrm{INT}(2D/\hbar\omega_{\mathrm{e}} - \tfrac{1}{2}). \tag{7.84}$$

If the molecule rotates as well as vibrates we expect the energy of equation (7.81) to be increased by approximately the rotational energy specified in equation (7.8):

$$E_l = \frac{l(l+1)\hbar^2}{2\mu r^2}. \tag{7.85}$$

Here we are assuming that the vibrational and rotational motions are completely independent, an assumption referred to as the Born–Oppenheimer approximation. It is a reasonable approximation since the energies involved differ considerably. It fails to the extent that the rotational motion causes the molecule to stretch, changing the potential function. A more subtle point is that, although harmonic vibration leaves the mean value of r unchanged, the mean value of $1/r^2$, which appears in equation (7.85), does depend on the amplitude of oscillation.

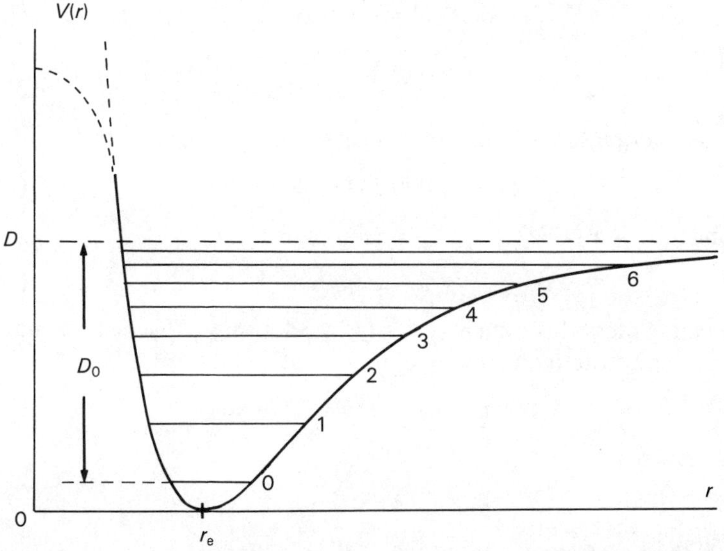

Figure 7.6 The Morse potential-energy curve and energy levels of a vibrating diatomic molecule. D_0 is the energy required to dissociate the molecule when it is in the ground vibrational level: r_e is the equilibrium separation of the atoms. The function is actually finite at the origin, although ideally it should be infinite.

Computer exercise 7.4: the Morse potential

In order to program the computer to investigate the Morse potential we require to evaluate the constants to be used in equation (7.80), and preferably to express them in dimensionless form. The example we will examine is that of molecular hydrogen, H_2. The required data for H_2 can be found in tabulations of molecular constants, and are

$$D_0 = 4.476 \text{ eV} \tag{7.86}$$

$$\hbar\omega_e = 0.5449 \text{ eV} \tag{7.87}$$

$$\mu = 0.504\,066 \text{ amu} \tag{7.88}$$

$$r_e = 0.074\,16 \text{ nm}. \tag{7.89}$$

In the Morse potential D represents the dissociation energy measured from the minimum of the curve. The tabulated value D_0 is relative to the lowest vibrational level, figure 7.6, so the value we require is

$$D = D_0 + (1 - \tfrac{1}{2}b)\hbar\omega_e/2. \tag{7.90}$$

Solving equations (7.90) and (7.83) we find

$$D = 4.745 \text{ eV} \tag{7.91}$$

and

$$b = 2.871 \times 10^{-2}. \tag{7.92}$$

We will adopt $\hbar\omega_e$ as our unit of energy, so

$$D = 4.745/0.5449 = 8.707 \text{ au}. \tag{7.93}$$

From equation (7.82)

$$a = \omega_e\sqrt{\mu/2D}. \tag{7.94}$$

We can express this dimensionlessly by adopting $\sqrt{\hbar/\mu\omega_e}$ as our unit of length, as before (equation (7.63)). We then find

$$a = \sqrt{\hbar\omega_e/2D} = 0.2396 \text{ au} \tag{7.95}$$

and similarly

$$r = 6.01 \text{ au}. \tag{7.96}$$

With these data equation (7.84) indicates that n_{max} should be 16. In practice, H_2 has 15 vibrational levels, so n_{max} is 14. A further discrepancy relates to the anharmonicity constant b, which is found experimentally to be 0.0268, as opposed to our 0.028 71. These inconsistencies simply reflect the impossibility of fitting four parameters (D_e, ω_e, r_e and b) by means of a three-parameter function!

We also note here that in dimensionless form equation (7.85) becomes

$$E_l = \frac{l(l+1)}{2r^2} \qquad r \text{ in au.} \tag{7.97}$$

Finally we have

$$V(r) = 8.707\{1 - \exp[-0.2396(r - 6.01)]\}^2. \tag{7.98}$$

Run the program 'General 3-D Potential (Num.)' and enter the potential of equation (7.98). (In BASIC this becomes: 8.707*(1 − EXP(−.2396*(R − 6.01)))^2.)

(*a*) Examine the ground-state wavefunction. We expect the energy to be approximately $\frac{1}{2}$ in dimensionless units. More precisely, equation (7.81) tells us that it will be $0.5 - 2.871 \times 10^{-2}/4 = 0.492\,82$. Use RI = 2, RF = 12, $l = 0$, $s = 0.1$ and $S = 10^6$. The program integrates the wavefunction starting at RI, with zero amplitude and a very small slope. After a few trials, or by using the eigenvalue-finding routine, it will be found that the actual eigenvalue is 0.492 748. The function has the shape of a slightly distorted Gaussian. Observe the form of the Morse potential curve. By altering 's' you can see it on various scales. Note that it is not in fact infinite at the origin, as it ideally should be in order to preclude the possibility of the two atoms being coincident. However, it is sufficiently large to reduce this probability to a vanishingly small value.

(*b*) Now set $l = 1$. According to equation (7.97) the energy should be increased by approximately $1/(6.01)^2$. Find the new eigenvalue and compare the shift with this estimate. Plot the wavefunction. It appears almost identical to the previous one. (You can compare them using the 'superimpose' option.) There is no limitation on the value l can take for a given n in this case, other than the fact that the energy must remain less than D.

(*c*) Now examine a highly excited vibration level. Hydrogen has 15 vibrational levels and experimentally the highest level occurs for $E = 8.467\,579$. The wavefunction penetrates further in towards $r = 0$ so you will need to start the integration earlier for this state. In the 'forbidden' region the wavefunction increases exponentially, so the earlier you begin the integration the smaller the scale factor you will need. The wavefunction now looks like a distorted Hermite–Gaussian function with 14 nodes (figure 7.7). The peaks are greater near the classical turning points. This becomes more evident when the radial density function $D(r)$ is plotted. This effect takes on a special significance in molecular spectroscopy when electronic transitions are considered. An electronically excited molecule can be considered as a different species, with different equilibrium separation r'_{2e}. Because electronic excitation is almost instantaneous compared to a vibrational period, the internuclear separation r does not change during the process. In the ground state, the most likely value of r is r_{1e}. However, in the excited state this separation may correspond to a stretched state of the molecule, so the molecule will most likely be found in an excited vibrational level of the excited electronic state, for which the peak in $D'(r)$ corresponds to r_{2e}. This effect is referred to as the Franck–Condon principle, and is illustrated in figure 7.8. A similar effect occurs in reverse on electronic de-excitation.

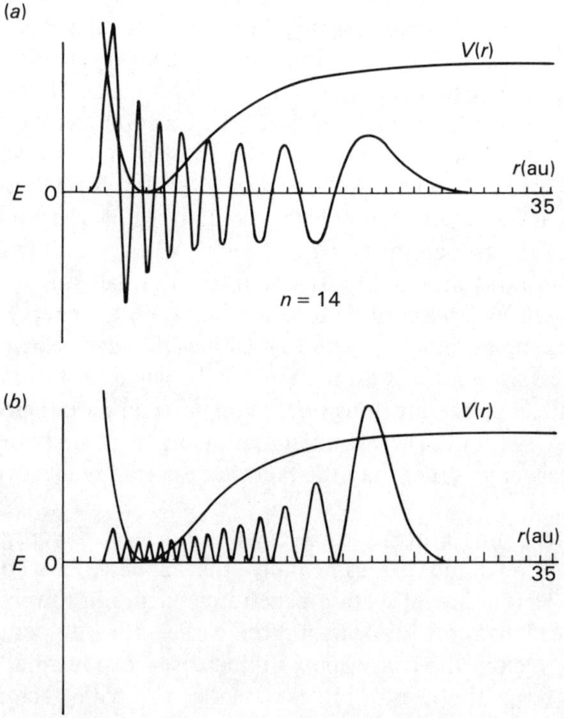

Figure 7.7 (*a*) The radial wavefunction and (*b*) the probability density distribution for the $n = 14$ vibrational state of molecular hydrogen, modelled by the Morse potential. The radial unit of length is equal to 0.0123 nm.

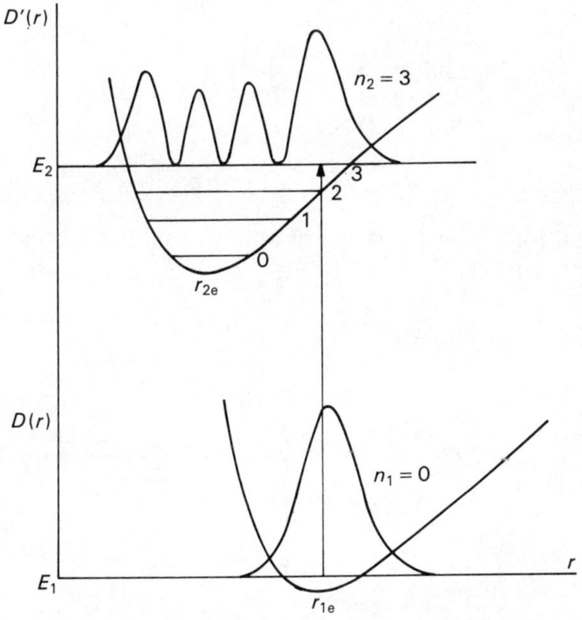

Figure 7.8 The Franck–Condon principle. On excitation from the ground state, where the atoms have only their zero-point motion, to an electronically excited upper state, the molecule is most likely to be found in a vibrational state for which the wavefunction is large at an interatomic separation equal to r_{1e}. This corresponds to the classical turning point, where the atoms are at rest.

〉7.7 The spherical square well

As a final example of a potential well with spherical symmetry we will study the three-dimensional analogue of the finite square-well potential which we examined in section 3.5. Here we consider the bound states of a particle confined in a spherical box defined by the potential function (figure 7.9)

$$V(r) = 0 \qquad r < a$$

$$V(r) = V \qquad r \geqslant a. \tag{7.99}$$

For this system the radial equation (7.20) can be written

$$\left(\frac{\hbar^2}{2m}\frac{\mathrm{d}^2}{\mathrm{d}r^2} + (E - V(r)) - \frac{l(l+1)\hbar^2}{2mr^2}\right)\chi(r) = 0 \tag{7.100}$$

or

$$\left(\frac{\mathrm{d}^2}{\mathrm{d}r^2} + k^2 - \frac{l(l+1)}{r^2}\right)\chi(r) = 0 \tag{7.101}$$

where

$$k^2 = 2m(E - V(r))/\hbar^2. \tag{7.102}$$

We note that for a bound state k will be imaginary in the region external to the well.

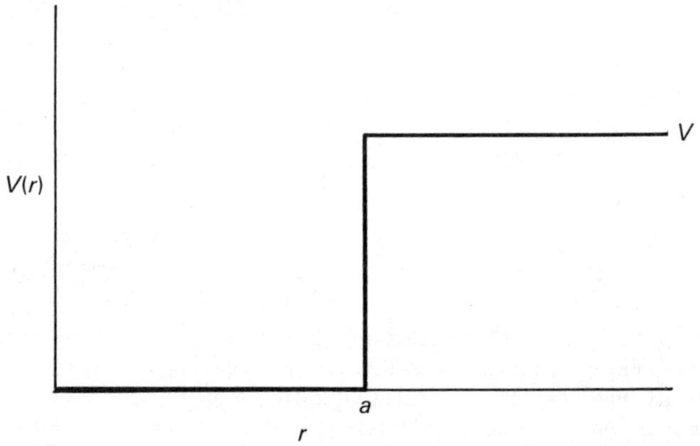

Figure 7.9 The potential energy function of a 'spherical square well'.

We will consider first the form of the solution when k is real. Further simplification of the equation occurs on changing the variable to ρ where

$$\rho = kr. \tag{7.103}$$

Hence

$$\left(\frac{\mathrm{d}^2}{\mathrm{d}\rho^2} + 1 - \frac{l(l+1)}{\rho^2}\right)\chi(\rho) = 0. \tag{7.104}$$

Let us suppose that we have found a solution of equation (7.104), $\chi_l(\rho)$ say. Then it can be verified by direct substitution that the function

$$\chi_{l+1}(\rho) = \frac{l+1}{\rho}\chi_l(\rho) - \frac{\mathrm{d}}{\mathrm{d}\rho}\chi_l(\rho) \tag{7.105}$$

is a solution of

$$\left(\frac{d^2}{d\rho^2} + 1 - \frac{(l+1)(l+2)}{\rho^2}\right)\chi(\rho) = 0. \tag{7.106}$$

Now the solution of equation (7.104) for $l = 0$ is easy to find since the equation becomes

$$\left(\frac{d^2}{d\rho^2} + 1\right)\chi(\rho) = 0 \tag{7.107}$$

which has independent solutions

$$\chi_0(\rho) = \sin\rho \tag{7.108}$$

and

$$\chi_0(\rho) = \cos\rho. \tag{7.109}$$

The general solution is a linear superposition of these.

Successive applications of equation (7.105) to the functions $\sin\rho$ and $\cos\rho$ yield two families of functions $\chi_l(\rho)$. The radial wavefunctions $R_l(\rho)$ are then obtained from

$$\chi_l(\rho) = \rho R_l(\rho). \tag{7.110}$$

Alternatively the functions $R_l(\rho)$ can be found directly from

$$R_{l+1}(\rho) = \frac{l}{\rho} R_l(\rho) - \frac{d}{d\rho} R_l(\rho) \tag{7.111}$$

which follows on substituting equation (7.110) into (7.105). In a similar way it can be shown that

$$R_{l-1}(\rho) = \frac{l+1}{\rho} R_l(\rho) - \frac{d}{d\rho} R_l(\rho) \tag{7.112}$$

and so combining (7.111) and (7.112) we find

$$R_{l+1}(\rho) = \frac{2l+1}{\rho} R_l(\rho) - R_{l-1}(\rho) \tag{7.113}$$

which is the simplest way of generating the solutions once the first two have been found.

The functions $R_l(\rho)$ so obtained are the spherical Bessel functions of the first and second kinds of order l. These are denoted $j_l(\rho)$ and $n_l(\rho)$ respectively. They are also referred to as spherical Bessel and spherical Neumann functions and are tabulated for $l \leqslant 4$ in table 7.5 and illustrated in figure 7.10. Both types of function are oscillatory, but display very different behaviour at the origin. The $j_l(\rho)$ are well behaved there but the $n_l(\rho)$ become infinite.

Table 7.5 Spherical Bessel and Neumann functions in terms of the variable $\rho = kr$.

l	$j_l(\rho)$	$n_l(r)$
0	$\dfrac{\sin\rho}{\rho}$	$-\dfrac{\cos\rho}{\rho}$
1	$\dfrac{\sin\rho}{\rho^2} - \dfrac{\cos\rho}{\rho}$	$-\dfrac{\cos\rho}{\rho^2} - \dfrac{\sin\rho}{\rho}$
2	$\left(\dfrac{3}{\rho^3} - \dfrac{1}{\rho}\right)\sin\rho - \dfrac{3\cos\rho}{\rho^2}$	$-\left(\dfrac{3}{\rho^3} - \dfrac{1}{\rho}\right)\cos\rho - \dfrac{3\sin\rho}{\rho^2}$
3	$\left(\dfrac{15}{\rho^4} - \dfrac{6}{\rho^2}\right)\sin\rho - \left(\dfrac{15}{\rho^3} - \dfrac{6}{\rho}\right)\cos\rho$	$-\left(\dfrac{15}{\rho^4} - \dfrac{6}{\rho^2}\right)\cos\rho - \left(\dfrac{15}{\rho^3} - \dfrac{6}{\rho}\right)\sin\rho$
4	$\left(\dfrac{105}{\rho^5} - \dfrac{45}{\rho^3} + \dfrac{1}{\rho}\right)\sin\rho - \left(\dfrac{105}{\rho^4} - \dfrac{10}{\rho^2}\right)\cos\rho$	$-\left(\dfrac{105}{\rho^5} - \dfrac{45}{\rho^3} + \dfrac{1}{\rho}\right)\cos\rho - \left(\dfrac{105}{\rho^4} - \dfrac{10}{\rho^2}\right)\sin\rho$

We will also need to consider the spherical Hankel functions of the first and second kinds, $h_l^{(1)}(\rho)$ and $h_l^{(2)}(\rho)$. They constitute an alternative pair of linearly independent solutions of equation (7.104), and bear the same relation to j_l and n_l as the functions $\exp(\pm i\phi)$ do to $\cos\phi$ and $\sin\phi$. They are defined by

$$h_l^{(1)}(\rho) = j_l(\rho) + in_l(\rho) \tag{7.114}$$

and

$$h_l^{(2)}(\rho) = j_l(\rho) - in_l(\rho) \tag{7.115}$$

and satisfy the relations (7.111)–(7.113)

We are now in a position to consider the nature of the radial wavefunction $R_l(\rho)$ in the internal ($\rho < ka$) and external ($\rho > ka$) regions of the potential well. Now a satisfactory $R_l(\rho)$ must be valid at $\rho = 0$, so that we can exclude $n_l(\rho)$ as a solution for the internal region, because it becomes infinite at the origin (see figure 7.6). For the external region both $j_l(\rho)$ and $n_l(\rho)$ are acceptable provided k is real, which it is in the unbound case. For bound states we require an external solution which approaches zero at large values of ρ. As already noted, k in this region is imaginary and we will

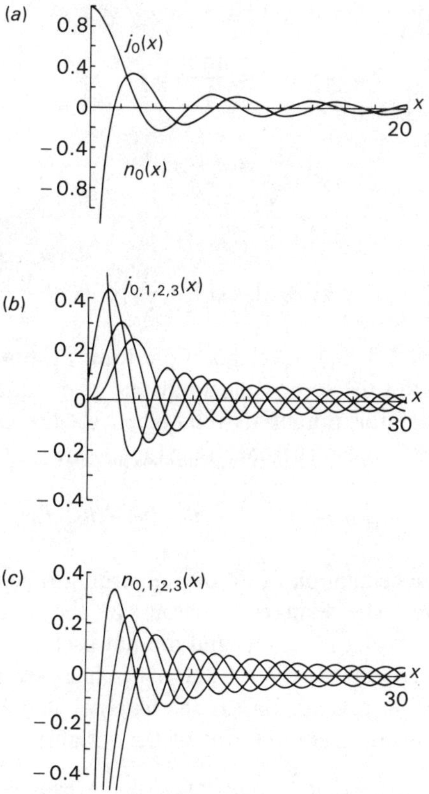

Figure 7.10 The spherical Bessel functions of the first and second kind, $j_n(x)$ and $n_n(x)$. The latter are infinite at the origin and therefore unsuitable as wavefunctions in that region. Note how, for large x, the functions settle down with successive $\frac{1}{4}\lambda$ relative phase shifts.

therefore redefine ρ here by letting

$$\rho = \varkappa r \tag{7.116}$$

where

$$\varkappa^2 = 2m(V - E)/\hbar^2 \qquad r \geqslant a \tag{7.117}$$

and $\varkappa$ is real. The effect of this is that the second term in equation (7.104) is reversed in sign or, equivalently, ρ is replaced by $\mathrm{i}\rho$. j_l, n_l, $h_l^{(1)}$ and $h_l^{(2)}$ now have imaginary arguments, and become functions of real exponentials. For instance, we have

$$h_0^{(1)}(\mathrm{i}\rho)=\frac{\sin \mathrm{i}\rho}{\mathrm{i}\rho}-\mathrm{i}\,\frac{\cos \mathrm{i}\rho}{\mathrm{i}\rho}$$

$$=-\frac{1}{\rho}\,\mathrm{e}^{-\rho}. \tag{7.118}$$

Similarly

$$h_1^{(1)}(\mathrm{i}\rho)=\mathrm{i}\left(\frac{1}{\rho}+\frac{1}{\rho^2}\right)\mathrm{e}^{-\rho}. \tag{7.119}$$

All the functions $h_l^{(1)}(\mathrm{i}\rho)$ have this monotonic, exponentially decaying character, and are thus acceptable solutions of equation (7.117) for the external region of the potential well, when we are considering bound states. They satisfy the recurrence relation

$$h_{l+1}^{(1)}(\mathrm{i}\rho)=\frac{2l+1}{\mathrm{i}\rho}\,h_l^{(1)}(\mathrm{i}\rho)-h_{l-1}^{(1)}(\mathrm{i}\rho) \tag{7.120}$$

which permits the generation of all the succeeding functions from the first two given above. The Hankel functions of the second kind lead to exponentially increasing functions and are not useful in this context.

The boundary conditions at $\rho=0$ and $\rho=\infty$ have led us to adopt $j_l(\rho)$ as the internal solution and $h_l^{(1)}(\mathrm{i}\rho)$ as the external solution. Collecting our results, and expressing them in terms of the variable r, our solution is

$$R_l(r)=Aj_l(kr) \qquad r<a \tag{7.121}$$

and

$$R_l(r)=Bh_l^{(1)}(\mathrm{i}\varkappa r) \qquad r\geqslant a \tag{7.122}$$

where

$$k^2=2mE\hbar^2 \tag{7.123}$$

and

$$\varkappa^2=2m(V-E)\hbar^2. \tag{7.124}$$

A and B are normalization constants. The eigenvalues are determined by equating the logarithmic derivatives of these functions at the boundary $r=a$:

$$\frac{kj_l'(ka)}{j_l(ka)}=\frac{\mathrm{i}\varkappa h_l'^{(1)}(\mathrm{i}\varkappa a)}{h_l^{(1)}(\mathrm{i}\varkappa a)}. \tag{7.125}$$

(Note that the symbol $'$ implies differentiation with respect to the *argument* of the function. We require the derivative with respect to r, hence the appearance of k and $\mathrm{i}\varkappa$ in the above equation.) For the case $l=0$ this reduces to

$$k\cot(ka) = -\varkappa \tag{7.126}$$

which is the condition (3.50) which had to be satisfied for the odd-parity eigenfunctions of the finite one-dimensional well studied in Chapter 3. This could have been anticipated since, with $l=0$, the TISE (7.101) is identical to the equation for the one-dimensional case, and with the same boundary condition at $r=a$. (It should be remembered, however, that equation (7.101) refers to $\chi(r)$ rather than to the wavefunction $R(r)$.) Negative values of r do not occur and the additional boundary condition ($\chi(0)=0$) which must be satisfied at the origin in three dimensions excludes the even-parity solutions. The $l=0$ eigenvalues can be determined graphically or numerically exactly as described in Chapter 3. The considerations which led to equation (3.57) for the number of bound levels tell us that in this case there will be $n_{\max}$ levels with $l=0$ where

$$n_{\max} = \mathrm{INT}\left(\frac{\gamma}{\pi}+\frac{1}{2}\right) \tag{7.127}$$

and

$$\gamma = \sqrt{2mV}a/\hbar. \tag{7.128}$$

Hence if

$$\sqrt{2mV}a/\hbar < \tfrac{1}{2}\pi \tag{7.129}$$

there will be no bound levels at all!

Equation (7.125) is much less tractable for $l \neq 0$ and as good a way as any to solve it is by computer integration of Schrödinger's equation as in the following exercise.

Computer exercise 7.5: the spherical square well

(*a*) As a preliminary measure, run the program '1-D Square-Well Eigenvalues', for a well of depth $V=10$ au and half-width $a=2$ au. Make a note of the eigenvalues.

(*b*) Load the program 'General 3-D Potential (RK)'. RK refers to the Runge–Kutta method of integration which, although slower than the Numerov, is better suited to the study of discontinuous potential functions. The default parameters provide a spherical well of radius 2 au and depth 10 au. Verify that the $l=0$ energy eigenvalues of this well are the same as those of the odd-parity eigenfunctions of the one-dimensional well. The functions you are plotting within the well are all portions of $j_0(kr)$, including one extra node with each energy increment, whilst external to the well they are all $h_0^{(1)}(\mathrm{i}\varkappa r)$. Compare the functions $\chi_{n,0}(r)$ and $R_{n,0}(r)$, $(K=0, 1)$. Note in particular the difference in behaviour at the origin.

(*c*) Set $l=1$ and find the lowest energy eigenvalue. This will correspond to an eigenfunction which is zero at the origin, and has no nodes. You will need to find the approximate energy and then use the eigenvalue-finding routine. In doing this you are solving numerically equation (7.125). When you plot the eigenfunction the part inside the well is a portion of $j_l(kr)$. Repeat this process for the next higher $l=1$ state, and the lowest $l=2$ state. These functions will be parts of $j_1(kr)$ and $j_2(kr)$ respectively. Note the energy eigenvalues. This will give you an impression of what the complete energy spectrum looks like.

Things to notice:

(i) We can define a principal quantum number n of the state in terms of the number of nodes, which is $n-l-1$.

(ii) For a given n, the energy goes down with l. Try to rationalize this in relation to the effective potential function, which is also plotted in each case.

(iii) Except near the top of the well, for a given n we have l values between $n-1$ and 0. Near the top of the well some of the lower l values do not occur, or rather, they are unbound.

(iv) In contrast to the Coulomb and isotropic oscillator cases, there is no degeneracy with respect to n, l. (The $(2l+1)$-fold m degeneracy remains of course.)

(*d*) Set the energy to some value larger than V, and plot the wavefunction. The internal functions are still spherical Bessel functions, but externally they are now superpositions of spherical Bessel and Neumann functions.

(*e*) You can examine more cycles of the spherical Bessel functions $j_l(kr)$ without the trouble of finding eigenvalues simply by setting RI and RF to some suitably large value, and plotting the wavefunctions for arbitrary energy.

〉7.8 General properties of spherically symmetric potential wells

For all spherically symmetric potentials eigenfunctions can be found which are products of a radial function $R_{n,l}(r)$ and a spherical harmonic (equation (7.1)). The eigenfunctions of the Coulomb potential, the isotropic oscillator and the spherical square well differ only in respect of the radial functions which as we have seen are distinct families of functions in each case. Closer inspection reveals certain common properties, which permit us to form a one-to-one correspondence between the functions of each well. The correspondence is between functions with common values of both l and number of nodes. The correspondence is obvious in the case of the Coulomb potential and the spherical square well, which each have $n - l - 1$ nodes, but whereas in the former case there is complete l degeneracy for a given n, there is no l degeneracy at all in the latter case. Naturally the correspondence will not extend to the infinity of states of the Coulomb potential unless we consider an infinitely deep square well.

The isotropic oscillator appears to be an exception, because of the rule that n and l must be both even or both odd, and it seems that some eigenfunctions must be 'missing'. That this is not the case can be seen in figure 7.11 where the appropriate correspondence is indicated. In this scheme we can see that a more logical labelling, in which n was determined by defining the number of nodes to be equal to $n - l - 1$, would show no l degeneracy for states with a given n, but would introduce an 'accidental' degeneracy between l values of the same parity but which belong to different families.

The fact that each potential well has corresponding wavefunctions is related to the fact that, as they are the eigenfunctions of their respective Hamiltonian operators, they must form complete sets, so that any arbitrary wavefunction may be expanded as a linear superposition of them. It is just as necessary for them all to be available in each case as it is for all the harmonics of $\cos(x)$ and $\sin(x)$ to be available for the Fourier analysis of an arbitrary waveform.

Characterization of the eigenfunctions in terms of l and the number of nodes becomes very useful when more complex potential wells are considered. For example, in atomic physics it is useful to model the motion of the single valence electron of an alkali metal in terms of a potential arising from the point-like positively charged nucleus plus a spherically symmetric distributed negative charge cloud resulting from all the other electrons. The fundamental features of the radial wavefunctions, their behaviour at the origin and the number of nodes remain the same as those of hydrogen, even though the scale and distribution of the functions may be very different, and this is enough to permit the unambiguous assignment of the quantum numbers n and l.

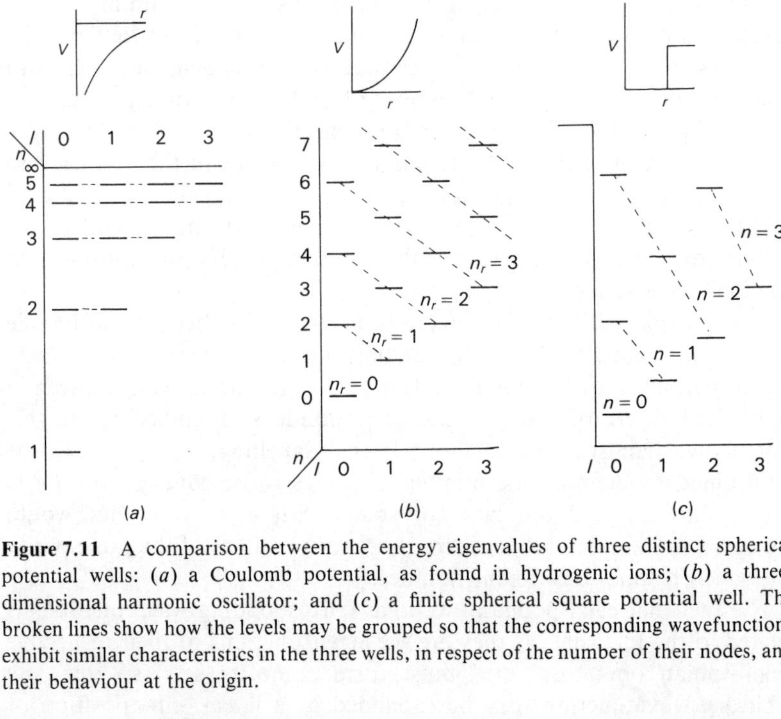

Figure 7.11 A comparison between the energy eigenvalues of three distinct spherical potential wells: (a) a Coulomb potential, as found in hydrogenic ions; (b) a three-dimensional harmonic oscillator; and (c) a finite spherical square potential well. The broken lines show how the levels may be grouped so that the corresponding wavefunctions exhibit similar characteristics in the three wells, in respect of the number of their nodes, and their behaviour at the origin.

〉 Problems

7.1 Perform the analysis leading from equation (7.18) to (7.28), by making the indicated substitutions.

7.2 Perform the analysis leading from equation (7.28) to (7.33), by making the indicated substitutions.

7.3 The eigenfunctions of the hydrogen atom u_{n,l,m_l} are all mutually orthogonal. Functions with l and m_l in common must rely on $R_{n,l}(r)$ for their orthogonality. Verify this for the pairs of eigenfunctions $(u_{1,0,0}, u_{2,0,0})$ and $(u_{3,1,1}, u_{2,1,1})$.

7.4 Calculate the probability that an electron is within a distance a_0 of the origin, for (a) a 2s electron and (b) a 2p electron. (This illustrates the much greater penetrating power of an s electron compared with a p electron.) (Hint: you can check your integrations by using the tabulation/normalization option in the hydrogen ion program. For each of two states first normalize the functions to one, after setting RF to a range large enough to include essentially the entire distribution. Then reduce the range RF to 1, and run the tabulation again. This will give you the integral over the range $0 \rightarrow a_0$.)

7.5 The following data relate to the diatomic molecule $^1H^{35}Cl$: $D_0 = 4.430$ eV, $\hbar\omega_e = 0.370\,668$ eV, $\mu = 0.979\,889$ amu, $r_e = 0.127\,46$ nm. Find the form of the Morse potential for this molecule, expressed in dimensionless units. What is the value of n_{max} for this potential? What is the energy of the $n = 0$ to 1 transition, in dimensionless and practical units? What is the spacing of the rotational lines in dimensionless and in practical units? (Hint: follow the method used in the computer exercise 7.4, and use the computer program to check your results.)

7.6 The wavefunctions of the isotropic harmonic oscillator are expressible in cartesian or spherical polar coordinates. Write down the explicit forms of the degenerate spherical eigenfunctions $u_{n,l,m_l} = R_{1,1}(r)\,Y_1^{1,0,-1}(\theta, \phi)$ (see tables 7.4 and 6.1), and the explicit forms of the degenerate cartesian functions with the same energy (see equation (7.59) and table 3.1). Show how the former can be expanded in terms of the latter. (Hint: this can be done by inspection. You will need to express x, y and z in terms of r, θ and ϕ, and recall de Moivre's theorem.)

7.7 The deuteron consists of a proton and a neutron loosely bound together by the strong nuclear force (with their spins parallel, i.e. in a triplet state). It is known that the potential well representing the interaction has a single level, bound by 2.22 MeV, and that $l = 0$.

It is also known that the range of the nuclear force is about 2×10^{-15} m. Find the depth of the potential on the assumption that it has the form of a spherical square well.

It is known from neutron–proton scattering experiments that the interaction between the particles when their spins are opposed (singlet state) can be modelled by a spherical square well of range 2.8×10^{-15} m and depth 11.9 MeV. Show that this well has *no* bound state. How deep would the well need to be for the singlet deuteron to be bound (with the given range)? (Hint: you need to solve equation (7.126) for the first part. You can use the program '1D Potential Well Eigenvalues' to do the work (review computer exercise 7.5), but you will need to use the 'atomic units' option. First find the appropriate reduced mass, then scale the atomic units of energy and distance accordingly.)

Chapter 8

Identical Particles

8.1 Two non-interacting particles: symmetry of wavefunctions

We have so far restricted our discussion to systems consisting of either a single particle moving in a fixed potential or else a system like the hydrogen atom, which can be reduced effectively to a one-particle system by means of the introduction of relative coordinates and the reduced mass of the system. (The analysis leading to this reduction is given in Appendix 3.) In this chapter we will consider the case of two particles, both moving in a fixed potential arising from an external source. The ideas introduced in the discussion are important because they form the basis for the development of quantum mechanics to cover many-particle systems such as atoms, molecules and solids.

We will make the plausible assumption that the Hamiltonian of the two-particle system may be written as a sum of terms representing respectively the kinetic energies of the two particles, their potential energies arising from the external potential, and initially we will also include the potential energy of the particles arising from their interaction with each other. Denoting the positions of the two particles by r_1 and r_2, and the position of particle one relative to particle two by r_{12} (see figure 8.1) the Hamiltonian takes the form

$$\hat{H}(r_1, r_2, p_1, p_2) = \frac{\hat{P}_1^2}{2m_1} + \frac{\hat{P}_2^2}{2m_2} + V_1(r_1) + V_2(r_2) + V(r_{12})$$

$$= -\frac{\hbar^2}{2m_1}\nabla_1^2 - \frac{\hbar^2}{2m_2}\nabla_2^2 + V(r_1) + V_1(r_2) + V_2(r_{12}) \quad (8.1)$$

where in the second equation we have introduced specific operators for the momenta of the two particles. The differential operators ∇_1 and ∇_2 operate only upon the coordinates of particles 1 and 2 respectively. The TISE for the system is then

$$\hat{H}(\boldsymbol{r}_1, \boldsymbol{r}_2)u(\boldsymbol{r}_1, \boldsymbol{r}_2) = Eu(\boldsymbol{r}_1, \boldsymbol{r}_2) \tag{8.2}$$

and the energy eigenfunctions will be functions of $\boldsymbol{r}_1$ and $\boldsymbol{r}_2$. (We have dropped the arguments $\boldsymbol{p}_1, \boldsymbol{p}_2$ in the Hamiltonian for the sake of brevity.) An example of the sort of system these equations apply to is the helium atom. Here the 'external' potentials V_1 and V_2 represent the Coulomb interactions between the two electrons and the nucleus, whilst $V(\boldsymbol{r}_{12})$ represents the Coulomb repulsion between the electrons. The first two terms in equation (8.1) represent the kinetic energies of the two electrons. (The small kinetic energy of the nucleus is ignored.)

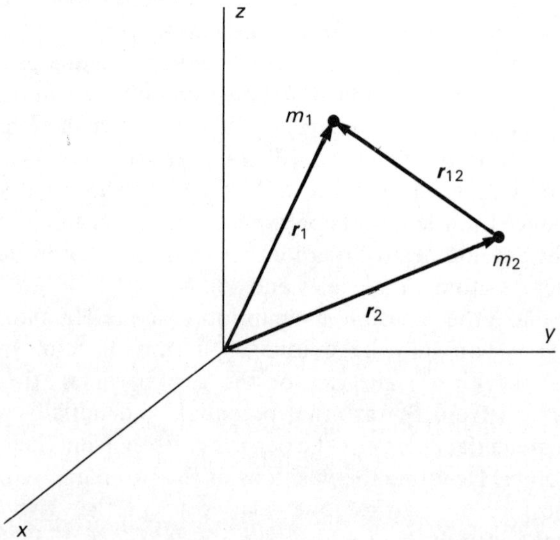

Figure 8.1 Position vectors for a two-particle system.

Exact analytical solutions of such systems are almost never possible; approximations must be employed. Here we will be content to examine the much simpler case where the two particles do not interact. The

Hamiltonian can then be written as

$$\hat{H}(r_1, r_2) = \frac{\hat{P}_1^2}{2m_1} + V_1(r_1) + \frac{\hat{P}_2^2}{2m_2} + V_2(r_2)$$
$$= \hat{H}_1(r_1) + \hat{H}_2(r_2) \tag{8.3}$$

where

$$\hat{H}_1(r_1) = \frac{\hat{P}_1^2}{2m_1} + V_1(r_1) \qquad \hat{H}_2(r_2) = \frac{\hat{P}_2^2}{2m_2} + V_2(r_2). \tag{8.4}$$

The important point to note here is that the Hamiltonian can now be split into two terms which are entirely independent of each other, since they depend only on properties of particles one and two respectively. The significance of this is that $\hat{H}_1(r_1)$ and $\hat{H}_2(r_2)$ and $\hat{H}(r_1, r_2)$ are commuting operators, and hence must have simultaneous eigenfunctions. Consider a function of the form

$$u(r_1, r_2) = u_a(r_1)u_b(r_2) \tag{8.5}$$

where $u_a(r_1)$ and $u_b(r_2)$ are eigenfunctions of $\hat{H}_1(r_1)$ and $\hat{H}_2(r_2)$ with eigenvalues E_a and E_b respectively. Applying the Hamiltonian of equation (8.3) to both sides of the equation we find

$$\hat{H}(r_1, r_2)u(r_1, r_2) = [\hat{H}_1(r_1) + \hat{H}_2(r_2)]u_a(r_1)u_b(r_2)$$
$$= [E_a + E_b]u_a(r_1)u_b(r_2)$$
$$= Eu_a(r_1)u_b(r_2) \tag{8.6}$$

where

$$E = E_a + E_b. \tag{8.7}$$

The second step in the above sequence depends on the fact that operating with $\hat{H}_1$ on $u_b(r_2)$ or with $\hat{H}_2$ on $u_a(r_1)$ has no effect, so that the product function is a simultaneous eigenfunction of $\hat{H}_1$ and $\hat{H}_2$. Equation (8.6) shows that $u(r_1, r_2)$ is also an eigenfunction of $\hat{H}(r_1, r_2)$, with eigenvalue $(E_a + E_b)$.

The analysis above can be extended to an arbitrary number of non-interacting particles. The eigenfunctions of the system are *products* of single-particle eigenfunctions, whilst the eigenvalues are *sums* of the single-particle eigenvalues.

The wavefunction $u(r_1, r_2)$ must now be interpreted as a *joint probability amplitude*. Thus the joint probability of particle 1 being in the volume element $d\tau_1$ at r_1 and particle 2 being in volume element $d\tau_2$ at

r_2 is given by

$$dP(r_1, r_2) = |u(r_1, r_2)|^2 \, d\tau_1 \, d\tau_2 = |u_a(r_1)|^2 |u_b(r_2)|^2 \, d\tau_1 \, d\tau_2. \quad (8.8)$$

Normalization of the wavefunction requires that

$$\int_{\substack{\text{All}\\\text{space}}} \int_{\substack{\text{All}\\\text{space}}} |u(r_1, r_2)|^2 \, d\tau_1 \, d\tau_2 = 1 \quad (8.9)$$

since the joint probability of the two particles being *somewhere* must be unity.

We have so far considered the two particles to be distinguishable in some way. They may perhaps have different masses, as suggested in equation (8.1). However, if both particles are for example electrons then all their properties are identical and there is no way in which we can follow the history of one of them in particular if their wavefunctions overlap. This undermines the preceding analysis since our labelling of the single-particle wavefunctions $u_a(r_1)$ and $u_b(r_2)$ implies that we know that it is specifically particle *one* which is in the state u_a and particle *two* which is in the state u_b. However, the result of any conceivable experiment could only tell us that one of the particles was in the state u_a and the other in u_b, but not which was which! The theory must be so formulated as to take this into account, and so far it does not. The situation can be remedied by noticing that, for identical particles, if $u_a(r_1)u_b(r_2)$ is a solution of the TISE for the system, then so also must be $u_a(r_2)u_b(r_1)$, in which the coordinates of the particles have been exchanged. Furthermore, any superpositions of these two solutions must themselves be solutions. Consider in particular the two functions:

$$u_S(r_1, r_2) = 2^{-1/2}[u_a(r_1)u_b(r_2) + u_a(r_2)u_b(r_1)] \quad (8.10)$$

$$u_A(r_1, r_2) = 2^{-1/2}[u_a(r_1)u_b(r_2) - u_a(r_2)u_b(r_1)]. \quad (8.11)$$

(The factor $\sqrt{2}$ is required in order to maintain normalization if the two component wavefunctions are separately normalized in the first place.) Neither of these functions implies an ability to identify the particles. To see this we simply exchange the symbols r_1 and r_2 on the right-hand sides of the equations. The effect of this is to leave the *symmetric* function u_S unchanged, whereas the *antisymmetric* function u_A is reversed in sign. This sign reversal lacks physical significance; the probability densities (and all other physical attributes) of both wavefunctions are unchanged under this operation of *particle exchange* and so either is an acceptable wavefunction.

The symmetric and antisymmetric functions which we have found exhibit some remarkable properties. Consider the antisymmetric function u_A of equation (8.11) when the two particles approach each other closely, so that $r_1 \to r_2$. Now the two terms in the square brackets become the same size and the value of the wavefunction approaches zero! We conclude that the probability of two identical particles being in the same place is zero if their wavefunction is antisymmetric, and we have reached this conclusion even though by hypothesis the particles *do not interact*! Conversely, the symmetric function of equation (8.10) describes a system in which the particles have an enhanced probability of being close together.

An example will illustrate these results. Consider two identical non-interacting particles confined in a one-dimensional infinite square potential well of width $2a$ which lies along the x axis. Suppose one particle is in the ground state whereas the other is in the first excited state. Making use of the single-particle wavefunctions of equation (3.20) the symmetric and antisymmetric wavefunctions of the system are

$$\begin{aligned} u_S(x_1, x_2) &= 2^{-1/2}a^{-1}[\cos(\pi x_1/2a)\sin(\pi x_2/a) + \cos(\pi x_2/2a)\sin(\pi x_1/a)] \\ u_A(x_1, x_2) &= 2^{-1/2}a^{-1}[\cos(\pi x_1/2a)\sin(\pi x_2/a) - \cos(\pi x_2/2a)\sin(\pi x_1/a)]. \end{aligned} \tag{8.12}$$

The energy of the system is, from equations (8.7) and (3.15),

$$E = E_a + E_b = \frac{\pi^2\hbar^2}{8ma^2}(1 + 4). \tag{8.13}$$

The probability densities corresponding to these wavefunctions are shown in figure 8.2. All possible combinations of x_1 and x_2 (which in real space are two points on the x axis) are shown in a two-dimensional plot with the probability density indicated by means of a contour map. It can be seen that for the antisymmetric case the probability is zero along the diagonal $x_1 = x_2$, but large in the two regions $(x_1, x_2) \simeq (\pm\frac{1}{2}a, \mp\frac{1}{2}a)$, where x_1 and x_2 differ widely. For the symmetric case on the other hand the probabilities have maxima near $(x_1, x_2) = (\pm\frac{1}{2}a, \pm\frac{1}{2}a)$, where x_1 and x_2 are equal.

Although the symmetric and antisymmetric wavefunctions (8.12) are degenerate this is only the case because the particles were assumed to be non-interacting. If we now imagine that the particles acquire identical electric charges but that the wavefunctions retain their symmetry characteristics (an assumption which we will shortly justify), then it is clear that the two types of function will represent states of different energy because the Coulomb repulsion will be much greater for the symmetric state than for the antisymmetric one. As a result the energy of the symmetric state will be higher. Such effects are crucial in accounting for the energies of atomic states in multi-electron atoms.

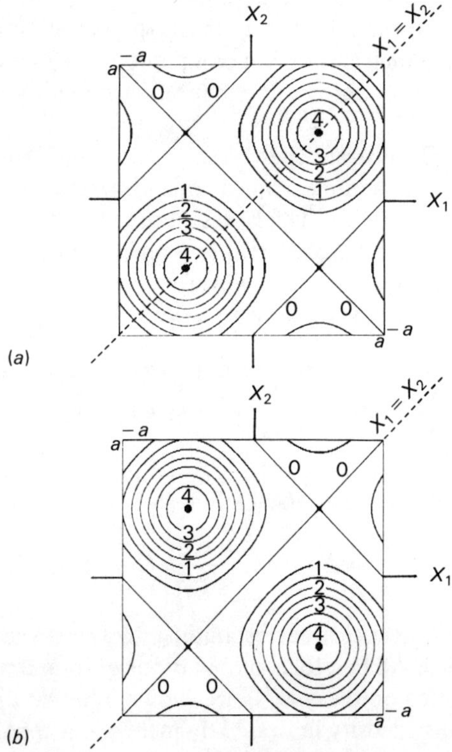

Figure 8.2 Contour plots of the probability density distribution for two non-interacting identical particles in a one-dimensional infinite potential well. One particle is in the lowest energy eigenstate, the other is in the first excited state. (*a*) Symmetric wavefunction. The probability is highest when both particles are in the same place. (*b*) Antisymmetric wavefunction. The probability is zero for the particles being found together.

Discussion of the symmetry properties of wavefunctions is facilitated by the introduction of a particle exchange operator $\hat{\mathscr{P}}$. This operator has the effect of interchanging the coordinates of the particles in any two-particle wavefunction, e.g.

$$\hat{\mathscr{P}}\psi(r_1, r_2) = \psi(r_2, r_1). \tag{8.14}$$

More generally, for a many-particle system, $\hat{\mathscr{P}}_{nm}$ interchanges the coordinates of the nth and mth particles.

We can show that $\hat{\mathscr{P}}$ is Hermitian by the following manipulations:

$$\begin{aligned}\int \psi_a^*(r_1, r_2)\hat{\mathscr{P}}\psi_b(r_1, r_2)\,d\tau_1\,d\tau_2 &= \int \psi_a^*(r_1, r_2)\psi_b(r_2, r_1)\,d\tau_1\,d\tau_2 \\ &= \int \psi_b(r_2, r_1)\hat{\mathscr{P}}^*\psi_a^*(r_2, r_1)\,d\tau_1\,d\tau_2 \\ &= \int \psi_b(r_1, r_2)\hat{\mathscr{P}}^*\psi_a^*(r_1, r_2)\,d\tau_1\,d\tau_2.\end{aligned} \tag{8.15}$$

The last step follows because the value of the integral cannot depend upon the choice of labels for the coordinates over which the integral is taken.

If $\hat{\mathscr{P}}$ operates twice we expect the wavefunction to return to its original form (at least to within a phase factor of unit magnitude):

$$\hat{\mathscr{P}}^2\psi(r_1, r_2) = \psi(r_1, r_2). \tag{8.16}$$

Now exactly the same argument that we used in section 5.12 in the case of the parity operator $\hat{\mathscr{P}}$ leads to the result that the eigenvalues of $\hat{\mathscr{P}}$ must also be ± 1. Hence, if $\psi(r_1, r_2)$ is actually an eigenfunction of $\hat{\mathscr{P}}$, then

$$\hat{\mathscr{P}}\psi(r_1, r_2) = \psi(r_2, r_2) = \pm\psi(r_1, r_2) \tag{8.17}$$

and we see that the symmetric and antisymmetric functions which we have been considering are eigenfunctions of $\hat{\mathscr{P}}$ with eigenvalues $+1$ and -1 respectively.

The Hamiltonian of a system consisting of identical particles cannot depend upon the labelling of the particles; it must therefore commute with the operator $\hat{\mathscr{P}}$:

$$[\hat{H}, \hat{\mathscr{P}}] = 0. \tag{8.18}$$

Hence energy eigenfunctions of definite symmetry can be found, even when a particle interaction term is present. Furthermore, Ehrenfest's theorem (section 5.11) tells us that the symmetry of simultaneous eigenfunctions of $\hat{H}$ and $\hat{\mathscr{P}}$ is a conserved property; no internal process can result in a change from a symmetric state to an antisymmetric state, nor vice versa.

〉8.2 Identical particles and spin: the Pauli exclusion principle

The full significance of the exchange symmetry of wavefunctions cannot be appreciated without taking into account the spins of the particles. The importance of the effects to be described can hardly be overstated, since the entire structure of matter depends upon them.

First we must consider the form to be expected for the total wavefunction of a particle with spin. Provided there is no interaction between the spin and translational motion of the particle a simple product of spatial and spin factors is appropriate, of the form

$$\psi(\boldsymbol{r}, \sigma) = \psi_{\mathrm{a}}(\boldsymbol{r})\chi_{\mathrm{a}'}(\sigma). \tag{8.19}$$

Here χ is the spin function and we are using σ to symbolize the 'spin coordinate'. The subscripts a and a$'$ signify the quantum numbers necessary to specify the state the particle is in. This works because spin operators such as $\hat{S}^2$ and $\hat{S}_z$ operate in a space which is entirely independent of ordinary three-dimensional space and so they automatically commute with space-only dependent operators.

As a specific example, an electron in a hydrogenic atom, with the quantum numbers n, l, m_l and $m_s = \frac{1}{2}$, has the total wavefunction

$$\psi(\boldsymbol{r}, \sigma) = u(\boldsymbol{r})_{n,l,m_l}\alpha(\sigma). \tag{8.20}$$

(In what follows, for spin-$\frac{1}{2}$ particles we will use α and β without subscripts to signify the states 'spin up' and 'spin down' in the z direction.) Considering now the case of two distinguishable particles, we expect the overall wavefunction to take the general form

$$\psi(\boldsymbol{r}_1, \sigma_1, \boldsymbol{r}_2, \sigma_2) = \psi_{\mathrm{a}}(\boldsymbol{r}_1)\chi_{\mathrm{a}'}(\sigma_1)\psi_{\mathrm{b}}(\boldsymbol{r}_2)\chi_{\mathrm{b}'}(\sigma_2). \tag{8.21}$$

However, if the particles are identical this must be taken into account in formulating the wavefunction. The process follows the lines of the discussion for the spatial function in the previous section. First we will widen the scope of the particle exchange operator $\hat{\mathscr{P}}$ so that it now exchanges both spatial and spin coordinates:

$$\hat{\mathscr{P}}\psi(\boldsymbol{r}_1, \sigma_1, \boldsymbol{r}_2, \sigma_2) = \psi(\boldsymbol{r}_2, \sigma_2, \boldsymbol{r}_1, \sigma_1). \tag{8.22}$$

Otherwise its properties remain the same; it is still a Hermitian operator with eigenvalues ± 1, so if ψ is an eigenfunction of $\hat{\mathscr{P}}$

$$\hat{\mathscr{P}}\psi(\boldsymbol{r}_1, \sigma_1, \boldsymbol{r}_2, \sigma_2) = \pm\psi(\boldsymbol{r}_1, \sigma_1, \boldsymbol{r}_2, \sigma_2). \tag{8.23}$$

These wavefunctions are either symmetric or antisymmetric under exchange of both space and spin coordinates. Observations show that these are the only types of total wavefunction which occur in nature!

There is more than one way of constructing overall wavefunctions of definite symmetry from single-particle functions. Our approach will be to look for satisfactory functions in which the spin and space parts remain in separate factors. We will consider only spin-$\frac{1}{2}$ particles, which include electrons, protons, neutrons, neutrinos, quarks and many other sub-nuclear particles. The possible single-particle spin states are α and β. Ignoring symmetry requirements the possible combined spin functions for two particles are

$$\chi(\sigma_1, \sigma_2) = \alpha(1)\alpha(2) \qquad \chi(\sigma_1, \sigma_2) = \beta(1)\beta(2) \tag{8.24}$$

$$\chi(\sigma_1, \sigma_2) = \alpha(1)\beta(2) \qquad \chi(\sigma_1, \sigma_2) = \beta(1)\alpha(2). \tag{8.25}$$

(We have simplified the notation by letting $\alpha(\sigma_1) \to \alpha(1)$, etc.) Of these four functions the upper two are symmetric under exchange of spin coordinates, but the lower two have no definite symmetry. They may however be used to form symmetric and antisymmetric combinations:

$$\chi(\sigma_1, \sigma_2) = 2^{-1/2}[\alpha(1)\beta(2) \pm \beta(1)\alpha(2)] \tag{8.26}$$

so that we have four spin functions of definite symmetry, of which three are symmetric and one antisymmetric. We can combine these with symmetric and antisymmetric spatial functions

$$\psi_S(r_1, r_2) = 2^{-1/2}[\psi_a(r_1)\psi_b(r_2) + \psi_a(r_2)\psi_b(r_1)] \tag{8.27}$$

$$\psi_A(r_1, r_2) = 2^{-1/2}[\psi_a(r_1)\psi_b(r_2) - \psi_a(r_2)\psi_b(r_1)] \tag{8.28}$$

to form eight overall wavefunctions of definite symmetry. It is a remarkable fact that four of these are never observed. For particles of half-integral spin, only systems whose wavefunctions are *antisymmetric* under exchange of spin and space coordinates are found in nature. The allowed functions under this scheme therefore combine space and spin factors of opposite symmetry:

$$\psi(r_1, \sigma_1, r_2, \sigma_2) = 2^{-1/2}[\psi_a(r_1)\psi_b(r_2) - \psi_a(r_2)\psi_b(r_1)]\alpha(1)\alpha(2) \tag{8.29}$$

$$\psi(r_1, \sigma_1, r_2, \sigma_2) = 2^{-1}[\psi_a(r_1)\psi_b(r_2) - \psi_a(r_2)\psi_b(r_1)][\alpha(1)\beta(2) + \alpha(2)\beta(1)] \tag{8.30}$$

$$\psi(r_1, \sigma_1, r_2, \sigma_2) = 2^{-1/2}[\psi_a(r_1)\psi_b(r_2) - \psi_a(r_2)\psi_b(r_1)]\beta(1)\beta(2) \tag{8.31}$$

$$\psi(r_1, \sigma_1, r_2, \sigma_2) = 2^{-1}[\psi_a(r_1)\psi_b(r_2) + \psi_a(r_2)\psi_b(r_1)][\alpha(1)\beta(2) - \alpha(2)\beta(1)]. \tag{8.32}$$

The discovery that the total wavefunction must be antisymmetric for particles of half-integral spin was made by Heisenberg in 1926. It is often expressed in a different way, namely that two identical particles in the same system cannot be in the same quantum state, i.e. having all quantum numbers the same. This is the famous Pauli exclusion principle, formulated in 1924. The two statements are equivalent, because if the two particles are in identical states an antisymmetric wavefunction vanishes, as can be verified for the four functions above. For the first three functions the spatial factor vanishes if a → b, whilst for the fourth the spin factor vanishes if $\alpha \to \beta$. The stability of matter depends crucially on this principle; without it all the electrons of an atom would fall into the ground state, with, to say the least, rather profound consequences for the chemical properties of the atom!

Examining the four spin functions it is easy to see that the three symmetric ones represent systems in which the z components of total spin angular momentum are respectively $\hbar$, 0 and $-\hbar$, whilst the antisymmetric one also represents a system with a zero z component. We can formalize this deduction by introducing the operator for the total z component of spin:

$$\hat{S}_z = \hat{S}_{1z} + \hat{S}_{2z}. \tag{8.33}$$

Applying this operator to the functions (8.26) for example we have

$$\begin{aligned}\hat{S}_z\chi(\sigma_1, \sigma_2) &= 2^{-1/2}(\hat{S}_{1z} + \hat{S}_{2z})[\alpha(1)\beta(2) \pm \alpha(2)\beta(1)] \\ &= 2^{-1/2}[(\tfrac{1}{2}\hbar - \tfrac{1}{2}\hbar) \pm (\tfrac{1}{2}\hbar - \tfrac{1}{2}\hbar)][\alpha(1)\beta(2) \pm \alpha(2)\beta(1)] \\ &= 0.\end{aligned} \tag{8.34}$$

Similarly

$$\hat{S}_z\alpha(1)\alpha(2) = \hbar\alpha(1)\alpha(2) \tag{8.35}$$

and

$$\hat{S}_z\beta(1)\beta(2) = -\hbar\alpha(1)\alpha(2) \tag{8.36}$$

so that the three symmetric spin functions are eigenfunctions of $\hat{S}_z$ with eigenvalues 0, $\pm\hbar$ and the antisymmetric function is an eigenfunction with eigenvalue 0. It is now natural to associate the three symmetric spin functions with the three possible orientations of a 'spin-1' system, and the antisymmetric spin function with a 'spin-0' system. These systems have total spin quantum numbers $S = 1$ and 0 respectively. The electron spins are behaving as if they are coupled, in the first case parallel to each other and in the second case anti-parallel.

Many of the ideas introduced in this section are realized in the helium atom which has two electrons. The He energy-level diagram is shown in figure 8.3. The features relevant to this discussion are:

(*a*) There are two essentially independent systems of energy levels. For one of these, referred to as the *singlet* system, $S = 0$, whilst for the other, the *triplet* system, $S = 1$. The principal reason that they are independent is that a transition from one system to the other would entail a change of symmetry of the spatial part of the wavefunction but, as we showed in section 8.1, that symmetry is conserved. Another reason is that such a transition would entail one of the electron spins flipping over, and the most common radiative process, electric dipole radiation, is incapable of accomplishing this.

(*b*) Within each system the level structure is similar to that of hydrogen, corresponding to the fact that only one of the electrons is excited at any time, and the states available to it are approximately hydrogenic in character. The unexcited electron occupies a 1s state, and so contributes nothing to the orbital angular momentum of the system.

(*c*) There is a state conspicuously absent in the triplet system. This state, which would be the lowest in the triplet system, is missing because both electrons would have to be in the same quantum state for it to exist, and this is forbidden by the Pauli exclusion principle. In fact the absence of this state was one of the principal pieces of evidence which led Pauli to formulate his principle. No restrictions arise in the singlet system, because there the spins are oppositely directed, so that the spatial parts of the wavefunctions are permitted to be the same for the two electrons. For all the other triplet states the spatial wavefunctions of the two electrons differ, so no difficulties arise there either.

(*d*) For pairs of corresponding states in the singlet and triplet systems the triplet state always lies below the singlet one. In the discussion of section 8.1 it was pointed out that antisymmetric space functions represented particles which avoided each other. Here we find that such functions must be associated with the triplet system. Conversely, symmetrical space functions are associated with the singlet system, and for these the particles tend to congregate together. Since charged electrons repel each other more the closer together they are, this tends to raise the potential energy of the system more for the singlets, and this explains their higher position.

(*e*) Although not directly relevant to the ideas we have been discussing, there is another feature of the energy levels shown in figure 8.3

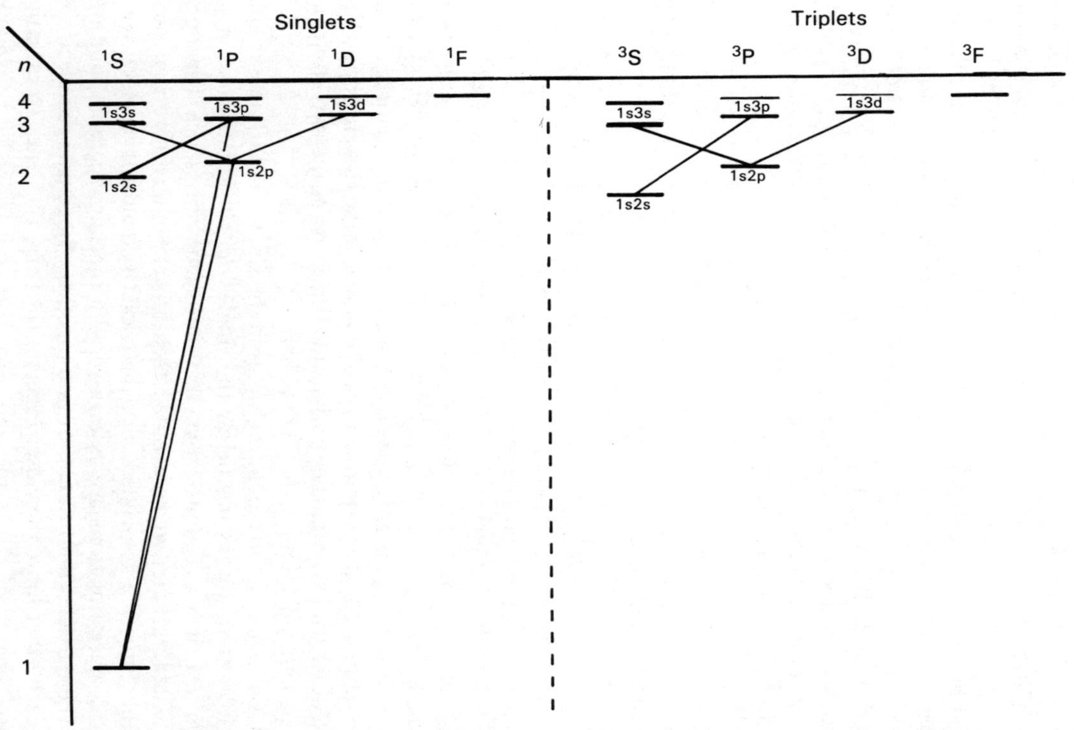

Figure 8.3 Energy levels of helium. Quantum numbers of the two single-particle electron states are shown for the lower levels.

which we can understand in a qualitative way. Whereas in hydrogen states of different l but the same n are degenerate, in helium they are not. The reason lies in the electrostatic screening which the unexcited 1s electron provides for the nucleus. This electron has a spherically symmetric charge distribution and while the other electron is well outside the bulk of this distribution it is shielded from the full effects of the nuclear charge. The extent to which the active electron penetrates this charge cloud depends upon its wavefunction. An s electron penetrates deeply, and is therefore more tightly bound than a p electron, and so on.

It is found that systems of two identical particles of integral or zero spin, such as photons and π mesons, have total wavefunctions which are symmetric under exchange of all coordinates. It follows that any number of such particles can be in the same quantum state, a fact which gives rise to profound differences of behaviour between such particles and particles of half-integral spin. Because the two classes of particles behave so differently they are given the generic names *fermions* (half-integral spin) and *bosons* (integral spin).

All the ideas we have discussed in this chapter with respect to two particles can be extended to systems of many identical particles. The symmetry requirements then apply under the exchange of space and spin coordinates of any pair of particles.

〉Problems

8.1 An alternative (and more direct) approach to the problem of constructing antisymmetric wavefunctions for two spin-$\frac{1}{2}$ particles leads to the set

$$\psi(r_1,\sigma_1,r_2,\sigma_2) = 2^{-1/2}[\psi_a(r_1)\alpha(1)\psi_b(r_2)\alpha(2) - \psi_a(r_2)\alpha(2)\psi_b(r_1)\alpha(1)]$$

$$\psi(r_1,\sigma_1,r_2,\sigma_2) = 2^{-1/2}[\psi_a(r_1)\alpha(1)\psi_b(r_2)\beta(2) - \psi_a(r_2)\alpha(2)\psi_b(r_1)\beta(1)]$$

$$\psi(r_1,\sigma_1,r_2,\sigma_2) = 2^{-1/2}[\psi_a(r_1)\beta(1)\psi_b(r_2)\alpha(2) - \psi_a(r_2)\beta(2)\psi_b(r_1)\alpha(1)]$$

$$\psi(r_1,\sigma_1,r_2,\sigma_2) = 2^{-1/2}[\psi_a(r_1)\beta(1)\psi_b(r_2)\beta(2) - \psi_a(r_2)\beta(2)\psi_b(r_1)\beta(1)].$$

Show that the functions (8.29)–(8.32) can be obtained from these by taking suitable linear combinations of them. (Hint: multiply equations (8.29)–(8.32) out and compare the terms with those in the equations above.)

8.2 Two particles are confined in a one-dimensional infinite square well extending from $-a$ to $+a$. The interaction between them is described by the δ-function potential $\lambda V\delta(x_1 - x_2)$, where x_1 and x_2 are the positions of the particles. (It represents a potential-energy function of magnitude V and range λ, where $V \to \infty$ and $\lambda \to 0$ in such a way that λV remains constant.) Assume that despite the interaction the wavefunctions of the particles are the ground- and first-excited-state eigenfunctions defined in equation (3.20). Calculate the expectation values of the interaction energy for the cases (a) the particles are distinguishable, (b) the particles are indistinguishable with half-integral spin, and (c) the particles are indistinguishable with zero spin. (Hint/discussion: the use of the original eigenfunctions despite the possibility that the interaction will modify them is justified in more advanced texts; the suggested calculation amounts to employing first-order perturbation theory. The form of the interaction is such that the particles are independent except when they are in the same place. Recall that the effect of a δ function under the integral sign is to pick out the value of the integrand at the point where the δ function is located.)

Chapter 9

Scattering in One Dimension

9.1 Introduction

We have so far concentrated almost entirely on solutions of Schrödinger's equation which describe bound states of matter. Whilst a great deal of physics has been learnt through the study of such systems it is probably true to say that even more has been learnt from scattering experiments. One of the earliest and most famous of all scattering experiments was performed by Rutherford, who directed a beam of α particles at a gold foil. Observations of the elastically scattered α particles provided the evidence for the nuclear model of the atom so familiar today. Scattering is almost the only experimental technique available to particle physicists, who employ high-energy accelerators to generate their probe beams. These may consist of a host of particle species from neutrinos to heavy ions, and the scattering is not merely inelastic but usually results in the transmutation of the colliding particles. The technique also plays an important part in the investigation of the much lower energy phenomena characteristic of atomic, solid state and surface physics. In these cases it is usually electrons which are employed. The common feature of all these experiments is that the particles of the beam and the outgoing collision products are not bound; they have positive net energy and propagate as travelling waves. Although we have occasionally mentioned the unbound solutions for the potential wells we have studied, they have appeared as standing waves, and their physical interpretation has been obscure. This deficiency will now be remedied. In this chapter we consider scattering in one dimension and also, because the analysis is closely related, a one-dimensional model of a solid. The subject of scattering in three dimensions, which requires rather different treatment, will be discussed in Chapter 10.

In the following treatment we will be representing the ingoing and outgoing particles by wavefunctions of the form $A\,e^{ikx}$. In section 2.8 we drew attention to the difficulty of normalizing such functions since they do not fall to zero at infinity, a circumstance which would seem to require the normalization constant A to be zero. Although techniques exist for dealing with this problem no difficulty arises in the following analysis since the physically meaningful quantities (reflection and transmission coefficients) will appear as ratios of amplitudes. An ideal treatment would describe the particles in terms of time-dependent wavepackets of the type in equation (2.47), satisfying the TDSE. Such wavefunctions are attractive both because they are normalizable and because they appeal to our intuitive feeling that if we are firing a particle at a target we must know *something* about its position, which ought to be time dependent! Whilst this approach is more satisfying much more mathematical complexity is introduced than is appropriate here. If we wish, we can think of the amplitude A above as referring to a steady state beam of particles, each with identical wavefunctions. There is then no difficulty associated with a finite density of flux filling the whole of space, since in that case there would have to be an infinite number of particles. It is important to bear in mind though that any interference effects come about as a result of the overlap of different parts of an individual particle's wavefunction. One particle does not affect another in this sense. These considerations allow us to use a time-independent approach and most of the experimentally verifiable predictions can be obtained in this way.

〉9.2 The potential step

One of the simplest scattering elements we can examine is the one-dimensional potential step, or semi-infinite potential barrier, as illustrated in figure 9.1. The potential is defined by

$$\begin{aligned} V(x) &= 0 \qquad x < 0 \\ V(x) &= V \qquad x \geqslant 0. \end{aligned} \tag{9.1}$$

For a particle of mass m moving along the x axis the TISE for the two regions takes the forms

$$-\frac{\hbar^2}{2m}\frac{d^2u(x)}{dx^2} = Eu(x) \qquad x < 0 \tag{9.2}$$

and

$$-\frac{\hbar^2}{2m}\frac{d^2u(x)}{dx^2} + V = Eu(x) \qquad x \geqslant 0. \tag{9.3}$$

The general solution of equation (9.2) can be written as

$$u_1 = A\,e^{ik_1x} + B\,e^{-ik_1x} \tag{9.4}$$

where

$$k_1 = (2mE)^{1/2}/\hbar. \tag{9.5}$$

A and B may be complex, and k_1 is real and positive.

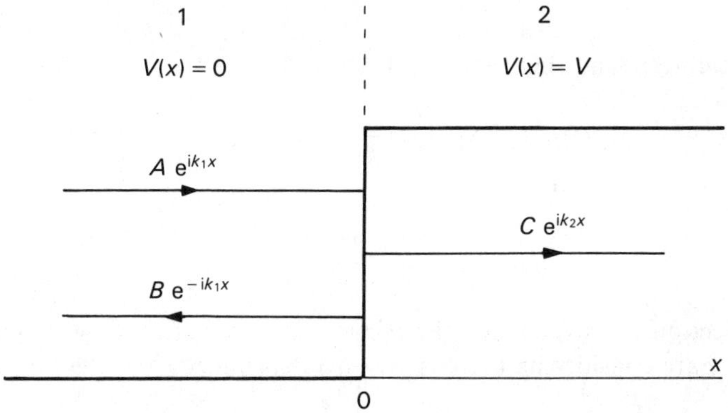

Figure 9.1 A potential step. A wave incident from the left is partially transmitted and partially reflected.

The expression (9.4) has been chosen because it represents a superposition of an incoming travelling wave from the left of amplitude A and an outgoing wave of amplitude B travelling towards the left. This is what we would expect for a scattering event in which the particle source is to the left and there is a possibility of the particle being reflected at the barrier.

The general solution of equation (9.3) for $x \geqslant 0$ is

$$u = C\,e^{ik_2x} + D\,e^{-ik_2x} \tag{9.6}$$

where

$$k_2 = [2m(E - V)]^{1/2}/\hbar. \tag{9.7}$$

The nature of this solution depends upon the sign of $E - V$. For the case $E < V$ the solution consists of real exponentials and we can immediately discard the positive one since it diverges at large x. If $E > V$ the solution consists of waves travelling to the right and left, but we can discard the wave travelling to the left since the right half-space contains no feature to generate such a wave. Hence we can analyse both cases together by writing the solution as

$$u_2 = C\,\mathrm{e}^{\mathrm{i}k_2x} \tag{9.8}$$

where k_2 may be real or imaginary.

Continuity of u and u' at $x = 0$ requires that

$$A + B = C \tag{9.9}$$

and

$$\mathrm{i}k_1A - \mathrm{i}k_1B = \mathrm{i}k_2C. \tag{9.10}$$

Eliminating C and then B we find

$$\frac{B}{A} = \frac{k_1 - k_2}{k_1 + k_2} \tag{9.11}$$

and

$$\frac{C}{A} = \frac{2k_1}{k_1 + k_2}. \tag{9.12}$$

In section 2.7 we defined the particle flux (equation (2.40)). For the case we are considering there is no time dependence and the flux is

$$j(x) = \frac{\mathrm{i}\hbar}{2m}\left(u\frac{\mathrm{d}u^*}{\mathrm{d}x} - u^*\frac{\mathrm{d}u}{\mathrm{d}x}\right). \tag{9.13}$$

From the form of $j(x)$ it is apparent that the flux will be zero unless u is complex. Since u and u' are continuous at the boundary so also is the flux. Inserting equations (9.4) and (9.8) into equation (9.13), we obtain the following expressions for the flux on either side of the boundary:

$$x < 0 \qquad j_1(x) = \frac{\hbar k_1}{m}(|A|^2 - |B|^2) \tag{9.14}$$

$$x \geqslant 0 \qquad j_2(x) = \frac{\hbar}{2m}|C|^2(k_2^* + k_2)\exp[\mathrm{i}(k_2 - k_2^*)]. \tag{9.15}$$

If k_2 is real

$$j_2(x) = \frac{\hbar k_2}{m} \mid C \mid^2 \tag{9.16}$$

and if k_2 is imaginary

$$j_2(x) = 0. \tag{9.17}$$

The interpretation of (9.14) is that the net flux is equal to the difference between the incident and reflected fluxes. It is therefore natural to define a reflection coefficient R which is the ratio of these two terms:

$$R = \frac{\mid B \mid^2}{\mid A \mid^2} = \left| \frac{k_1 - k_2}{k_1 + k_2} \right|^2. \tag{9.18}$$

Similarly, if k_2 is real we can define a transmission coefficient T which is the ratio of the transmitted flux to the incident flux:

$$T = \frac{k_2}{k_1} \frac{\mid C \mid^2}{\mid A \mid^2} = \frac{4k_1 k_2}{(k_1 + k_2)^2}. \tag{9.19}$$

Hence we find

$$R + T = 1. \tag{9.20}$$

If k_2 is imaginary $j_2(x)$ is zero, and by flux continuity so is $j_1(x)$. We conclude from equation (9.14) that $\mid A \mid^2 = \mid B \mid^2$, and so $R = 1$, $T = 0$, and equation (9.20) remains true.

Equation (9.15) indicates that, although $j_2(x) = 0$ if k_2 is imaginary, there is no requirement that $C = 0$. Writing

$$k_2 = \mathrm{i}\varkappa \qquad \varkappa \text{ real} \tag{9.21}$$

equations (9.11) and (9.12) become

$$\frac{B}{A} = \frac{k_1 - \mathrm{i}\varkappa}{k_1 + \mathrm{i}\varkappa} \tag{9.22}$$

and

$$\frac{C}{A} = \frac{2k_1}{k_1 + \mathrm{i}\varkappa}. \tag{9.23}$$

Equation (9.22) shows that

$$\mid A \mid = \mid B \mid \tag{9.24}$$

and therefore

$$R = 1 \tag{9.25}$$

in agreement with the result obtained by considering the flux.

Equation (9.23) shows that $|C|/|A| \neq 0$ showing that the wave penetrates the classically forbidden region $x > 0$, even though the particle flux is zero. Such a disturbance is referred to as an evanescent or non-propagating wave. It should be remembered that we are considering a steady state; there must of course have been an initial flux to set up the probability density in this region when the beam was switched on. This behaviour is reminiscent of the total internal reflection which occurs when light is incident at greater than the critical angle upon an interface at which the refractive index decreases. Evanescent waves are launched into the second medium in that case also.

It should be noted that, for $E < V$, B/A and C/A are complex, corresponding to the phase differences which exist between the incident, reflected and transmitted waves. It is therefore not possible to choose the constants A, B and C to be all real. We can choose the amplitude and phase of any one of the three, then the other two complex quantities are determined by equations (9.22) and (9.23).

The choice of any particular phase is actually equivalent to a choice of time zero. To see this we recall (section 2.4) that the time-dependent wavefunctions in the two regions are

$$u_1 \exp(-\mathrm{i}\omega t) = A \exp[\mathrm{i}(k_1 x - \omega t)] + B \exp[-\mathrm{i}(k_1 x + \omega t)] \tag{9.26}$$

and

$$u_2 \exp(-\mathrm{i}\omega t) = C \exp[-\mathrm{i}(\varkappa x + \omega t)] \tag{9.27}$$

where

$$\omega = E/\hbar \qquad |A| = |B|. \tag{9.28}$$

Moving time zero to t_0 has the effect of multiplying all the coefficients by $\mathrm{e}^{-\mathrm{i}\omega t_0}$, a constant phase factor. This means that it is always possible to pick a time to view the waves at which any particular component is real. In this example, if we choose C to be real and $t = 0$, then u_2 is real and so also must be u_1 in order to satisfy the boundary conditions at $x = 0$, but the incident and reflected components of u_1 are complex (with equal and opposite phase angles). The phase difference at $x = 0$ between the incident and scattered wave can be found from equation (9.22), and is given by

$$\delta = \frac{2k_1\varkappa}{k_1^2 - \varkappa^2}. \tag{9.29}$$

It is a general property of scattering events that they introduce such phase shifts, and their value is characteristic of the particular form of the scattering potential. Much use is made of this property in the analysis of scattering in three dimensions.

The fact that the wavefunction can be chosen to be real is a property held in common with the bound states of the potential wells we have studied in earlier chapters. This is not surprising since conditions at the step are identical to those at one of the boundaries of the finite square well analysed in section 3.5, and the wavefunctions we have found for the two cases are similar. In the case of the step the energy eigenvalues form a continuum, whereas they are discrete for the well. However, we can regard the step as one side of an infinitely wide well, for which the energy spectrum would become continuous.

Examining equations (9.11) and (9.12) we find that if $E > V$ (k_2 real) there are finite probabilities for both reflection and transmission. This contrasts with the behaviour of a classical particle, which would certainly be transmitted. Figure 9.2 shows the transmission T as a function of $\varepsilon = E/V$. The transmission asymptotically approaches 1 at very large energies. This is also the case when V is negative; reflection occurs at

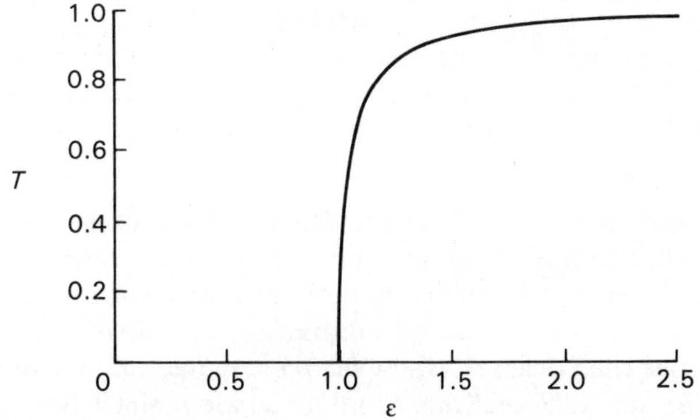

Figure 9.2 The transmission of a potential step as a function of $\varepsilon = E/V$. The step is of infinite width so no transmission occurs until $\varepsilon > 1$.

downhill steps as well as uphill ones! Again these properties are typical of wave motion. The partial reflection and transmission of light at a boundary between two different media is a familiar phenomenon.

When $E > V$ the flux is finite on both sides of the step, so the functions u_1 and u_2 must be complex. On the right the solution is a pure travelling wave; on the left it is a superposition of standing and travelling waves. Equations (9.11) and (9.12) make it clear that the amplitudes A, B and C can differ in phase only by 0 or π, since their ratios are real. It is therefore possible for them all to be chosen to be real. Even so, the complex natures of u_1 and u_2 make it impossible to represent the solution in a single graph and we must plot both real and imaginary parts.

The probability density function u^*u for a steady state travelling wave is simply a constant, independent of x. It is useful to consider the form of u^*u in a region where waves are propagating in either direction. Consider

$$u^*u = |A\,e^{ik_1x} + B\,e^{-ik_1x}|^2 \tag{9.30}$$

$$= |A|^2 + |B|^2 + 2|A||B|\cos(k_1x + \delta). \tag{9.31}$$

δ is the phase difference between A and B. Clearly

$$\frac{u_1^*u_1(\max)}{u_2^*u_2(\min)} = \frac{(|A| + |B|)^2}{(|A| - |B|)^2} = \frac{(1 + R^{1/2})^2}{(1 - R^{1/2})^2}. \tag{9.32}$$

The last equality assumes, as in our case, that the wave B is a reflection of wave A. This relation will allow us to determine R from the probability density.

Computer exercise 9.1: the potential step
Load the program '1-D Potential Step'. The default parameters illustrate a step of height 10 eV, on which an electron (or beam of electrons) is incident. For correct interpretation of what the computer shows you it is essential to appreciate the following points:

(i) Despite the fact that we are describing scattering, the computer integration, which proceeds across the screen in the usual way, bears no relation whatever to the motion of the scattered particle. As it happens, it is usually more convenient to integrate in the direction towards the incident particle, because it is simpler to set up the required initial values.

(ii) The solutions of Schrödinger's equation which we have found analytically, and which the computer will plot for us, are only solutions when taken in their entirety. The incident and reflected waves are not in themselves solutions which satisfy the boundary conditions; only their sum is a solution, and this is what we see on the screen. Thus in the case of total reflection we will see a standing wave, not two waves travelling in opposite directions.

(iii) In order to represent the complete complex solutions to the equation the computer integrates the equation twice, once for the real part and once for the imaginary part, and plots both of these, as well as the probability density. It is therefore necessary to provide four initial values: the real and imaginary amplitudes and their derivatives. It can be quite inconvenient to set up the precise conditions to produce a pure travelling wave, which consists of real and imaginary parts with a $\frac{1}{2}\pi$ phase difference. If required the program will do this automatically and it is then only necessary to specify the required amplitude. The comments in (i) and (ii) apply to the real and imaginary parts independently.

(*a*) Running the program with the default parameters shows the wavefunction for a totally reflected particle with $E = 5$ eV. The imaginary part of the function has been made zero throughout by the choice of initial conditions. In order to produce a decaying exponential, the integration is started on the far right with a zero amplitude and small negative slope, a technique introduced in connection with the asymmetrical well in computer exercise 3.6. Note the standing-wave character of the wave on the left. Now introduce an imaginary component to the solution. First reduce the slope of the real function to half of its value, then alter the imaginary slope from zero to the same magnitude, but of opposite sign. Note that the amplitude and standing-wave nature of the probability density are completely unchanged. Changing the phase of the wavefunction is equivalent to changing the instant at which we 'view' the wave. Since the wavefunctions are stationary states there should be no observable time dependence, and this is what we find.

(*b*) Now have a look at the wavefunction for a particle with $E > V$, which is partially transmitted. Choose the 'Pure Travelling Wave' option, and set $E = 20$ eV and $A = 0.5$. A is the amplitude of the transmitted wave on the right which, as we have seen, must be a pure travelling wave. Note that the probability density on the right is constant as expected for a travelling wave; although the real and imaginary parts are oscillatory they differ in phase by $\frac{1}{2}\pi$ and combine to give a uniform probability density. On the left the probability has a constant and an oscillating component, corresponding to a combination of travelling and standing waves or, equivalently, incident and reflected waves of unequal amplitude. The phase of the wavefunction can conveniently be changed in this case by altering XI, to 7 say. The integration then starts with the same values but at a different x position. Again this is equivalent to a time-zero shift. The real and imaginary parts on the left will now look quite different, but the probability density will remain unchanged.

(*c*) Change E to 12 eV. The situation is much as before but now there is a much larger reflected component, so the dips in the probability density are much deeper. With $E = 11$ eV this is even more exaggerated. Note the large increase in wavelength on the right, corresponding to the much lower momentum in this region. An initially surprising observation, given that the particle is most probably reflected, is the larger mean probability density on the right than on the left. This is explained by the much lower particle velocity on the right. The net flux is identical on each side, but the incident and reflected fluxes separately are much greater than the transmitted flux.

(*d*) Using the tabulation option, and the same conditions as in (*c*), make a note of the ratio $u^*u(\max)/u^*u(\min)$, for several particle energies (equation (9.32)). Now change the starting point of XI $= -8$. The result will represent a particle incident on a downhill step from the right. Measure the standing-wave ratio on the right for the same energies as before. You should find that the values are the same; the reflection probability is the same in either direction. Now, however, the probability density is much lower in the transmitted wave, because it has the greater velocity. Calculate R from equation (9.32). Does it agree with the value predicted by equation (9.18)?

(*e*) It is a well known property of waves that the reflection at a discontinuity is a function of the abruptness of the transition. The program allows us to change the step either to a linear ramp or to a ramp which blends smoothly into the regions of constant potential. (Analytic solutions for the linear ramp can readily be found in terms of the Airy functions met with in section 3.12. The smooth ramp is actually sine in form.) Run the program for ramps of several different lengths $2a$, both linear and smooth. You should discover that long ramps cause much less reflection than short ones and, for a given length, the reflection is less for the smoothly blended ramp. In this context a long ramp is one which is much longer than the particle wavelength, and in the limit $a \gg 2\pi/k_2$ the transmission becomes ≈ 1. This behaviour corresponds to that of a classical particle which will certainly be transmitted provided $E > V$. This is to be expected since any macroscopic particle will have an infinitesimal wavelength compared with the width of any potential step it might encounter.

〉9.3 The rectangular potential barrier

We now consider a barrier of finite width, figure 9.3, defined by the potential

$$\begin{aligned} V(x) &= 0 \qquad & |x| \geqslant b \\ V(x) &= V \qquad & -b < x < b. \end{aligned} \tag{9.33}$$

We can immediately write down the general solutions for the three regions:

$$u_1 = A\,e^{ik_1x} + B\,e^{-ik_1x} \qquad x \leqslant -b \tag{9.34}$$

$$u_2 = P\,e^{ik_2x} + Q\,e^{-ik_2x} \qquad -b < x < b \tag{9.35}$$

$$u_3 = C\,e^{ik_1x} + D\,e^{-ik_1x} \qquad x \geqslant b \tag{9.36}$$

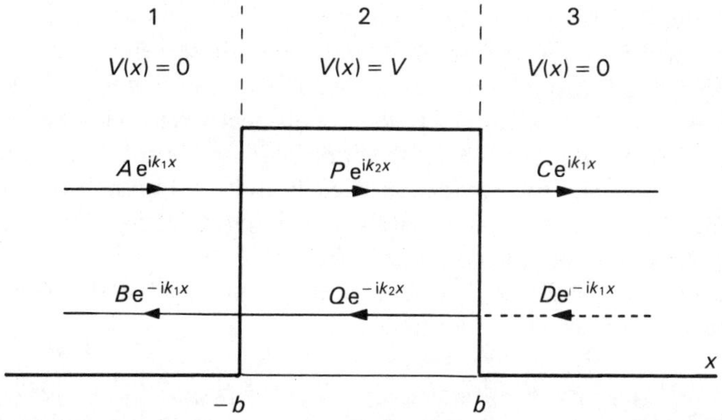

Figure 9.3 A potential barrier. A wave incident from the left is partially reflected and partially transmitted at both potential steps. The component De^{-ik_1x} will not be present unless another source or reflector exists to the right of the figure.

where

$$k_1 = (2mE)^{1/2}/\hbar \tag{9.37}$$

and

$$k_2 = [2m(E-V)]^{1/2}/\hbar. \tag{9.38}$$

For a particle flux incident from the left we anticipate that D will be zero. However, for the analysis which follows it is convenient to retain it in order to maintain the generality and symmetry of the equations. In the barrier itself we expect waves travelling in either direction because there will be partial reflection at the second boundary even when the flux is incident from one side only. k_2 will be real or imaginary as the sign of $E-V$ is positive or negative. The analysis also covers the case of negative V, corresponding to scattering at a square potential well.

Matching u and u' at both boundaries gives the following four equations:

$$A\ \mathrm{e}^{-\mathrm{i}k_1b}+B\ \mathrm{e}^{\mathrm{i}k_1b}=P\ \mathrm{e}^{-\mathrm{i}k_2b}+Q\ \mathrm{e}^{\mathrm{i}k_2b} \tag{9.39}$$

$$\mathrm{i}k_1A\ \mathrm{e}^{-\mathrm{i}k_1b}-\mathrm{i}k_1B\ \mathrm{e}^{\mathrm{i}k_1b}=\mathrm{i}k_2P\ \mathrm{e}^{-\mathrm{i}k_2b}-\mathrm{i}k_2Q\ \mathrm{e}^{\mathrm{i}k_2b} \tag{9.40}$$

$$P\ \mathrm{e}^{\mathrm{i}k_2b}+Q\ \mathrm{e}^{-\mathrm{i}k_2b}=C\ \mathrm{e}^{\mathrm{i}k_1b}+D\ \mathrm{e}^{-\mathrm{i}k_1b} \tag{9.41}$$

$$\mathrm{i}k_2P\ \mathrm{e}^{\mathrm{i}k_2b}-\mathrm{i}k_2Q\ \mathrm{e}^{-\mathrm{i}k_2b}=\mathrm{i}k_1C\ \mathrm{e}^{\mathrm{i}k_1b}-\mathrm{i}k_2D\ \mathrm{e}^{\mathrm{i}k_1b}. \tag{9.42}$$

If we set $D=0$ we have four equations with five variables, which allows us to solve for the ratio of any pair of coefficients. As with the potential step we will be interested in B/A and C/A, these being related to the reflection and transmission coefficients. The algebra involved is straightforward but long and tedious. It is more instructive to attack the problem using matrix techniques. This is the approach used in advanced scattering theory, and is also similar to that used in matrix optics. We treat each interface individually, find a matrix describing its action, then multiply the two matrices for the combined effect.

Eliminating first B and then A from equations (9.39) and (9.40) we can express the two equations in matrix form by

$$\begin{bmatrix}A\\B\end{bmatrix}=\frac{1}{2}\begin{bmatrix}(1+k_2/k_1)\exp[\mathrm{i}(k_1-k_2)b] & (1-k_2/k_1)\exp[\mathrm{i}(k_1+k_2)b]\\(1-k_2/k_1)\exp[-\mathrm{i}(k_1+k_2)b] & (1+k_2/k_1)\exp[-\mathrm{i}(k_1-k_2)b]\end{bmatrix}\begin{bmatrix}P\\Q\end{bmatrix}. \tag{9.43}$$

Similarly, equations (9.41) and (9.42) can be written

$$\begin{bmatrix}P\\Q\end{bmatrix}=\frac{1}{2}\begin{bmatrix}(1+k_1/k_2)\exp[\mathrm{i}(k_1-k_2)b] & (1-k_1/k_2)\exp[-\mathrm{i}(k_1+k_2)b]\\(1-k_1/k_2)\exp[\mathrm{i}(k_1+k_2)b] & (1+k_1/k_2)\exp[-\mathrm{i}(k_1-k_2)b]\end{bmatrix}\begin{bmatrix}C\\D\end{bmatrix}. \tag{9.44}$$

Substituting (9.44) in (9.43), and introducing

$$\alpha=k_2/k_1+k_1/k_2 \qquad \beta=k_2/k_1-k_1/k_2 \tag{9.45}$$

we find

$$\begin{bmatrix} A \\ B \end{bmatrix} = \begin{bmatrix} [\cos 2k_2b - (\tfrac{1}{2}\mathrm{i}\alpha)\sin 2k_2b]\exp(2\mathrm{i}k_1b) & -(\tfrac{1}{2}\mathrm{i}\beta)\sin 2k_2b \\ (\tfrac{1}{2}\mathrm{i}\beta)\sin 2k_2b & [\cos 2k_2b + (\tfrac{1}{2}\mathrm{i}\alpha)\sin 2k_2b]\exp(-2\mathrm{i}k_1b) \end{bmatrix} \begin{bmatrix} C \\ D \end{bmatrix}. \tag{9.46}$$

If at this point we specify $D = 0$ we readily obtain the following expressions for the ratios we require:

$$\frac{B}{A} = \frac{\mathrm{i}\beta \sin 2k_2b \exp(-2\mathrm{i}k_1b)}{2[\cos 2k_2b - (\tfrac{1}{2}\mathrm{i}\alpha)\sin 2k_2b]} \tag{9.47}$$

$$\frac{C}{A} = \frac{\exp(-2\mathrm{i}k_1b)}{[\cos 2k_2b - (\tfrac{1}{2}\mathrm{i}\alpha)\sin 2k_2b]}. \tag{9.48}$$

Inserting α and β, the squared magnitudes of these expressions are

$$R = \frac{|B|^2}{|A|^2} = \left(1 + \frac{4k_1^2k_2^2 \operatorname{cosec}^2 2k_2b}{(k_1^2 - k_2^2)^2}\right)^{-1} \tag{9.49}$$

$$T = \frac{|C|^2}{|A|^2} = \left(1 + \frac{(k_1^2 - k_2^2)^2 \sin^2 2k_2b}{4k_1^2k_2^2}\right)^{-1}. \tag{9.50}$$

If k_2 is imaginary we can obtain the reflection and transmission coefficients directly from equations (9.49) and (9.50) by making the substitution

$$k_2 = \mathrm{i}\varkappa \tag{9.51}$$

obtaining

$$R = \frac{|B|^2}{|A|^2} = \left(1 + \frac{4k_1^2\varkappa^2 \operatorname{cosech}^2 2\varkappa b}{(k_1^2 + \varkappa^2)^2}\right)^{-1} \tag{9.52}$$

$$T = \frac{|C|^2}{|A|^2} = \left(1 + \frac{(k_1^2 + \varkappa^2)^2 \sinh^2 2\varkappa b}{4k_1^2k_2^2}\right)^{-1}. \tag{9.53}$$

Hence

$$R + T = 1 \tag{9.54}$$

in both cases. Note that the ks do not appear in the definitions of R and T above, equations (9.49) and (9.50) (cf equation (9.19)); the potential and hence the particle speed are now the same on either side of the barrier, unlike the case of the potential step, and so the fluxes and probability densities have the same proportionality for each of the three travelling waves external to the barrier. Equations (9.50) and (9.53) can be written in simpler form by introducing the dimensionless parameters

$$\gamma = (2m\,|\,V\,|)^{1/2}\,b/\hbar \qquad \varepsilon = E/|\,V\,|. \tag{9.55}$$

We have used $|\,V|$ so that the expressions remain valid for either a potential barrier or a well; γ and ε are positive since E is assumed positive for scattering. γ is a measure of the strength of the potential structure. For a well we found in section 3.5 that it determined the number of energy eigenstates. Substitution of γ and ε results in

$$T = \left(1 + \frac{\sinh^2 2\gamma\sqrt{1-\varepsilon}}{4\varepsilon(1-\varepsilon)}\right)^{-1} \qquad \varepsilon < 1 \tag{9.56}$$

$$T = \left(1 + \frac{\sin^2 2\gamma\sqrt{\varepsilon-1}}{4\varepsilon(\varepsilon-1)}\right)^{-1} \qquad \varepsilon > 1\,. \tag{9.57}$$

The physical content of equations (9.56) and (9.57) is best conveyed by means of graphs. Figure 9.4 shows T, the transmission of the barrier, as a function of ε for several different values of γ. R is obtained from the graph with the help of equation (9.54). The oscillatory nature of T when $\varepsilon > 1$ is apparent in equation (9.50), which indicates that the transmission is unity whenever

$$2k_2 b = n\pi \qquad n = 1, 2, 3, \ldots\,. \tag{9.58}$$

This is the condition that the barrier width is equal to an integral number of half wavelengths. What is happening is that the multiple partial reflections occurring within the barrier are interfering destructively to give no reflected wave at all. Conversely the multiple partial transmissions are interfering constructively to give total overall transmission. This is precisely analogous to the behaviour of light in a Fabry–Perot interferometer. This consists of two plane parallel mirrors. Light incident normally is totally transmitted if the mirror spacing is an integral number of half waves. This remains true even when the reflectivities of the mirrors approach unity! The reason is that the greater the reflectivity the greater the effective number of interfering beams there are to sum.

When $\varepsilon < 1$ we find that there is still a finite probability of transmission. This phenomenon is usually referred to as 'tunnelling' since it directly violates the classical prediction for a particle, which is that it will be quite unable to surmount the barrier. Again however it is a well known property of waves, and in the case of light it is manifested in the phenomenon of

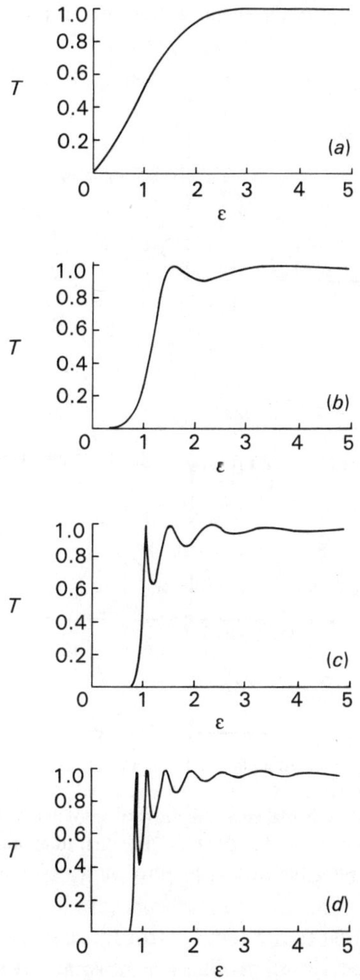

Figure 9.4 The transmission of a barrier as a function of ε, for various barrier strengths γ. $\gamma = 1$ (*a*), 2 (*b*), 4 (*c*), 6 (*d*). As the barrier strength increases the transmission tends towards the classical 'all or nothing' characteristic.

frustrated total internal reflection. This occurs in the sort of situation shown in figure 9.5. Light incident internally at greater than the critical angle on the face of the prism sets up an evanescent wave in the second medium. This wave decays exponentially with distance. If a second prism is brought close to the first, a propagating wave is launched into it, and partial transmission occurs.

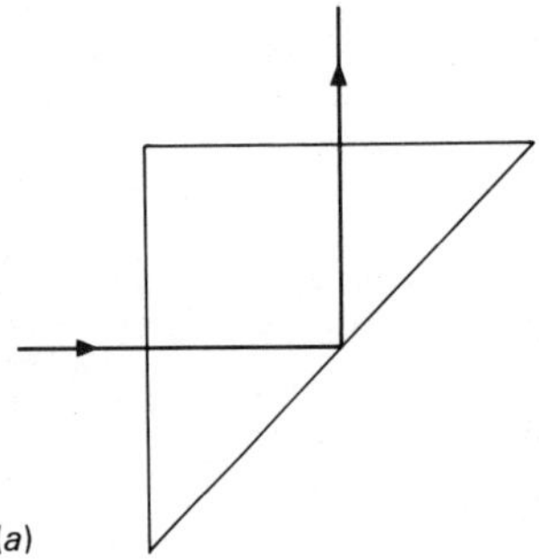

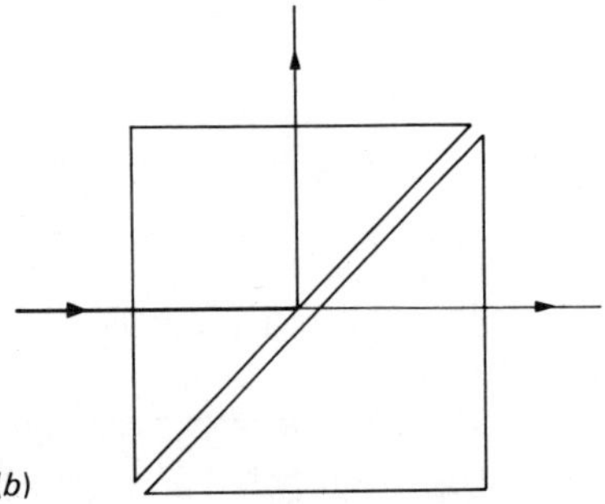

Figure 9.5 'Tunnelling' in a classical system. (*a*) Total internal reflection in a prism. (*b*) Frustrated total internal reflection and partial transmission occur when a second surface is placed within a fraction of a wavelength of the first. The evanescent wave to the right of the first diagonal surface decays exponentially until it enters the second prism.

For those who feel uncomfortable with the idea of a particle *penetrating* a potential barrier quantum mechanics provides an alternative picture, based upon the uncertainty principle, in which the particle *surmounts* it. The idea is that the energy of a particle fluctuates, with fluctuations of magnitude ΔE occurring on a timescale $\Delta t \sim \hbar/\Delta E$. Obviously a sufficiently large fluctuation ($E + \Delta E > V$) will enable the particle to pass over the top of the barrier. The width of the barrier is significant because the particle has a limited range of motion in the time interval Δt, determined by its speed. These ideas can be made quantitative but we will not pursue them here. Whether this picture is any more comfortable than the tunnelling one is a matter of taste!

Examination of figure 9.4 shows that as γ, the strength of the barrier, is increased the probability of transmission for $\varepsilon < 1$ decreases rapidly, the onset of transmission near $\varepsilon = 1$ becomes very abrupt, and the oscillations become very rapid. In fact the curve approaches that of figure 9.2 for the potential step for very large values of γ since we then have a barrier of infinite width.

When $\varepsilon < 1$ the wavefunction u_2 within the barrier, equation (9.35), becomes a sum of rising and falling real exponentials since k_2 is then imaginary. It is interesting to examine the wavefunction within the barrier for this case in more detail since we have noted that the flux will be zero unless the wavefunction is complex. (We exclude constant overall phase factors.) Equation (9.44) with $D = 0$ provides expressions for P and Q in terms of C. We then find

$$u_2 = \tfrac{1}{2}C \exp(\mathrm{i}k_1 b)\{(1 - \mathrm{i}k_1/\varkappa)\exp[\varkappa(b - x)] + (1 + \mathrm{i}k_1/\varkappa)\exp[-\varkappa(b - x)]\}. \quad (9.59)$$

This shows that the function is indeed intrinsically complex, and that this is only the case because of the inclusion of exponentials of both signs. We had to exclude the positive exponential in the case of the potential step, since the wavefunction would have diverged for large x. Here the range of validity of this function is limited to $|x| < b$, so the question of divergence does not arise. The term is required in this case in order to satisfy all the boundary conditions and, equivalently, to provide the continuity of flux which we require. It may be verified that the flux (equation (9.13)) carried by this wavefunction is $\hbar k_1 |C|/m$, which is equal to the flux carried by the travelling wave to the right of the barrier.

In many cases of physical importance the height and width of the barrier are such that the probability of transmission is very small. This will be the case provided $\varkappa b \gg 1$, because then the width of the barrier is much greater than the characteristic length of the exponential wavefunction components within it. In that case we can approximate T by neglecting the one in the denominator of equation (9.53) and keeping only the positive exponential in the sinh term:

$$T \approx \frac{16k_1^2\varkappa^2}{(k^2 + \varkappa^2)} \mathrm{e}^{-4\varkappa b} = 16\varepsilon(1 - \varepsilon)\,\mathrm{e}^{-4\varkappa b}. \quad (9.60)$$

Computer exercise 9.2: the rectangular potential barrier

Load the program '1-D Potential Barrier'. This follows the same pattern as the potential-step program, in that the real and imaginary parts of the wavefunction and the probability density are all plotted. The potential represents a rectangular barrier or a potential well (and refers to the half-width of either as a).

(a) Accepting the default values results in a barrier of height 5 eV and half-width $b = 0.05$ nm ($\gamma = 0.573$). The initial values correspond to an electron incident from the left with an energy of 3 eV.

Although the probability density on the left shows large oscillations it never reaches zero. Also, there is an obvious phase shift between the real and imaginary parts, so the density is not that of a pure standing wave; there is a travelling-wave component. On the right the density is that of a pure travelling wave, constant with x. Analysis has shown that the flux is constant throughout and, since the propagation constant k_2 is the same on either side of the barrier, we might expect the dips on the left to be at the same level as the constant density on the right. Intuition plays us false here; the levels are clearly different, a result of interference between the incident and reflected waves. Examination of equation (9.32) shows that the minima in the probability density have the value $(|A| - |B|)^2$ whereas the flux in this region has the value $\hbar k_1 (|A|^2 - |B|^2)/m$, so the two are not proportional! Within the barrier the functions are sums of positive and negative real exponentials; note the curvature away from the x axis. Run the program several times, incrementing XI by 0.05 nm each time. You are viewing the wave at different times; note how the real and imaginary parts of the function change radically while the probability density remains exactly the same.

(b) With the same parameters as for (a), use the tabulation option to obtain an accurate value for the transmission T. Compare your result with the value predicted by equation (9.56).

(c) Increase the energy to 5 eV ($\varepsilon = 1$). This is the classical threshold for transmission of the particle but, according to equations (9.50) and (9.57), the probability of transmission is now given by

$$T = (1 + k_1 b^2)^{-1} = (1 + \gamma^2)^{-1} \qquad (9.61)$$

since $k_2 = 0$. Use this equation to find T and hence R, and compare this with the transmission you observe on the screen, making use of equation (9.32) and the tabulation option to get an accurate value. Notice that the wavefunction within the barrier is linear; it is a wave of infinite wavelength.

(d) Equation (9.58) shows that $T = 1$ when the barrier width equals an integral number n of half wavelengths. Use this relation and equation

(9.38) to find the energies for which this condition will be satisfied. Verify your results by plotting the wavefunctions for $n = 1$ and 2.

〉9.4 Non-rectangular barriers

For all but the simplest shaped barriers exact calculations of transmission coefficients become quite intractable and we must either use numerical methods or approximations. For the latter the general approach is to treat the barrier as if it consisted of an array of consecutive rectangular barriers of varying height, as in figure 9.6. The transmission of a typical element is calculated and integrated over the barrier to obtain the overall transmission. This works well provided that the potential is smoothly varying and that the barrier height and width are such that the transmission is small. The effect of this is that strong local reflections do not occur within the barrier. Furthermore, if the wave is strongly attenuated, we do not expect there to be much reflection from the rear part of the barrier, since the amplitude there is small. This is illustrated, for the rectangular barrier, in equation (9.59), which shows that the forward and backward travelling amplitudes contain the factors $e^{\varkappa b}$ and $e^{-\varkappa b}$ respectively. We shall therefore assume that over most of the barrier there is almost no backward travelling disturbance. With this approximation it is simple to calculate the attenuation of the wave on traversing

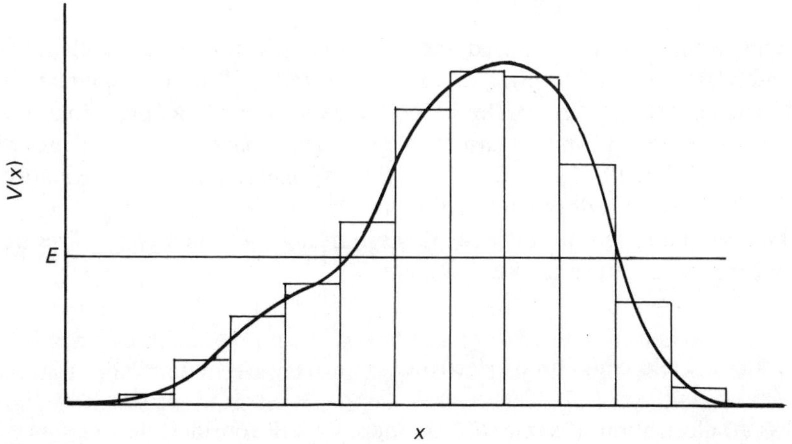

Figure 9.6 A barrier of arbitrary shape may be modelled as a series of elementary rectangular barriers.

one of the elementary rectangular barriers, because the wavefunction is then a decaying exponential, and the decay constant must be $\varkappa(x)$, where

$$\varkappa(x) = [2m(V(x) - E)]^{1/2}/\hbar. \tag{9.62}$$

For the element at x therefore

$$u(x + \Delta x) = u(x)\exp(-\varkappa(x)\Delta x). \tag{9.63}$$

Expanding both sides of this equation and retaining only first-order terms, we find

$$u(x) + \left(\frac{\mathrm{d}u(x)}{\mathrm{d}x}\right)_x \Delta x \simeq u(x)(1 - \varkappa(x)\Delta x)$$

or

$$\frac{\mathrm{d}u(x)}{u(x)} \simeq -\varkappa(x)\,\mathrm{d}x. \tag{9.64}$$

Integrating this over the barrier, between the turning points x_1 and x_2 (for which $E = V$),

$$u(x_2) \simeq u(x_1)\exp\left(-\int_{x_1}^{x_2} \varkappa(x)\,\mathrm{d}x\right) \tag{9.65}$$

and so the transmission is given by

$$T = \left|\frac{u(x_2)}{u(x_1)}\right|^2 = \exp\left(-2\int_{x_1}^{x_2} \varkappa(x)\,\mathrm{d}x\right). \tag{9.66}$$

This result, crudely derived here, can be justified in a much more rigorous fashion by using a method introduced in this context by Wentzel, Kramers and Brillouin, and known as the WKB approximation. Note that for the special case of a rectangular barrier $\varkappa(x)$ is constant and equation (9.66) yields $e^{-4\varkappa b}$, the exponential factor in equation (9.60). This is the dominant factor, since it is strongly dependent upon E, but the failure to reproduce the slowly varying coefficient emphasizes the approximate nature of the treatment.

〉9.5 Field emission of electrons

As an illustration of some of these ideas we will consider the emission of electrons from a conductor subjected to an intense electric field, a process referred to as *field* or *cold emission*. The electrons in a conductor are retained within it by potential barriers at the surfaces, of electrostatic origin. For our purpose we will regard them as potential steps of height V and infinite width, as in figure 9.7(a). The electrons themselves have a range of energies with a fairly well defined maximum value. We will suppose that a uniform field exists perpendicular to the surface. This could be achieved by arranging a pair of conductors at different potentials as in a parallel plate capacitor. The potential barrier at the negative surface then becomes tilted, as in figure 9.7(b). An electron within the metal is now presented with a triangular barrier of finite width and barrier penetration can be expected to occur. If the surface is defined by $x=0$, and the electric field is $\mathscr{E}$, the potential can be expressed as

$$V(x)=0 \qquad x<0$$
$$V(x)=V-e\mathscr{E}x \qquad x\geqslant 0. \tag{9.67}$$

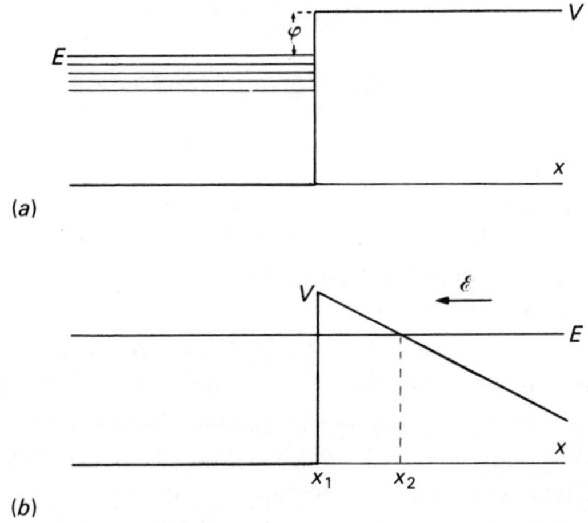

Figure 9.7 Field emission of electrons. (a) The electrons in a conductor are restrained by the infinitely wide potential barrier of height equal to the work function φ. (b) An electric field reduces the barrier to a triangular one of finite width, through which the electrons can be transmitted.

For an electron of energy E the barrier extends from $x_1=0$ to x_2, where

$$E=V-e\mathscr{E}x_2. \tag{9.68}$$

The electrons with the greatest probability of emission are those of maximum energy, and for these

$$\varphi = V - E \tag{9.69}$$

where φ is the work function. For such electrons the barrier extends to

$$x_2 = \varphi / e\mathscr{E}. \tag{9.70}$$

Equations (9.68)–(9.70) enable us to express the barrier transmission, (9.66), in terms of the electric field $\mathscr{E}$ and the barrier width x_2:

$$\begin{aligned} T &= \exp\left(-2 \int_0^{x_2} \{ [2me\mathscr{E}(x_2 - x)]^{1/2} / \hbar \} \, \mathrm{d}x \right) \\ &= \exp\left(\frac{-4(2me\mathscr{E})^{1/2}}{3\hbar} x_2^{3/2} \right). \end{aligned} \tag{9.71}$$

In terms of the work function φ the transmission is

$$T = \exp\left(\frac{-4\sqrt{2m}}{3\hbar e} \frac{\varphi^{3/2}}{\mathscr{E}} \right). \tag{9.72}$$

Evaluating the constants in this expression we obtain

$$T = \exp\left(-6.83 \times 10^9 \frac{\varphi^{3/2}}{\mathscr{E}} \right) \tag{9.73}$$

where the work function φ is to be expressed in electron volts. For most metals this quantity is between 2 and 6 eV, so it is apparent that fields of the order of 10^{10} V m^{-1} will be required to obtain high transmission. Electric fields of this magnitude cannot easily be obtained with the geometry described here but can be achieved in the neighbourhood of a very sharp point. By this means the relation (9.72) has been tested experimentally and the dependence upon φ and $\mathscr{E}$ verified. The absolute value of the transmission, however, is usually higher than predicted because of surface roughness. The effect of this is to cause local electric fields which are much greater than the mean. This effect was capitalized upon by Mueller in 1937 who used it to form a greatly magnified image of a surface according to the facility with which it emitted electrons, a technique known as field emission microscopy. The related technique of scanning tunnelling microscopy also relies on this effect.

A very similar process to that described above is exploited in various electronic devices which rely on the tunnelling mechanism for their operation. In these devices the electric field is set up in a slab of insulating material situated between two conductors. The slab can be made extremely thin so that very high fields can be attained with modest potential differences.

Computer exercise 9.3: field emission
Run the program '1-D Potential Step'. This has an option ('B' in item 4 of the edit screen) to plot a triangular barrier rather than a step.

(a) Set $V = 10$ eV, $E = 5$ eV and $a = 1$ so that $\varphi = 5$ eV and $\mathscr{E} = 5 \times 10^{10}$ V m^{-1}. Set up a travelling wave of amplitude 0.2 on the left. Measure the transmission of the barrier using equations (9.20) and (9.32) and the tabulation option, and compare this with the value predicted by equation (9.73). You should find that the two values agree within a factor of about two, even though the barrier can hardly be said to satisfy the conditions for the WKB approximation to apply, and in fact the wavefunctions (real and imaginary) within the barrier are not simply monotonically decreasing functions, as suggested in the discussion of section 9.4.

(b) Make similar measurements for more opaque barriers, by progressively increasing a to 20. Note that as the barrier becomes thicker the wavefunctions (real and imaginary) become monotonically decreasing functions. For very opaque barriers determination of T using equation (9.32) becomes impractical. However, in these cases the wave on the right is almost a pure standing wave and so the squared amplitude of the incoming wave is $|A|^2 = \frac{1}{4}(u^*u)_{\max}$. C is the value you set for the amplitude of the travelling wave on the left which will need to be very small ($\sim 10^{-6}$ for $a = 20$), and $T = |C|^2/|A|^2$. Use your values to verify the $\mathscr{E}$ dependence indicated in equation (9.73), by plotting $\ln(T)$ against $\mathscr{E}^{-1}$.

(c) Similarly measure T for a range of values of $\varphi = V - E$, and verify the dependence predicted by equation (9.73). Note the remarkable strength of this dependence.

〉9.6 α-particle emission

An early major success of the quantum theory was the convincing explanation of the characteristics of α-particle emission. In this process an energetic α particle is spontaneously ejected from a nucleus. The time of emission is random but different species of emitting nuclei have well defined characteristic mean lifetimes. An inverse correlation between energy and lifetime was noted by Rutherford as early as 1907, and the strength of the dependence is remarkable. Whereas the energies of the particles emitted by different species range over a factor of about 2 (from 4 to 9 MeV), the half lives of the species range from less than a microsecond to more than 10^{11} years, a factor of 10^{24} or so!

According to classical physics α emission is impossible! α-particle scattering experiments by Rutherford using 8.7 MeV α particles on uranium demonstrated that a repulsive Coulomb potential barrier existed which was at least 8.7 MeV in height. How then is it possible to account for the fact that uranium was known to emit 4.2 MeV α particles? Any particle surmounting the barrier would be observed to have an energy of at least 8.7 MeV. Assuming for the moment that α particles have a separate existence in the nucleus, we must conclude that, within a certain radius, which we can identify with the radius of the nucleus, there is an attractive force, which can be represented by a potential well. Combining this with the repulsive Coulomb barrier we arrive at the schematic form of potential function for the α-particle–nucleus system shown in figure 9.8. It is now clear that the emission of an α particle can only be accomplished by quantum mechanical tunnelling, a fact first recognized in 1928 by Gamow and independently by Gurney and Condon.

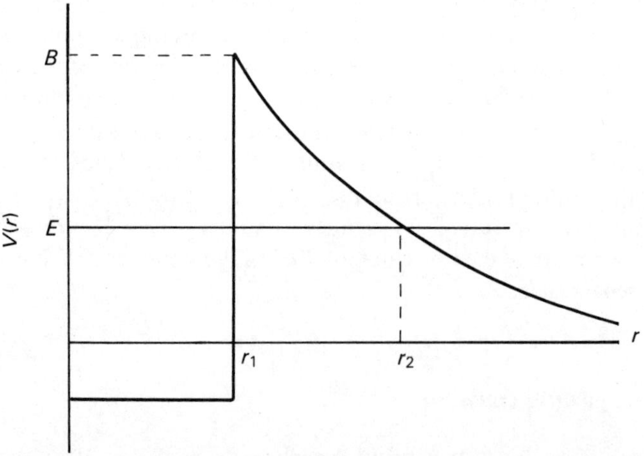

Figure 9.8 A schematic diagram of the potential well experienced by an α particle in an atomic nucleus. In α-particle emitters the barrier is of finite width (i.e. $E > 0$) and the particle can escape.

The naive model we visualize consists of an α particle rattling about in a spherical well, bouncing off the inside face of the barrier many times until eventually it is successful in tunnelling through it. Although the process is essentially a three-dimensional one, it can be reduced to one dimension by the same process that we used in our earlier studies of spherical potential wells. Specifically, the TISE for the α particle will be equation (7.20):

$$\left(-\frac{\hbar^2}{2\mu}\frac{\mathrm{d}^2}{\mathrm{d}r^2}+\frac{l(l+1)\hbar^2}{2\mu r^2}+V(r)\right)\chi(r)=E\chi(r). \tag{9.74}$$

For simplicity we will assume that the α particle bounces radially, so that $l=0$. We will also assume that the potential rises abruptly to B at a sharply defined nuclear boundary at r_1. Thereafter it falls according to Coulomb's law. We therefore have

$$V(r)=Br_1/r \qquad r\geqslant r_1. \tag{9.75}$$

$B=V(r_1)$ represents the electrostatic interaction energy between the α particle of charge $2e$ and the daughter nucleus of charge $(Z-2)e$ when the α particle is at the surface of the nucleus. Hence

$$B=\frac{2(Z-2)e^2}{4\pi\varepsilon_0 r_1}. \tag{9.76}$$

We also note that for an α particle of energy E the barrier extends to r_2, where

$$E=Br_1/r_2. \tag{9.77}$$

Since equation (9.74) is identical to the one-dimensional TISE we can use the one-dimensional approximation for the transmission of the barrier, equation (9.66), substituting r for x:

$$\begin{aligned} T&=\exp\left(-2\int_{r_1}^{r_2}\varkappa(r)\,\mathrm{d}r\right)\\ &=\exp\left[-\frac{2\sqrt{2\mu}}{\hbar}\int_{r_1}^{r_2}\left(\frac{r_1}{r}B-E\right)^{1/2}\mathrm{d}r\right]\\ &=\exp\left[-\frac{2\sqrt{2\mu E}}{\hbar}\int_{r_1}^{r_2}\left(\frac{r_2}{r}-1\right)^{1/2}\mathrm{d}r\right]=\mathrm{e}^{-2G} \end{aligned} \tag{9.78}$$

where we have used equation (9.62) for $\varkappa$ and (9.77) for B. G is referred to as the Gamow factor.

The integral is readily performed by means of the substitution

$$r_2/r=\operatorname{cosec}^2\theta \tag{9.79}$$

and yields the result

$$G = -\frac{\sqrt{2\mu E}}{\hbar} r_2\left[\frac{\pi}{2} - \sin^{-1}\left(\frac{r_1}{r_2}\right)^{1/2} - \left(\frac{r_1}{r_2}\right)^{1/2}\left(1 - \frac{r_1}{r_2}\right)^{1/2}\right]. \quad (9.80)$$

If $E \ll B$, then $r_1 \ll r_2$ and this expression can be approximated by

$$G \simeq -\frac{\sqrt{2\mu E}}{\hbar} r_2\left[\frac{\pi}{2} - 2\left(\frac{r_1}{r_2}\right)^{1/2}\right]. \quad (9.81)$$

Hence

$$T = \exp\left[-\frac{\sqrt{2\mu}}{\hbar}\left(\frac{\pi B r_1}{\sqrt{E}} - 4\sqrt{B r_1}\right)\right] \quad (9.82)$$

or, inserting B from equation (9.76),

$$T = \exp\left[-\frac{\sqrt{\mu}(Z-2)e^2}{\sqrt{2}\hbar\varepsilon_0}\frac{1}{\sqrt{E}} + \frac{4e}{\hbar}\left(\frac{\mu(Z-2)}{\pi\varepsilon_0} r_1\right)^{1/2}\right]. \quad (9.83)$$

Evaluating the constants for a typical α emitter with $Z = 92$ and $r_1 = 9.2 \times 10^{-15}$ m we find

$$T = \exp(-355E^{-1/2} + 85) \quad (9.84)$$

where E is expressed in MeV. We can now account for the extraordinary energy sensitivity of α-decay lifetimes. A variation in E from 4 to 9 MeV introduces a factor of $e^{355/6} \simeq 5 \times 10^{25}$! The form of the dependence on E indicated in equation (9.83) has been extensively tested experimentally and shows remarkable agreement, considering the simplicity of the model. The strong sensitivity to r_1 provides a powerful method of determining nuclear radii.

In order to calculate the actual decay constant it only remains to estimate the rate at which the α particle collides with the barrier. If the speed of the particle inside the nucleus is v, then the collision rate is v/r_1 and hence the decay constant is

$$\lambda = \frac{v}{r_1} T. \quad (9.85)$$

The speed can be estimated from the energy of the particle on emission (neglecting any contribution from the internal potential). For a 6 MeV particle $v \simeq 1.7 \times 10^7$ m s^{-1}. Hence the α particle makes perhaps 10^{21} collisions with the barrier each second. Evaluating (9.84) for an energy of 6 MeV we find $T \simeq 9.4 \times 10^{-27}$ so that $\lambda \sim 1.75 \times 10^{-5}$ s^{-1} which is very similar to experimental values for 6 MeV α particles. In reality it is found that α particles have rather less than unit probability of existing ready formed within the nucleus, so that an additional factor needs to be included in equation (9.85).

〉9.7 Scattering by a potential well

The analysis of section 9.3 remains valid for a potential well; we merely need to take V as negative in all the equations. We will also write a for b to be consistent with earlier notation for a well. With the substitutions of equation (9.55), for a well equation (9.50) becomes

$$T = \left(1 + \frac{\sin^2 2\gamma\sqrt{\varepsilon + 1}}{4\varepsilon(\varepsilon + 1)}\right)^{-1} \qquad \varepsilon > 1. \tag{9.86}$$

Figure 9.9 shows T as a function of ε for potential wells of several strengths. Classically a particle would be transmitted for any positive E, but quantum mechanically the situation is much as for the case of the barrier; there are finite probabilities of reflection and transmission for all energies except those special values which result in an exact number of half waves fitting within the width of the well (equation (9.58)). These energies are identical with the energy eigenvalues of a potential well the same width as the actual well but of infinite depth and we can regard the effect as the result of temporary entrapment of the particle as it bounces back and forth within the well. Figure 9.9 shows that the sharpness of the peaks in transmission, referred to as resonances, increases with the strength parameter of the well. This is again analogous to a Fabry–Perot interferometer, where the sharpness of the transmission peaks increases with the reflectivity of the mirrors.

Although conditions exist for which $T = 1$, equation (9.49) shows that apart from the trivial case of $k_1 = 0$ it is never the case that $R = 1$. Nevertheless under certain circumstances the reflectivity can be quite high, as can be seen in figure 9.9. The analogy with optics leads us to expect that maximum reflection will occur when the width of the well encompasses an odd number of quarter waves. From equation (9.86) we observe that R will be maximized (T minimized) when

$$\frac{1}{\varepsilon(\varepsilon + 1)} \sin^2 2\gamma\sqrt{\varepsilon + 1} \tag{9.87}$$

has a maximum. Differentiating we find that this occurs when

$$\tan 2\gamma\sqrt{\varepsilon + 1} = \frac{2\gamma\varepsilon\sqrt{\varepsilon + 1}}{2\varepsilon + 1}. \tag{9.88}$$

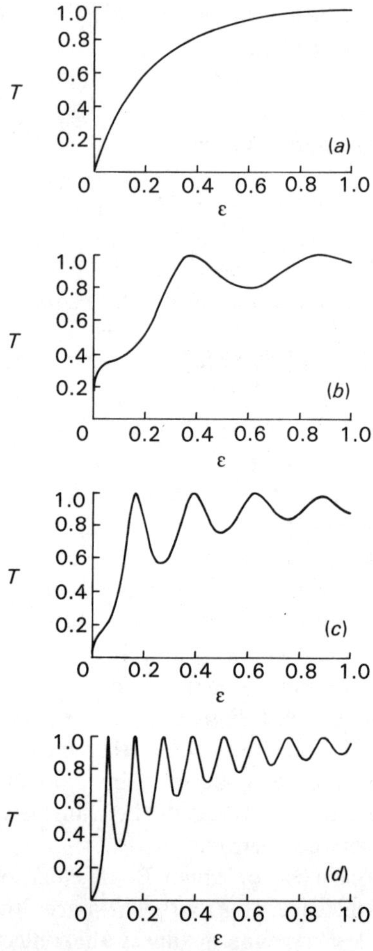

Figure 9.9 The transmission of a potential well as a function of ε, for various well strengths γ. $\gamma = 1$ (a), 8 (b), 16 (c), 32 (d). As γ increases the transmission tends towards the classical characteristic, which exhibits a transmission of unity for all positive energies.

For large ε the right-hand side is large, and the equation has roots

$$2\gamma\sqrt{\varepsilon + 1} = 2k_2 a = (2n + 1)\pi/2 \tag{9.89}$$

in agreement with expectations. However, for small ε the minima are significantly displaced from these values.

Computer exercise 9.4: scattering by a rectangular potential well
Run the program '1-D Potential Barrier' again. Enter $V = -9$ and $a = 1$ for a potential well 9 eV deep and 0.2 nm wide ($\gamma = 1.54$).

(*a*) Observe a typical scattering event, for example by selecting XI = 2, XF = 5, $E = 10$ and the travelling-wave option. Note the similarity in general features to scattering from a barrier.

(*b*) Change the energy to one of the eigenvalues of a well of width 0.2 nm and infinite depth. (The '1-D Square-Well Eigenvalues' program will tell you what these are. Remember that the energy eigenvalue is referred to the bottom of the potential well.) You should find that the transmission $T = 1$ for any of these values. Note the relation between the wavelength and the well width in each case.

(*c*) In order to obtain a high reflection coefficient (other than by letting $k_1 \to 0$), figure 9.9 indicates that we require a fairly strong well, and an energy which corresponds to the first minimum in the curve. Setting $a = 2$ and $V = -34.29$ will result in a well for which $\gamma = 6$. Figure 9.4 indicates a minimum in T near $\varepsilon = 0.3$; in fact equation (9.88) has a root at $\varepsilon = 0.319$, or $E = 10.77$ eV. Plot the wavefunction and observe the fairly strong reflection. Measure the value of R with the help of equation (9.32) and the tabulation option. Does your value agree with figure 9.9? Notice that the well includes slightly less than 2.25λ. Verify that the energy given does correspond to the maximum reflection by making small (10%) variations in E.

⟩ 9.8 Bound states of the potential well

The final case to be considered occurs when $0 > E > V$. The solutions are of course just the bound-state solutions of the finite potential well discussed in section 3.5, but it is interesting to see how the analysis of section 9.3 leads to the same results for the eigenvalues. First we recognize that it is now k_1 which is imaginary, and so we write

$$k_1 = \mathrm{i}\varkappa. \tag{9.90}$$

Examining equations (9.34) and (9.36) we find that A and D must be zero in order to have decaying exponentials in the two outer regions. Equation (9.46) then yields the result

$$0 = (2\cos 2k_2 a - \mathrm{i}\alpha \sin 2k_2 a) C\, \mathrm{e}^{-2\varkappa a}. \tag{9.91}$$

Inserting the expression for α, equation (9.45), we obtain

$$\tan 2k_2a = \frac{2\varkappa/k_2}{1-\varkappa^2/k_2^2}. \tag{9.92}$$

With the help of the identity

$$\tan 2\theta = \frac{2\tan\theta}{1-\tan^2\theta} \tag{9.93}$$

we find the roots of equation (9.92) to be

$$\tan k_2a = \varkappa/k_2 \tag{9.94}$$

and

$$-\cot k_2a = \varkappa/k_2. \tag{9.95}$$

These are identical to equations (3.49) and (3.50) and their solutions give the eigenvalues found in section 3.5. Inserting equations (9.94) and (9.95) in turn into equation (9.44) with $D=0$ shows that $P=Q$ and $P=-Q$ for the two cases, indicating that the eigenvalues relate to the even- and odd-parity eigenfunctions respectively.

〉9.9 Periodic potentials: the Bloch theorem

In a crystalline solid the atoms (or ions) form a regular array or lattice. The behaviour of the electrons in the solid can be modelled in terms of their response to the three-dimensionally periodic potential presented to them by the lattice atoms. One of the most important phenomena exhibited by the electrons in solids, the band structure of their allowed energies, persists even when the model is drastically simplified to a one-dimensional periodic array of alternating square wells and potential barriers, as in figure 9.10. This model was first studied in 1930 by R de L Kronig and W G Penney and is named after them.

In attempting to find the energy eigenfunctions for the potential illustrated in figure 9.10, we will assume that the number of wells and barriers is very large indeed, as would be the case for a real solid. We can in principle adopt and extend the technique that we used for the single potential barrier in section 9.3 but the large number of boundaries to consider makes the calculation intractable unless we can somehow exploit the periodicity. The Bloch theorem provides the formalism to do so.

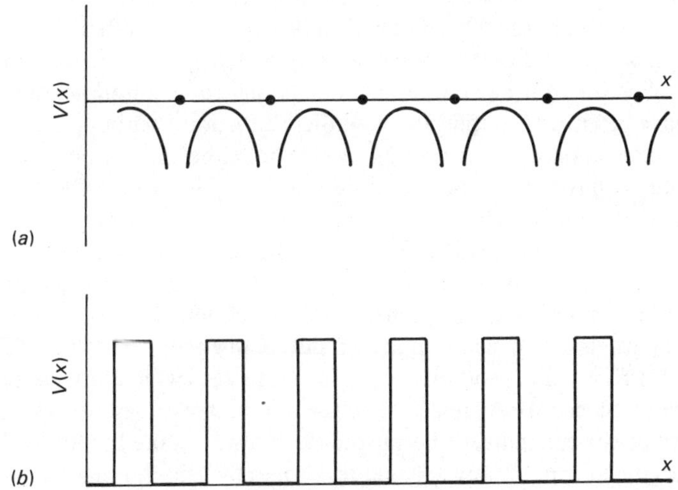

Figure 9.10 The Kronig–Penney one-dimensional model of a solid. (*a*) A semi-realistic representation of the potential arising from a linear chain of positive ions. (*b*) An idealization of (*a*) representing the potential as an array of rectangular wells and barriers.

First we must discuss the boundary conditions to be satisfied by the eigenfunctions we intend to find. Because the number of atoms is very large, over most of the array the effects of the boundaries at the surfaces of the crystal become negligible, and we would like to eliminate these effects rigorously from our model. We could accomplish this by considering an infinite array but dealing with this presents other problems. What we require is a finite array with no boundaries! This ideal would actually be realized in the case of a circular chain of atoms. If there were N atoms with separation d the wavefunction would then be required to satisfy the periodic boundary condition (PBC):

$$u(x+Nd)=u(x) \tag{9.96}$$

since the points x and $x+Nd$ would be one and the same. In this model there are no boundaries and so travelling-wave solutions to the wave equation can be employed as an alternative to standing-wave solutions. The latter are required when boundaries are present since the solutions must fall to zero there. Although a circular chain of potential wells might appear of limited applicability we can consider a finite array of N atoms and simply impose the PBC of equation (9.96). This implies that the wave leaving one end of the finite array exactly matches the wave entering the other. We can continue to employ travelling-wave solutions and these can have arbitrary amplitude in either direction. Although standing waves also consist of travelling waves in either direction, they are required to have the same amplitude and in this respect the PBC is a weaker requirement. The motivation for using PBCs is one of mathematical convenience. The justification is: (a) the belief that the bulk properties of a material cannot depend upon conditions at the surface; (b) the fact that the number of independent eigenfunctions is identical whether standing waves or PBCs are imposed; and (c) in either case the eigenfunctions form a valid complete set in terms of which any arbitrary wavefunction can be expressed. The use of a finite N does not restrict the validity of the calculations because we can always take the limit $N \to \infty$ at the conclusion of any calculation, thereby eliminating any effects introduced by a lattice of restricted extent.

In order to prove the Bloch theorem we will introduce a non-Hermitian translation operator $\hat{T}(x)$ which increases the argument of a function by d, the period of the lattice, so that

$$\hat{T}(x)f(x) = f(x+d). \tag{9.97}$$

The Hamiltonian for the system has the form

$$\hat{H}(x) = -\frac{\hbar^2}{2m}\frac{\mathrm{d}^2}{\mathrm{d}x^2} + V(x) \tag{9.98}$$

where $V(x)$ is periodic in x, with period d. This means that the Hamiltonian itself is periodic in x,

$$\hat{H}(x) = \hat{H}(x+d) \tag{9.99}$$

and therefore commutes with $\hat{T}(x)$:

$$\hat{T}(x)\hat{H}(x)\psi(x) = \hat{H}(x+d)\hat{T}(x)\psi(x) = \hat{H}(x)\hat{T}(x)\psi(x). \tag{9.100}$$

It follows that simultaneous eigenfunctions of $\hat{H}(x)$ and $\hat{T}(x)$ must exist. Let $u(x)$ be such a function. We now determine the eigenvalues t of $\hat{T}(x)$, from the eigenvalue equation

$$\hat{T}(x)u(x) = tu(x). \tag{9.101}$$

Then

$$u(x+d) = tu(x) \tag{9.102}$$

and if we apply $\hat{T}(x)$ N times we arrive at the equation

$$\hat{T}^N(x)u(x) = u(x+Nd) = u(x) = t^N u(x) \tag{9.103}$$

where for the second equality we have invoked the PBC, equation (9.96). This shows that

$$t^N = 1 \tag{9.104}$$

so that the N distinct eigenvalues of $T(x)$ are the Nth roots of unity:

$$t = 1^{1/N} = \exp(2\mathrm{i}m\pi/N) \tag{9.105}$$

where

$$m = 0, \pm 1, \pm 2, \ldots, \pm \begin{cases} \frac{1}{2}(N-1) & N \text{ odd} \\ \frac{1}{2}N & N \text{ even.} \end{cases} \tag{9.106}$$

(Note that there are not really $N+1$ roots in the case N even as the roots with $m = \pm\frac{1}{2}N$ are then identical.) Now any eigenfunction of $\hat{T}(x)$ must satisfy

$$\hat{T}(x)u(x) = u(x+d) = \exp(2\mathrm{i}m\pi/N)u(x) \tag{9.107}$$

and so $u(x)$ must take the form

$$u(x) = w(x)\mathrm{e}^{ikx} \tag{9.108}$$

where

$$k = 2m\pi/Nd \tag{9.109}$$

and $w(x)$ is any function with period d in x.

A function satisfying equation (9.108) is called a Bloch function and has the properties that incrementing x by d merely introduces the phase factor $\exp(2\mathrm{i}m\pi/N)$ in accordance with equation (9.107), whilst increasing x by Nd leaves it unchanged, in accordance with the PBC, equation (9.96). Once the wavefunction is known in any cell of the lattice, equation (9.107) can be used to find it in any other. The form of the periodic function $w(x)$ is determined by the precise nature of the periodic potential. Remembering that the time dependence of an energy eigenfunction takes the form of a factor $\mathrm{e}^{-\mathrm{i}\omega t}$ we can see that the Bloch functions combine features of both travelling and standing waves. They can be thought of as travelling waves modulated by a stationary envelope.

〉9.10 The Kronig–Penney model

We must now discover the particular Bloch functions which satisfy the TISE for the potential of figure 9.10. Much effort can be saved by using the analysis of section 9.3 in which we studied the transmission of a rectangular barrier. To maintain continuity with that analysis for the time being we will set the origin in the centre of a barrier. The widths of the wells and barriers are defined as $2a$ and $2b$ respectively and so the period of the lattice is $d = 2(a + b)$. Figure 9.11 illustrates two cells of the lattice situated on either side of the origin.

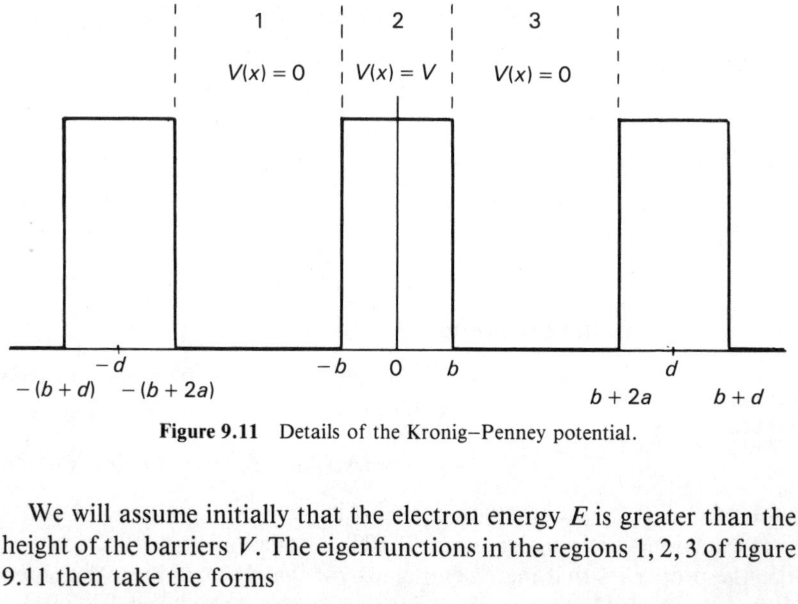

Figure 9.11 Details of the Kronig–Penney potential.

We will assume initially that the electron energy E is greater than the height of the barriers V. The eigenfunctions in the regions 1, 2, 3 of figure 9.11 then take the forms

$$u_1 = A\,\mathrm{e}^{\mathrm{i}k_1x} + B\,\mathrm{e}^{-\mathrm{i}k_1x} \qquad -b-2a < x < -b \qquad (9.110)$$

$$u_2 = P\,\mathrm{e}^{\mathrm{i}k_2x} + Q\,\mathrm{e}^{-\mathrm{i}k_2x} \qquad -b < x < b \qquad (9.111)$$

$$u_3 = C\,\mathrm{e}^{\mathrm{i}k_1x} + D\,\mathrm{e}^{-\mathrm{i}k_1x} \qquad b < x < b+2a. \qquad (9.112)$$

k_1 and k_2 are given by

$$k_1 = (2mE)^{1/2}/\hbar \qquad k_2 = [2m(E-V)]^{1/2}/\hbar. \qquad (9.113)$$

As we have shown, the wavefunctions must take the form of Bloch functions, equation (9.108), in order to satisfy the requirements of periodicity. This requires that the wavefunction in one cell of the lattice differs from the function in the equivalent position in the adjacent cell only by the phase factor e^{ikd}. Thus for example

$$u_3(x+d) = u_1(x)\exp(\mathrm{i}kd) \tag{9.114}$$

or explicitly

$$\begin{aligned} &C\exp[\mathrm{i}k_1(x+d)] + D\exp[-\mathrm{i}k_1(x+d)] \\ &\qquad = [A\exp(\mathrm{i}k_1x) + B\exp(-\mathrm{i}k_1x)]\exp(\mathrm{i}kd). \end{aligned} \tag{9.115}$$

This equality must hold for all values of x and so, on collecting terms in $\exp(\mathrm{i}k_1x)$ and $\exp(-\mathrm{i}k_1x)$, we find

$$A = C\exp[-\mathrm{i}(k-k_1)d] \qquad B = D\exp[-\mathrm{i}(k+k_1)d]$$

or

$$\begin{bmatrix} A \\ B \end{bmatrix} = \begin{bmatrix} \exp[-\mathrm{i}(k-k_1)d] & 0 \\ 0 & \exp[-\mathrm{i}(k+k_1)d] \end{bmatrix} \begin{bmatrix} C \\ D \end{bmatrix}. \tag{9.116}$$

On the other hand we have already found in equation (9.46) the relationship between the wavefunctions on either side of the barrier necessary in order to satisfy the local boundary conditions at $\pm b$. It is

$$\begin{bmatrix} A \\ B \end{bmatrix} = \begin{bmatrix} [\cos 2k_2b - (\tfrac{1}{2}\mathrm{i}\alpha)\sin 2k_2b]\exp(2\mathrm{i}k_1b) & -(\tfrac{1}{2}\mathrm{i}\beta)\sin 2k_2b \\ (\tfrac{1}{2}\mathrm{i}\beta)\sin 2k_2b & [\cos 2k_2b + (\tfrac{1}{2}\mathrm{i}\alpha)\sin 2k_2b]\exp(-2\mathrm{i}k_1b) \end{bmatrix} \begin{bmatrix} C \\ D \end{bmatrix} \tag{9.117}$$

where

$$\alpha = k_2/k_1 + k_1/k_2 \qquad \beta = k_2/k_1 - k_1/k_2. \tag{9.118}$$

Subtracting these two matrix equations and making use of the fact that $d = 2(a+b)$, we find

$$\begin{bmatrix} \{\cos 2k_2b - (\tfrac{1}{2}\mathrm{i}\alpha)\sin 2k_2b - \exp[-\mathrm{i}(kd-2k_1a)]\}\exp(2\mathrm{i}k_1b) & -(\tfrac{1}{2}\mathrm{i}\beta)\sin 2k_2b \\ (\tfrac{1}{2}\mathrm{i}\beta)\sin 2k_2b & \{\cos 2k_2b + (\tfrac{1}{2}\mathrm{i}\alpha)\sin 2k_2b - \exp[-\mathrm{i}(kd+2k_1a)]\exp(-2\mathrm{i}k_1b)\} \end{bmatrix} \begin{bmatrix} C \\ D \end{bmatrix} = 0. \tag{9.119}$$

This equation represents two homogeneous equations in C and D, and in order for these to be consistent the determinant of the square matrix must equal zero. Multiplying out the determinant, converting the exponentials to sines and cosines and inserting the values of α and β, we obtain the condition

$$\cos kd = \cos 2k_1a \cos 2k_2b - \frac{k_1^2 + k_2^2}{2k_1k_2} \sin 2k_1a \sin 2k_2b. \quad (9.120)$$

The condition for the case $E < V$ is readily obtained from this on making the substitution

$$k_2 = \mathrm{i}\varkappa \quad (9.121)$$

so that, from (9.113),

$$\varkappa = [2m(V - E)]^{1/2}/\hbar. \quad (9.122)$$

Hence

$$\cos kd = \cos 2k_1a \cosh 2\varkappa b - \frac{k_1^2 - \varkappa^2}{2k_1\varkappa} \sin 2k_1a \sinh 2\varkappa b. \quad (9.123)$$

We have found that k can take only certain values, given by equation (9.109), and for each one of these there will be a set of values of E which satisfy equations (9.120) and (9.123). These are the energy eigenvalues for the system. Before we discuss these solutions we will verify that equation (9.123) reduces as it should to the standard conditions (3.49) and (3.50) for the energy eigenvalues of an isolated potential well, if we let the well separation become very large. In that case $b \to \infty$ and

$$\sinh 2k_2b \to \cosh 2k_2b \to \infty. \quad (9.124)$$

Equation (9.123) then becomes

$$0 = \cos 2k_1a - \frac{k_1^2 - \varkappa^2}{2k_1\varkappa} \sin 2k_1a \quad (9.125)$$

or

$$\tan 2k_1a = \frac{2k_1\varkappa}{k_1^2 - \varkappa^2}. \quad (9.126)$$

We have already shown that this equation is equivalent to equations (3.49) and (3.50) in the discussion following equation (9.92).

Equations (9.120) and (9.123) can both be written in the form

$$\cos kd = F \cos(2k_1a + \delta). \quad (9.127)$$

When $E > V$

$$F = \left[1 + \left(\frac{k_1^2 + k_2^2}{2k_1k_2}\right)^2 \sin^2 2k_2b\right]^{1/2} \tag{9.128}$$

and

$$\tan \delta = \frac{k_1^2 + k_2^2}{2k_1k_2} \tan 2k_2b \tag{9.129}$$

whereas if $E < V$

$$F = \left[1 + \left(\frac{k_1^2 - \varkappa^2}{2k_1\varkappa}\right)^2 \sinh^2 2k_2b\right]^{1/2} \tag{9.130}$$

and

$$\tan \delta = \frac{k_1^2 - \varkappa^2}{2k_1k_2} \tanh 2\varkappa b. \tag{9.131}$$

Examining equation (9.127) we note that the left-hand side takes values which are bounded by ± 1. However, the right-hand side takes greater excursions, because the amplitude of the cosine function F is always greater than one, as can be seen from equations (9.128) and (9.130). This implies that some values of E, which determines F through k_1, k_2 and $\varkappa$, must be forbidden. This is made clear in figure 9.12 which shows the general form of the 'Kronig–Penney' function (the right-hand sides of equations (9.120) and (9.123)). Also shown are the limits ± 1. The eigenvalues are determined by the intersection of the curve with vertical lines indicating the N discrete values taken by cos kd, which lie between these limits. When the number of wells N becomes very large the eigenvalues form quasi-continuous bands separated by 'forbidden' gaps. It will be observed that there are actually less than N intersections within each band because cos kd is an even function and so $\pm m$ lead to the same energy; all states are doubly degenerate except the one for which $m = 0$, and, in the case N even, when $m = \pm\frac{1}{2}N$, since, as already noted, these refer to the same root of unity. The pairs of degenerate eigenfunctions consist of travelling waves in opposite directions. The non-degenerate functions at the edges of the bands are standing waves. These results can be deduced from the form of equation (9.119), which can be expressed

$$\begin{bmatrix} p + q\mathrm{e}^{-\mathrm{i}kd} & r \\ r^* & p^* + q^*\mathrm{e}^{-\mathrm{i}kd} \end{bmatrix} \begin{bmatrix} C \\ D \end{bmatrix} = 0. \tag{9.132}$$

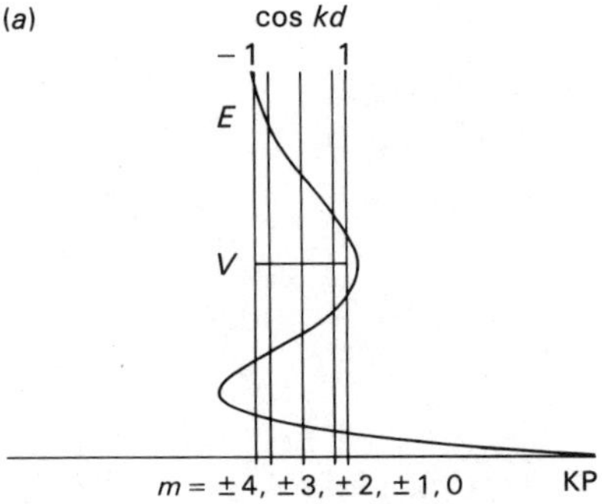

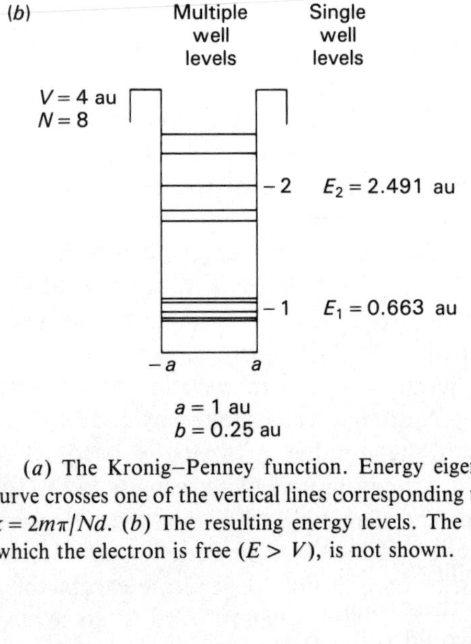

Figure 9.12 (*a*) The Kronig–Penney function. Energy eigenvalues occur where the curve crosses one of the vertical lines corresponding to the allowed values of $k = 2m\pi/Nd$. (*b*) The resulting energy levels. The third band of levels, for which the electron is free ($E > V$), is not shown.

Hence

$$\frac{C}{D} = \frac{-r}{p + qe^{-ikd}} \qquad \frac{C}{D} = -\frac{p^* + q^* e^{-ikd}}{r^*}$$

so that

$$\left|\frac{C}{D}\right|^2 = \frac{p + qe^{ikd}}{p + qe^{-ikd}}. \tag{9.133}$$

Now the permitted values of k are

$$k = 2\pi m/Nd \qquad m = 0,\ \pm 1,\ \pm 2, \ldots, \pm \tfrac{1}{2}N \tag{9.134}$$

so if $m = 0$

$$\left|\frac{C}{D}\right|^2 = \frac{p+q}{p+q} = 1. \tag{9.135}$$

If $m = \pm \tfrac{1}{2}N$, $k = \pm \pi/d$ and in either case $e^{ikd} = -1$. Then

$$\left|\frac{C}{D}\right|^2 = \frac{p-q}{p-q} = 1. \tag{9.136}$$

Equation (9.116) shows that equivalent expressions hold for A and B, and so the band-edge wavefunctions are indeed standing waves throughout the lattice. However, equation (9.133) shows that in general

$$\left|\frac{C}{D}\right|_{+k} \neq \left|\frac{C}{D}\right|_{-k} \neq 1. \tag{9.137}$$

The solutions for $\pm k$ are quite independent of each other and are combinations of travelling and standing waves. In fact their character becomes more nearly that of a pure travelling wave as the centre of the band is approached, since it is there that C/D differs most from unity.

Because of the relation between the propagation constant k and momentum, and the fact that we do not expect there to be bounds upon the possible values of the latter, it is sometimes useful to think of k as extending over an unlimited range. We are free to do this because $\cos kd$ is periodic, and so the energy eigenvalues are unaffected if $kd \rightarrow kd + 2l\pi$ (l integer). If we adopt this viewpoint for a moment we find that band edges occur whenever $kd = n\pi$ or, since $k = 2\pi/\lambda$, when

$$d = \tfrac{1}{2}n\lambda \qquad n \text{ integer}. \tag{9.138}$$

This is precisely the condition for Bragg reflection at normal incidence (the only possibility in one dimension). This gives us some physical insight into why these values of k correspond to standing waves. Any travelling wave would suffer partial reflection at each boundary, but if this condition is satisfied all such reflections reinforce each other. Eventually waves propagating in either direction achieve equal amplitudes, and a standing wave is formed. This mechanism is obviously effective whether we are dealing with bound ($E < V$) or free ($E > V$) eigenfunctions, and so in a periodic potential even the 'continuum' states show band structure. (Of course for a finite lattice all the eigenvalues are discrete, including those for $E > V$.)

The form of the Bloch function (equation (9.108)) tells us a little more about the character of the standing waves. If $m = 0$ then

$$u(x) = w(x) \tag{9.139}$$

and the wavefunction, like $w(x)$, is periodic with period d. If $m = \pm\frac{1}{2}N$ then

$$u(x) = w(x)\exp(\pm i\pi x/d) \tag{9.140}$$

either of which is periodic with period $2d$. In the first case the wavefunction is identical in every cell. In the second case the wavefunction alternates in sign in successive cells but is otherwise identical.

If the origin is placed symmetrically in the centre of a barrier or of a well then the potential and hence the Hamiltonian are even functions of x and so the standing wavefunctions, which are non-degenerate, must have definite parity. Whether they are even or odd in any particular case depends upon the choice of origin. If $E < V$ it is convenient to move the origin to the centre of a well since we can then compare the functions with the bound states of an isolated well, which we know to alternate in parity. Each band of the multiple-well system originates from one of these bound states, and the band-edge standing waves retain the same parity. Consequently we find an alternation in parity of the standing waves on either side of a band gap. This will become clearer on inspection of figure 9.13, which shows a set of eigenfunctions for a lattice consisting of eight cells. The wells in isolation support just two bound states of opposite parity. It should be noted that only the real part of the function is shown, but this does not affect the band-edge functions, which, being standing waves, can be expressed as real functions.

Computer exercise 9.5: multiple potential wells

The program 'Multiple Well Eigenvalues' contains two energy-eigenvalue-finding routines. The first will find the eigenvalues for a single finite square well and is identical to the earlier program. The second calculates the eigenvalues for a periodic lattice consisting of an arbitrary number of wells, by numerical solution of the Kronig–Penney equations, (9.120) and (9.123). Solutions can be found within any specified energy range. The Kronig–Penney function can be plotted, as can the energy bands for any well parameters. Once the eigenvalues have been found the corresponding eigenfunctions can be plotted using the program 'Multiple Potential Wells'.

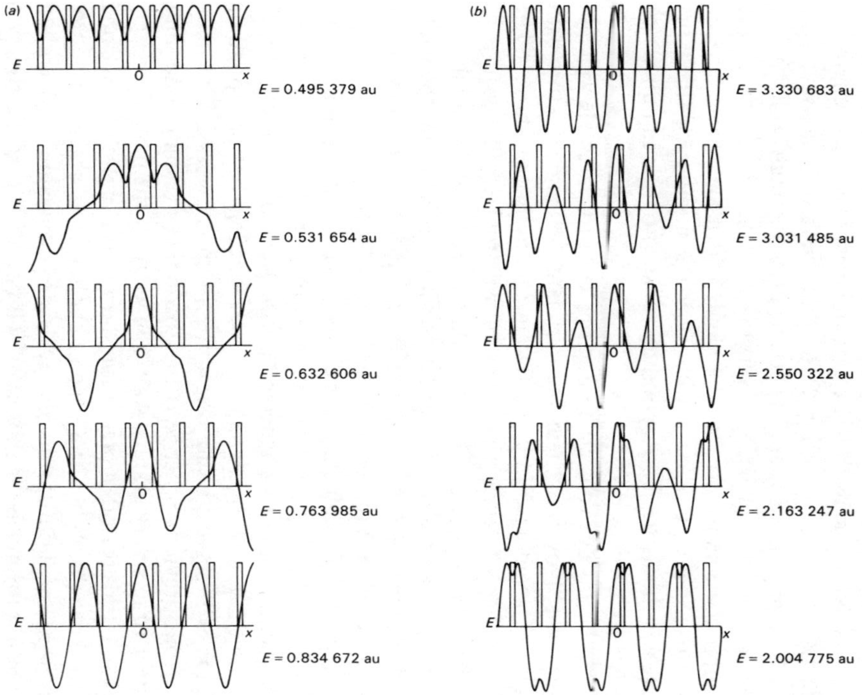

Figure 9.13 Kronig–Penney wavefunctions for a system of eight potential wells. The wavefunctions of (*a*) the first band and (*b*) the second band shown in figure 9.12(*b*). Only the band-edge functions are unique. The others have been selected to exhibit even (*a*) and odd (*b*) parities.

(*a*) First accept the default parameters for a lattice of eight wells, of depth 4 au, width 2 au and barrier width 1 au. Plot the Kronig–Penney function. The horizontal axis represents the value of the function; the vertical axis represents energy. Allowed values of cos kd lie between the vertical lines at ± 1 and are indicated at the bottom of the plot. The program automatically sets an energy range of twice the well depth, so that cases $E > V$ and $E < V$ are illustrated. Try the effect of varying the well parameters. Note that as the barrier width is increased the amplitude of the function becomes very large, so that the bands become very narrow; they asymptotically approach the discrete single-well values. Conversely, as the barriers become very narrow the amplitude of the function approaches unity, so that the bands widen and ultimately the band gaps close. Varying V or a will affect the number of times the function crosses the figure, corresponding to the variation in number of discrete single-well levels.

(*b*) Use the eigenvalue routines to find the levels of the single well and the multiple wells. Now plot them and correlate the band structure with the Kronig–Penney plots. Vary the parameters as in (*a*) above. (The maximum number of levels the program will find is 50 because of computer memory limitations.) Notice that increasing the number of wells merely increases the number of levels within a band, but the width of the band remains the same. (Only if the number of wells is even will there be levels at both band edges.)

(*c*) Revert to the default parameter values and make a note of the eigenvalues for that potential. Now run the program 'Multiple Potential Wells'. The default lattice parameters are the same as for the previous program with the lowest eigenvalue entered. Plotting the corresponding eigenfunction will serve to show how the program works. The main departure from previous practice is that periodic boundary conditions are implemented in a very graphic way. The width of the plot is precisely equal to eight lattice periods. Integration starts in the centre of a well and proceeds to the right. When the edge of the screen is reached the plot continues from the left. If the energy is an eigenvalue, and correct starting conditions have been chosen, it will be found that the function precisely repeats itself when the starting point is reached. Verify that this works for several of the eigenvalues you have noted. Now enter an energy lying in one of the band gaps. The wavefunction will be found to diverge, at a rate which depends upon the position of the chosen energy in the gap.

(*d*) Now enter a value of energy which is not one of the eigen-energies but which lies within an allowed band. Although it does not satisfy the periodic boundary condition and match itself after one circuit of the lattice neither does it diverge. After a few circuits it should become apparent that eventually it will match. What this means is that any energy within the band is an eigen-energy of a lattice of some length. Recall that the band becomes a continuum for an infinite lattice.

(*e*) Although the routine plots only the real parts of the wavefunctions the different character of the band-edge (standing-wave) and band-centre (mixed standing- and travelling-wave) functions can nevertheless be illustrated. It has been pointed out that the standing-wave functions have definite parity with respect to a well centre. Those for the lowest band must be even (starting conditions $y = 1$, $y' = 0$) and those for the second band must be odd ($y = 0$, $y' = 1$). Verify that this is necessary. Now plot an eigenfunction from the interior of the band. It will be found that satisfactory functions are obtained for a whole range of initial conditions. Not only can even or odd parity conditions be chosen, but conditions corresponding to functions with mixed parity can be chosen also. One way to do this is to start with $y = 1$, $y' = 0$ but then systematically to vary XI. It will be found that the function repeats for almost any initial conditions! Because the two travelling waves are degenerate and independent, a wide range of initial conditions can be accommodated by taking appropriate linear combinations of them. As always, the program plots the entire solution, i.e. the superposition required to satisfy the initial conditions.

〉 Problems

9.1 Show that the reflection coefficient for a potential step ($E > V$) is independent of the direction of the incident wave.

9.2 Calcul .e the probability that an electron of energy E incident norr ally on the surface of a metal will be reflected. The potential enei gy of the electron inside the metal is $-V$, outside it is zero. Model the problem as a potential step. Evaluate your result for $V = 15$ eV (e.g. Ni) and energies $E = 30$, 3 and 0.1 eV. What restrictions must apply to the electron energy for your analysis to be valid if the width of the step is comparable to atomic dimensions?

9.3 A rectangular potential structure is defined as follows:

$$\phi(x) = 0 \qquad |x| > 0.1 \text{ nm}$$
$$\phi(x) = +10 \text{ V} \qquad |x| \leqslant 0.1 \text{ nm}.$$

Calculate the transmission coefficient for (a) an electron, (b) a positron and (c) a proton, all with energies 5 eV.

What thickness would the barrier need to have for the transmission of the proton to equal the original positron transmission?

9.4 Perform the analysis relating equations (9.39)–(9.42) to (9.46).

9.5 Perform the analysis relating equation (9.78) to (9.80).

9.6 Perform the analysis relating equation (9.119) to (9.120).

9.7 The surface of an electrode passes through the origin, at which point the normal to the surface lies along the x axis. The potential outside the electrode varies with x parabolically:

$$V(x) = V(1 - x^2/a^2) \qquad 0 \leqslant x \leqslant a$$
$$V(x) = 0 \qquad x < 0,\ x > a.$$

Show that the transmission of this barrier for an electron of energy $E < V$ is given approximately by

$$T = \exp\left[-\frac{(2mV)^{1/2}}{\hbar}\frac{\pi a}{2}\left(1 - \frac{E}{V}\right)\right].$$

9.8 A double potential step is defined by

$$V(x) = 0 \qquad x < -b$$
$$V(x) = V_2 \qquad |x| \leqslant b \qquad V_3 > V_2.$$
$$V(x) = V_3 \qquad x > b$$

Show that the conditions for zero reflection of a particle of mass m and energy $E > V_3$ are

$$2b = \tfrac{1}{4}\lambda_2 \qquad k_2 = (k_1 k_3)^{1/2}$$

where

$$k_1 = (2mE)^{1/2}/\hbar \qquad k_2 = [2m(E - V_2)]^{1/2}/\hbar$$
$$k_3 = [2m(E - V_3)]^{1/2}/\hbar$$

and

$$\lambda_2 = 2\pi/k_2.$$

(Hint/discussion: this is the quantum mechanical equivalent of the $\frac{1}{4}\lambda$ anti-reflection coating used in optics and the $\frac{1}{4}\lambda$ matching section used in electromagnetic transmission lines. The calculation is best performed by adapting the analysis of the barrier in section 9.3, changing k_1 to k_3 at the second interface. The required condition is found by setting $B = D = 0$. You can see the effect in action with the aid of the 1D potential step program.)

9.9 Show that in the limit $E \gg V$, the Kronig–Penney dispersion relation (equation (9.120)) approaches the free-particle relation, i.e. $k = (2mE)^{1/2}/\hbar$.

〉 Chapter 10

〉 Scattering in Three Dimensions

〉 10.1 Classical scattering

Most scattering events of significance involve motion in three dimensions and this adds considerably to the complexity of the analysis. A typical experimental arrangement is shown schematically in figure 10.1. A collimated beam of particles of mass m_1 is incident upon a scattering object of mass m_2.

It is useful to consider first of all what happens during a scattering event according to classical mechanics. In that case the particle is deflected, and the scatterer, initially at rest, recoils. Both move along curved trajectories as indicated in the diagram and if the force law is known the precise form of these trajectories can be calculated. As always in two-body interactions the analysis is considerably simplified by separating the relative and centre of mass motions. In the frame in which the centre of mass remains at rest (the CM frame) both particles move towards each other initially, but the analysis is identical (see Appendix 3) to the case of a fictitious particle of reduced mass $\mu = m_1 m_2/(m_1 + m_2)$ moving under the influence of a fixed (i.e. very massive) force centre, as in figure 10.2. It is then a matter of straightforward kinematics to relate the motions in the laboratory frame to the relative motion so calculated. If the mass of the scatterer is very large compared with that of the incident particle the two descriptions will be almost identical, since the centre of mass will remain essentially stationary at the scatterer throughout the encounter. We will assume scattering from a fixed centre in all that follows.

It is apparent that the angular deflection for any particular particle will be determined by its impact parameter, b, the distance of closest approach that would occur if there were no forces operating. In all practical cases the incident beam has a width large compared with the scatterer, so that particles of essentially all impact parameters are present. The result is a flux of scattered particles, the angular distribution of which is characteristic of the force law operating. It is the object of scattering experiments to deduce the nature of the force from the observed distribution.

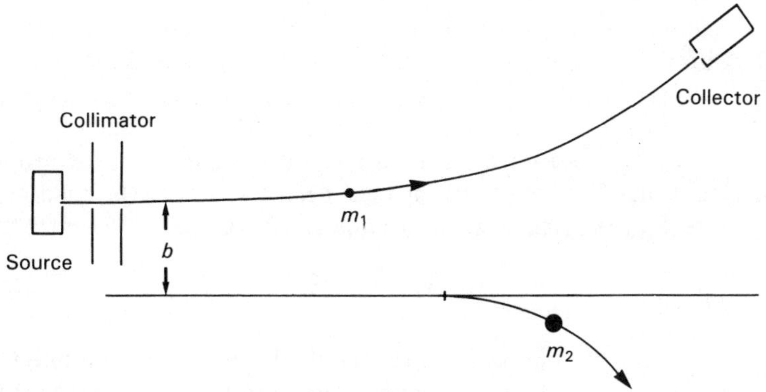

Figure 10.1 A scattering experiment, showing the impact parameter and the recoil of the target.

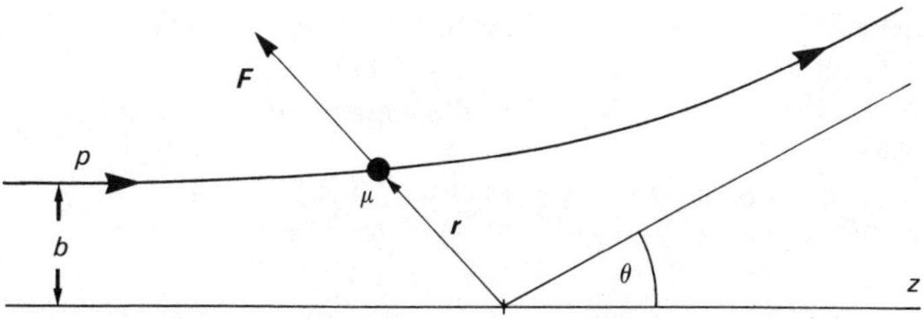

Figure 10.2 The scattering event of figure 9.1 transformed by the introduction of a particle of reduced mass μ, scattered at a recoilless centre of force. $\boldsymbol{r}$ is the relative coordinate and θ is the final scattering angle.

If the force is a conservative one, so that it can be expressed in terms of a potential, then the scattering will be elastic; the kinetic energy of the particle will be the same before and after the collision. If in addition the force is central the angular momentum of the particle about the force centre, $\boldsymbol{L}$, will be conserved throughout the encounter. The magnitude of $\boldsymbol{L}$ is determined by the impact parameter, together with the initial momentum of the particle, p, so that $|\boldsymbol{L}| = pb$. Another feature of a central force is that axial symmetry prevails. Allowing the beam direction to define the z axis, then the scattered distribution will be a function of θ but not of ϕ, where θ and ϕ are the usual spherical polar angles.

〉10.2 Scattering cross sections

The results of scattering experiments are usually expressed in terms of differential and total cross sections. Assuming that the detector of the scattered particles is located at a distance which is very large compared with the dimensions of the scattering zone, in the direction (θ, ϕ), then the rate at which scattered particles are collected, dN, will be proportional to j, the flux density of incident particles, and to the solid angle subtended by the detector at the scatterer, dΩ. Hence

$$\mathrm{d}N = j\,\mathrm{d}\Omega\,\frac{\mathrm{d}\sigma}{\mathrm{d}\Omega}. \tag{10.1}$$

The constant of proportionality, written dσ/dΩ, is called the differential cross section. Its value depends upon θ and also upon the energy of the incident particle. It is clear from this equation that dσ must have the dimensions of area. Classically this area can be identified with the elementary area (far ahead of the interaction region) through which particles of the incident beam must pass in order to be deflected into the detector, as shown in figure 10.3. By including dΩ in its definition the differential cross section is made independent of the parameters of the detector.

If we integrate the differential cross section over all solid angles we obtain the total cross section:

$$\sigma = \int \frac{\mathrm{d}\sigma}{\mathrm{d}\Omega}\,\mathrm{d}\Omega. \tag{10.2}$$

σ represents the area through which a particle in the incident beam must pass in order to be scattered through any angle greater than zero.

Referring to figure 10.3 we can see that

$$\mathrm{d}\sigma = b\,\mathrm{d}b\,\mathrm{d}\phi. \tag{10.3}$$

Also, the elementary solid angle at (θ, ϕ) is

$$\mathrm{d}\Omega = \sin\theta\,\mathrm{d}\theta\,\mathrm{d}\phi. \tag{10.4}$$

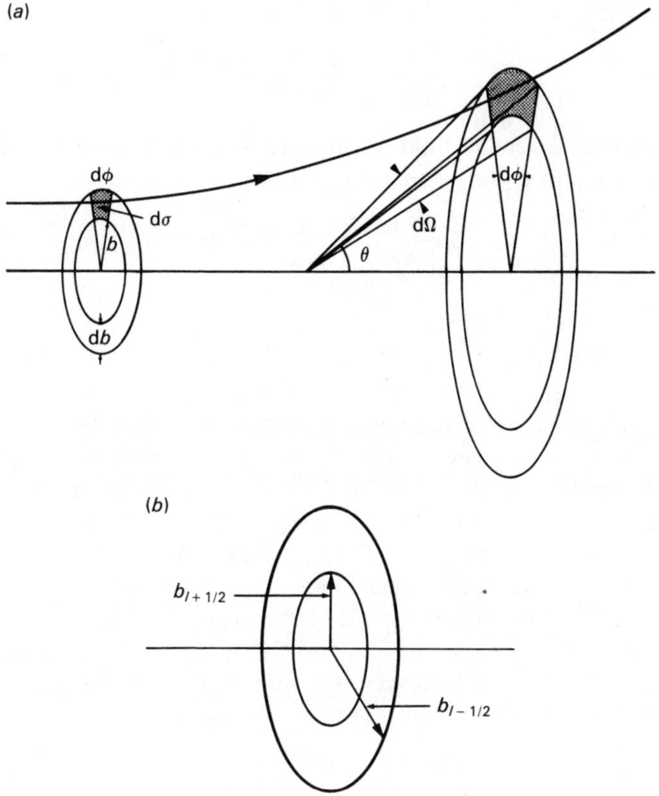

Figure 10.3 (*a*) Particles scattered into a solid angle dΩ at (θ, ϕ) have all passed through the shaded area dσ. (*b*) The annular region representing the cross section for scattering into the lth partial wave.

Hence the differential cross section can be written

$$\frac{\mathrm{d}\sigma}{\mathrm{d}\Omega} = \frac{b}{\sin\theta}\frac{\mathrm{d}b}{\mathrm{d}\theta}. \tag{10.5}$$

Analysis of the impact according to classical mechanics provides an expression for $\mathrm{d}b/\mathrm{d}\theta$ and hence for the cross section. As an example, we may quote the well known result for Rutherford (Coulomb) scattering, in which α particles of charge $2e$ and energy E are scattered from nuclei of charge Ze. It is found that

$$b = \frac{Ze^2}{E}\cot\tfrac{1}{2}\theta \tag{10.6}$$

and therefore, from equation (10.5),

$$\frac{d\sigma}{d\Omega} = \frac{Z^2 e^4}{4E^2 \sin^4 \frac{1}{2}\theta}. \tag{10.7}$$

Integrating this over the solid angle lying between θ and π (i.e. for all angles of scattering greater than θ) we find

$$\sigma_\theta = \frac{Z^2 e^4}{4E^2} \int_0^\pi \frac{2\pi \sin\theta}{\sin^4 \frac{1}{2}\theta} \, d\theta$$

$$= \frac{2\pi Z^2 e^4}{E^2} \frac{1}{\sin^2 \frac{1}{2}\theta}. \tag{10.8}$$

Evidently this diverges as $\theta \to 0$. Since the Coulomb force has an infinite range, it is clear that particles will be deflected to some extent for any impact parameter, however large. In practice, the Coulomb field of a charge is usually screened by other charges beyond some finite distance. This is the case when the target is an atomic nucleus; the atomic electrons shield the field of the nucleus at distances greater than the atomic radius. In addition the incident beam must have some finite angle of divergence, and scattering within this range cannot be distinguished.

Experimental verification of equation (10.7) by Rutherford established that the force followed an inverse square law down to distances of the order 10^{-14} m, thereby establishing the existence of atomic nuclei.

〉10.3 The wave mechanics of scattering

Scattering can only be described classically provided that the wavelength of the particle is much smaller than the dimensions of the target, in which case diffraction effects are negligible. If this condition is not satisfied we cannot speak in terms of particle trajectories. For instance, to define an impact parameter we must know the lateral position of the particle relative to the target, but this introduces an uncertainty in the transverse momentum. The description of a particle trajectory requires the precise simultaneous knowledge of both these quantities, and this would violate the uncertainty principle. Accordingly we must resort to a quantum mechanical description.

As in our treatment of scattering in one dimension, we may again use a time-independent approach, in which we consider a steady flux of particles incident parallel to the z axis. The particles may then be represented by the wavefunction for a plane wave:

$$u_{\text{in}} = \mathrm{e}^{\mathrm{i}kz}. \tag{10.9}$$

In practice the wave will be restricted both laterally and longitudinally by the experimental arrangements; however, we will ignore the longitudinal limitation. To take it into account we would need to adopt a wavepacket approach which is unnecessarily complicated for our purposes. On the other hand, we need to recognize that the incident beam has a limited lateral dimension which, whilst being very large compared with the dimension of the scatterer, is sufficiently small that it does not encompass the detector. This is accomplished by means of collimating slits, as shown in figure 10.1. The effect of the slits is to introduce diffraction so that the wave is no longer exactly plane but, since it is much wider than the scatterer, this will have no significant effect on the scattering process. The choice of unit amplitude for the incident wave is for convenience. The results of the calculations are expressed as ratios of fluxes so no loss of generality results.

We will model the scatterer by means of a spherically symmetric potential, centred at the origin, which falls to zero as $r \to \infty$. The overall wavefunction can be written

$$u = \mathrm{e}^{\mathrm{i}kz} + f(\theta)\frac{\mathrm{e}^{\mathrm{i}kr}}{r} \qquad r \to \infty. \tag{10.10}$$

The first term represents the incoming beam. The second term, which represents the scattered particles, has the form of an outgoing spherical wave, with amplitude $f(\theta)$. Since there is axial symmetry, the amplitude $f(\theta)$, which describes the angular distribution, is a function of θ only. The r in the denominator is required for conservation of flux as r varies. Looking back at the classical picture, we see that the trajectory followed by the outgoing particle is not precisely radial, although at large distances the discrepancy is negligible. The same applies in the wave picture; the wavefunction of equation (10.10) is the asymptotic wavefunction and applies only at large r. For the same reason the constant $k = \sqrt{2\mu E}/\hbar$ is the same in both terms. Its value changes during the encounter but since we are dealing with elastic scattering it has the same value before and after. Because the detector is necessarily situated at a very large distance compared to the size of an atomic or nuclear scatterer, it is the asymptotic wavefunction which governs the observations.

We will now show how $f(\theta)$ is related to the differential scattering cross section. To do this we first calculate the flux density in the incoming and outgoing waves. In general we would expect there to be interference effects between the two components of equation (10.10). The justification for treating them separately is that they can be measured independently. For the outgoing spherical wave we can place the detector outside the collimated beam. For the collimated beam we can place the detector at such a large distance ahead of the scatterer that the $1/r$ dependence reduces the spherical wave to negligible amplitude.

Recalling equation (2.38) the flux density is

$$j = \mathrm{Re}\,\frac{\hbar}{\mathrm{i}\mu}\,u^*\nabla u. \tag{10.11}$$

Applying this to the first term in equation (10.10) we find

$$j_{\mathrm{in}} = \mathrm{Re}\,\frac{\hbar}{\mathrm{i}\mu}\,\mathrm{e}^{-\mathrm{i}kz}\,\mathrm{i}k\,\mathrm{e}^{\mathrm{i}kz} = \frac{\hbar k}{\mu} = v \tag{10.12}$$

where v is the speed of the particles.

For the second term

$$j_{\mathrm{out}} = \mathrm{Re}\,\frac{\hbar}{\mathrm{i}\mu}\,f^*(\theta)\,\frac{\mathrm{e}^{-\mathrm{i}kr}}{r}\,\nabla\left(f(\theta)\,\frac{\mathrm{e}^{\mathrm{i}kr}}{r}\right). \tag{10.13}$$

In taking the gradient only the radial function contributes to the real part. Hence we find

$$\begin{aligned} j_{\mathrm{out}} &= \mathrm{Re}\,\frac{\hbar}{\mathrm{i}\mu}\,|\,f(\theta)\,|^2\,\frac{\mathrm{e}^{-\mathrm{i}kr}}{r}\,\mathrm{e}^{\mathrm{i}kr}\left(\frac{\mathrm{i}k}{r}-\frac{1}{r^2}\right) \\ &= \frac{\hbar k}{\mu}\,\frac{1}{r^2}\,|\,f(\theta)\,|^2. \end{aligned} \tag{10.14}$$

The number $\mathrm{d}N$ of particles per unit time entering a detector of area $\mathrm{d}A$ at a distance r is therefore

$$\mathrm{d}N = \frac{\hbar k}{\mu}\,|\,f(\theta)\,|^2\,\frac{\mathrm{d}A}{r^2}. \tag{10.15}$$

In this equation $\hbar k/\mu$ is the incident flux density and $\mathrm{d}A/r^2$ is the solid angle subtended by the detector, so comparison with equation (10.1) reveals that

$$\frac{\mathrm{d}\sigma}{\mathrm{d}\Omega} = |\,f(\theta)\,|^2 \tag{10.16}$$

which is the desired relationship.

〉10.4 Partial-wave analysis

Our task now is to relate $f(\theta)$ to the potential function which is the cause of the scattering. There are several ways of approaching this problem, but the only one we will consider here is the method of partial-wave analysis. This method was invented by Rayleigh in connection with the diffraction of sound waves, and was applied in the analysis of quantum mechanical scattering in 1927 by Faxen and Holtsmark. Although the method provides exact results in principle, they are in a form which is only useful for low-energy scattering. For high-energy scattering, perturbation methods are available such as the Born approximation but the reader is referred elsewhere for these.

In order to proceed we must solve the TISE (section 6.4) for the particle moving in the potential $V(r)$ of the scatterer. The potential may be positive or negative but we will assume that $V(r) \to 0$ as $r \to \infty$ more rapidly than $1/r$, which is to say faster than a Coulomb potential. When we have found the solution, comparison of its asymptotic form with equation (10.10) will then permit the determination of $f(\theta)$.

Now any solution can be expressed as a superposition of functions of the form

$$u(r,\theta,\phi) = \sum_{l,m} R_l(r)\, Y_l^m(\theta,\phi) \tag{10.17}$$

since the $R_l(r)\, Y_l^m(\theta,\phi)$ form a complete set. However, in this case it is clear that the angular momentum of the scattered particle about the origin can have no z component since its initial velocity is in that direction, and so the sum will only contain the functions $Y_l^0(\theta,\phi)$, which differ only in normalization from the Legendre polynomials $P_l(\cos\theta)$. Accordingly we write

$$u(r,\theta) = \sum_{l} R_l(r) P_l(\cos\theta). \tag{10.18}$$

As can be deduced from equations (6.48) and (6.49), the Legendre polynomials are orthogonal, and normalized so that

$$\int_{-1}^{1} P_l(\cos\theta) P_{l'}(\cos\theta)\, \mathrm{d}(\cos\theta) = \frac{2\delta_{ll'}}{2l+1}. \tag{10.19}$$

The functions $R_l(r)$ must satisfy the radial equation (7.18), but as usual it is convenient to make the substitution $\chi_l(r) = rR_l(r)$, which must satisfy the simpler equation (7.20):

$$\left(-\frac{\hbar^2}{2\mu}\frac{d^2}{dr^2}+\frac{l(l+1)\hbar^2}{2\mu r^2}+V(r)\right)\chi_l(r)=E\chi_l(r). \qquad (10.20)$$

In this case E is positive and has a continuous spectrum. Its value is chosen by the experimenter. For this reason R_l and χ_l carry only the angular momentum label l, but they do of course depend upon E.

Let us suppose that $V(r)$ becomes negligible beyond say $r = a$. In this external region then, the equation to be solved is

$$\left(-\frac{\hbar^2}{2\mu}\frac{d^2}{dr^2}+\frac{l(l+1)\hbar^2}{2\mu r^2}\right)\chi_l(r)=E\chi_l(r) \qquad r>a. \qquad (10.21)$$

We have already dealt with this equation in connection with the spherical square well (section 7.7) where it was shown that satisfactory solutions for $\chi_l(r)/r = R_l(r)$ are the spherical Bessel and Neumann functions, $j_l(kr)$ and $n_l(kr)$, tabulated in table 7.5 and illustrated in figure 7.10. Hence the general solution for a particular value of l is

$$R_l(r) = A_l j_l(kr) + B_l n_l(kr) \qquad (10.22)$$

where

$$k = (2\mu E)^{1/2}/\hbar. \qquad (10.23)$$

Although the spherical Neumann functions are singular at the origin there is no need to exclude them here since the solution is only applicable for $r > a$. Equation (10.22) can be simplified if we replace j_l and n_l by their asymptotic forms. This is permissible because the function (10.10) which is to be expanded is itself an asymptotic form. The asymptotic form of the spherical Bessel functions can be found by noting that the leading terms (for large r) of successive js in table 7.5 follow the repeating sequence:

$$\frac{\sin kr}{kr}, \frac{-\cos kr}{kr}, \frac{-\sin kr}{kr}, \frac{\cos kr}{kr}, \ldots$$

which are equivalent to

$$\frac{\sin kr}{kr}, \frac{\sin(kr-\frac{1}{2}\pi)}{kr}, \frac{\sin(kr-\pi)}{kr}, \frac{\sin(kr-\frac{3}{2}\pi)}{kr}, \ldots.$$

In general

$$j_l(kr) = \frac{\sin(kr-\frac{1}{2}l\pi)}{kr} \qquad r\to\infty. \qquad (10.24)$$

Similarly we find that

$$n_l(kr) = \frac{-\cos(kr - \frac{1}{2}l\pi)}{kr} \qquad r \to \infty. \tag{10.25}$$

This behaviour is clearly illustrated in figure 7.10, where it can be seen that each successive Bessel function lags behind its predecessor by $\frac{1}{2}\pi$ at large r.

Inserting equations (10.24) and (10.25) into (10.22) we obtain

$$R_l(r) = A_l \frac{\sin(kr - \frac{1}{2}l\pi)}{kr} - B_l \frac{\cos(kr - \frac{1}{2}l\pi)}{kr}$$

$$= C_l \frac{\sin(kr - \frac{1}{2}l\pi + \delta_l)}{kr} \tag{10.26}$$

where

$$C_l^2 = A_l^2 + B_l^2 \qquad \delta_l = -\tan^{-1} B_l/A_l. \tag{10.27}$$

The asymptotic expansion (10.18) now reads

$$u(r, \theta) = \sum_l C_l \frac{\sin(kr - \frac{1}{2}l\pi + \delta_l)}{kr} P_l(\cos\theta). \tag{10.28}$$

This equation is valid for r large and $> a$, where a is the radius beyond which the potential becomes negligible. In many cases of interest the potential function approaches zero only asymptotically. Analysis shows that the asymptotic form (10.28) is valid even in those cases, provided $V(r)$ approaches zero more rapidly than r^{-1}, which is what we assumed initially. Unfortunately this excludes the important case of Coulomb scattering, which as we have already noted leads to an infinite cross section in the classical calculation; special techniques have been developed to deal with this case. The potential falls so slowly that the asymptotic region is never reached, and the phase shift in equation (10.28) never settles down to a steady value, but continues to increase logarithmically.

The significance of the phases in equation (10.28) begins to emerge if we express the incident plane wave in a similar expansion. This wave-function, $u = \exp(ikz)$, is a solution of equation (10.21), the TISE equation in the absence of the scattering potential. Now, however, the range of validity covers $0 \leqslant r \leqslant \infty$, and so in the expansion we must exclude the Neumann functions of equation (10.22), which are singular at the origin, and write

$$\exp(ikz) = \exp(ikr\cos\theta) = \sum_l a_l j_l(kr) P_l(\cos\theta). \tag{10.29}$$

The coefficients a_l may be determined by the standard technique introduced in section 5.6. The well known result is

$$\exp(\mathrm{i}kz) = \sum_l (2l+1)\mathrm{i}^l j_l(kr) P_l(\cos\theta). \tag{10.30}$$

This equation, which expresses the decomposition of a plane wave into a linear superposition of spherical waves, has a simple interpretation. The position of a particle whose wavefunction is a plane wave is completely indeterminate, although its momentum $\hbar k$ is quite definite. (Plane waves are simultaneous eigenfunctions of momentum and energy if no forces are acting.) This means that its angular momentum (with respect to the origin) is also indeterminate. Consequently its expansion must include components with all possible angular momentum eigenvalues, $L^2 = l(l+1)\hbar^2$. These components are the terms of the sum in equation (10.30), and they are referred to as partial waves. Now a particular value of l will, since the momentum is fixed, correspond to a particular discrete impact parameter, through the relation $|L| = \hbar kb$. Since we expect b to be a continuous variable this appears to be an absurd result, but we should remember that components of $\boldsymbol{r}$ and $\hat{\boldsymbol{L}}$ do not commute so that a determination of l could only restrict b to within certain limits. As a rough estimate let us suppose that the range of impact parameters for a particular value of l runs from $b_{l-1/2}$ to $b_{l+1/2}$ (see figure 10.3), where

$$\begin{aligned} \hbar k b_{l+1/2} &= [(l+\tfrac{1}{2})(l+\tfrac{3}{2})]^{1/2}\hbar \\ \hbar k b_{l-1/2} &= [(l-\tfrac{1}{2})(l+\tfrac{1}{2})]^{1/2}\hbar. \end{aligned} \tag{10.31}$$

Hence the annular area through which the particle must pass in order to qualify as a particle of angular momentum quantum number l is equal to

$$\pi(b_{l+1/2}^2 - b_{l-1/2}^2) = (2l+1)/k^2 \tag{10.32}$$

and this explains the weighting factor in the expansion (10.30). The partial waves have a standing-wave nature, but we can decompose each into a superposition of outgoing and ingoing spherical waves. These components are the spherical Hankel functions of the first and second kind respectively, introduced in section 7.7. The first few are given in table 10.1.

Adding equations (7.114) and (7.115) we find that the spherical Bessel functions can be expressed as

$$j_l(kr) = \tfrac{1}{2}h_l^{(1)}(kr) + \tfrac{1}{2}h_l^{(2)}(kr). \tag{10.33}$$

We will need the asymptotic form of the plane-wave expansion and this is readily found by inserting equation (10.24) in (10.30):

$$\exp(\mathrm{i}kz) = \sum_l (2l+1)\mathrm{i}^l \frac{\sin(kr - \frac{1}{2}l\pi)}{kr} P_l(\cos\theta) \qquad r \to \infty. \quad (10.34)$$

The picture we have developed so far is as follows. In the absence of a scattering potential the incident plane wave can be analysed into partial waves of specific angular momentum. Each of these comprises an ingoing and outgoing spherical travelling wave. We can picture a spherical wave converging on the origin and then diverging from the origin. Naturally the flux is the same in each part: particles are neither created nor annihilated at the origin! If a scattering potential is introduced centred on the origin, the wavefunction for large r, equation (10.28), has the same general form as (10.34) apart from the introduction of a phase shift. This is the accumulated effect of the variation in k which takes place as the wave propagates through the region of the potential. At large distances a phase shift is in fact the only possible change that the potential could introduce, because of all the restraints which must be satisfied. These are: (i) energy is conserved and so k is the same before and after. (ii) Angular momentum is conserved and so l is the same before and after. The partial waves retain their identity. (iii) The ingoing and outgoing fluxes must be equal for particle conservation. This means that the amplitudes of the ingoing and outgoing components must be the same with or without the potential. Hence the effect of the scattering potential is entirely contained in the phase shifts of the partial waves.

Table 10.1 Spherical Hankel functions of the first and second kind.

l	$h_l^{(1)}(kr)$	$h_l^{(2)}(kr)$
0	$-\dfrac{\mathrm{i}}{kr}\mathrm{e}^{\mathrm{i}kr}$	$\dfrac{\mathrm{i}}{kr}\mathrm{e}^{-\mathrm{i}kr}$
1	$-\left(\dfrac{1}{kr}+\dfrac{\mathrm{i}}{k^2r^2}\right)\mathrm{e}^{\mathrm{i}kr}$	$-\left(\dfrac{1}{kr}-\dfrac{\mathrm{i}}{k^2r^2}\right)\mathrm{e}^{-\mathrm{i}kr}$
2	$\left(\dfrac{\mathrm{i}}{kr}-\dfrac{3}{k^2r^2}-\dfrac{3\mathrm{i}}{k^3r^3}\right)\mathrm{e}^{\mathrm{i}kr}$	$-\left(\dfrac{\mathrm{i}}{kr}-\dfrac{3}{k^2r^2}-\dfrac{3\mathrm{i}}{k^3r^3}\right)\mathrm{e}^{-\mathrm{i}kr}$

The next step is to obtain the relation between the phase shifts and the scattering amplitude $f(\theta)$. This is done by substituting equation (10.28), the asymptotic expansion of the complete wavefunction u, into the left-hand side of equation (10.10), and substituting equation (10.30), the asymptotic expansion of the plane wave, into the right-hand side of equation (10.10). The result is

$$\sum_l C_l \sin(kr - \tfrac{1}{2}l\pi + \delta_l)P_l(\cos\theta) = \sum_l (2l+1)\mathrm{i}^l \sin(kr - \tfrac{1}{2}l\pi)P_l(\cos\theta) + f(\theta)\frac{\mathrm{e}^{\mathrm{i}kr}}{r} \qquad r \to \infty. \tag{10.35}$$

Replacing the sines by their exponential forms and collecting terms we find

$$\frac{\mathrm{e}^{\mathrm{i}kr}}{2\mathrm{i}kr}\left(\sum_l [C_l\,\mathrm{e}^{\mathrm{i}\delta_l} - (2l+1)\mathrm{i}^l]P_l(\cos\theta) - 2\mathrm{i}kf(\theta)\right) + \frac{\mathrm{e}^{-\mathrm{i}kr}}{2\mathrm{i}kr}\left(\sum_l [C_l\,\mathrm{e}^{-\mathrm{i}\delta_l} - (2l+1)\mathrm{i}^l]P_l(\cos\theta)\right) = 0. \tag{10.36}$$

Now this equation must be true for all values of r provided they are large, and so the expressions in large brackets must vanish. In addition, the equation must be true for all θ, and this means that the individual terms of the second summation must vanish separately. The second requirement determines all the coefficients C_l:

$$C_l = (2l+1)\mathrm{i}^l\,\mathrm{e}^{\mathrm{i}\delta_l}. \tag{10.37}$$

Inserting this into the first summation and setting the contents of the first large brackets to zero we find

$$f(\theta) = \frac{1}{2\mathrm{i}k}\sum_l (2l+1)\mathrm{i}^l(\mathrm{e}^{2\mathrm{i}\delta_l} - 1)P_l(\cos\theta) = \frac{1}{k}\sum_l (2l+1)\mathrm{i}^l\,\mathrm{e}^{\mathrm{i}\delta_l}\sin\delta_l P_l(\cos\theta). \tag{10.38}$$

From this and equation (10.16) we can immediately obtain the differential scattering cross section:

$$\frac{\mathrm{d}\sigma}{\mathrm{d}\Omega} = \left|\frac{1}{k}\sum_l (2l+1)\mathrm{i}^l\,\mathrm{e}^{\mathrm{i}\delta_l}\sin\delta_l P_l(\cos\theta)\right|^2. \tag{10.39}$$

When this expression is evaluated cross products such as $P_{l'}(\cos\theta) \times P_l(\cos\theta)$ appear. These represent interference between the various partial waves and give the scattering its characteristic angular distribution. The total scattering cross section is obtained on integrating equation (10.39) over all solid angles according to equation (10.2). In this operation the orthogonality of the $P_l(\cos\theta)$ comes into play (equation (10.19)) and all the cross products integrate to zero. We therefore find

$$\sigma = \int_0^\pi |f(\theta)|^2 \, 2\pi \sin\theta \, d\theta$$

$$= \frac{2\pi}{k^2} \sum_l (2l+1)^2 \sin^2\delta_l \frac{2}{2l+1}$$

$$= \frac{4\pi}{k^2} \sum_l (2l+1)\sin^2\delta_l. \qquad (10.40)$$

We can proceed no further without specifying the precise form of the potential, since of course this determines the actual phase shifts. Note that, although equation (10.40) is exact, the calculation of a total cross section entails the evaluation of the phase shifts of, in principle, an infinite number of partial waves. At first sight this appears to render the method useless, but in many practical cases only the first few phase shifts need be calculated. The reason is that, as indicated in the discussion prior to equation (10.31), the impact parameter corresponding to the lth partial wave is of order $b = [l(l+1)]^{1/2}/k$. If the range of the potential, a, is much less than this the particle will not interact with the scatterer. It follows that the lth partial-wave phase shift will vanish, and the corresponding term and all succeeding ones will not contribute to the sum in equation (10.40). Since b is inversely proportional to $k = (2\mu E)^{1/2}/\hbar$ we find that the lower the energy, the smaller the number of partial waves that contribute. For this reason the method is principally applicable to low-energy scattering. If it happens that only the $l = 0$ partial wave contributes, then the scattering is referred to as s wave. If $l = 1$ contributes, this is referred to as p-wave scattering and the total scattering amplitude will be the sum of the s- and p-wave scattering amplitudes, and so on.

〉10.5 Hard-sphere scattering

The first example that we will consider is the case of particles impinging upon a totally impenetrable sphere of radius a. It is fairly easy to show that classically this would result in an isotropic distribution, apart from a geometrical shadow behind the sphere. Quantum mechanically the sphere can be modelled by the potential (see figure 10.4):

$$V(r) = \infty \qquad r < a$$
$$V(r) = 0 \qquad r \geqslant a. \tag{10.41}$$

In order to calculate the phase shifts δ_l we must solve equation (10.20) for this potential. In the internal region $R_l(r) = 0$ since $V(r) = \infty$. In the external region $V(r) = 0$, and the solutions of the equation have already been determined (equation (10.22)), and are

$$R_l(r) = A_l j_l(kr) + B_l n_l(kr). \tag{10.42}$$

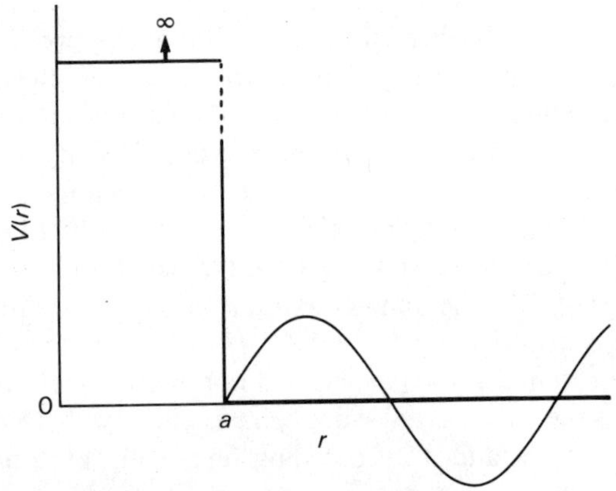

Figure 10.4 A 'hard sphere' potential. The wavefunction must fall to zero at the boundary.

In this case the boundary condition

$$R_l(a) = 0 \tag{10.43}$$

must be satisfied and this determines the phase shifts. Inserting equation (10.43) in (10.42) we find

$$\frac{j_l(ka)}{n_l(ka)} = -\frac{B_l}{A_l} = \tan\delta_l \tag{10.44}$$

where we have used equation (10.27) for the second equality. For very-low-energy scattering, such that $ka \ll 1$, we need consider only the s wave. For $l = 0$ equation (10.44) becomes

$$-\tan ka = \tan\delta_0$$

or

$$\delta_0 = -ka. \tag{10.45}$$

The differential s-wave scattering cross section (equation (10.39)) is then

$$\frac{d\sigma_0}{d\Omega} = \frac{1}{k}\sin^2 ka \tag{10.46}$$

since $P_0(\cos\theta) = 1$. Hence s-wave scattering is independent of angle, i.e. isotropic. The total s-wave cross section is obtained by integrating equation (10.46) over all angles, and is thus

$$\sigma_0 = \frac{4\pi}{k^2}\sin^2 ka. \tag{10.47}$$

In the limit of very-low-energy scattering $\sin ka \to ka$ and the total cross section becomes $4\pi a^2$. This is just four times the classical cross section for the scattering of particles by an impenetrable sphere. This should not concern us unduly since we expect wave rather than particle phenomena to dominate the behaviour in this limit.

Exceptionally, for hard-sphere scattering it is possible to obtain an expression for the total cross section using the partial wave analysis even in the limit of high energy. In equation (10.44), for large ka we can use the asymptotic forms of the spherical Bessel and Neumann functions (equations (10.24) and (10.25)) to obtain

$$-\tan(ka - \tfrac{1}{2}l\pi) = \tan\delta_l$$

and hence

$$\sin\delta_l = -\sin(ka - \tfrac{1}{2}l\pi). \tag{10.48}$$

Inserting this in (10.39) we find that the total cross section is

$$\sigma = \frac{4\pi}{k^2}\sum_{l=0}^{ka}(2l+1)\sin^2(ka - \tfrac{1}{2}l\pi). \tag{10.49}$$

The sum is truncated at $l = ka$ since this is the highest order partial wave which will contribute. The $\sin^2$ term is a rapidly oscillating function in the sum and can be approximated by its mean value, which is $\frac{1}{2}$. Hence

$$\begin{aligned}\sigma &\simeq \frac{2\pi}{k^2}\sum_{l=0}^{ka}(2l+1) \simeq \frac{2\pi}{k^2}(1 + 2ka + 1)\frac{ka}{2} \qquad ka \gg 1\\ &\simeq 2\pi a^2.\end{aligned} \tag{10.50}$$

Surprisingly, even in the high-energy limit, we find a cross section which is still twice the classical one. The explanation is that one half of the cross section arises from diffraction, and in the high-energy limit this diffraction results in a very sharply peaked forward spike in the intensity distribution. Since it is so narrow it is nearly unobservable. The remainder of the scattering cross section gives rise, in this limit, to the isotropic distribution which is normally detected. The forward spike is a typical wave phenomenon and can be seen in the centre of the diffraction pattern formed when light is diffracted by, for example, a ball bearing. In that context it is known as Poisson's spot.

〉10.6 Scattering at a spherical square well

As a second example we will consider the case of a spherical square well. The bound states of such a system were studied in section 7.7. For scattering it is more convenient to define the potential to be zero at infinity, so that

$$\begin{aligned} V(r) &= -V \qquad & r < a \\ V(r) &= 0 \qquad & r \geqslant a \end{aligned} \tag{10.51}$$

as in figure 10.5.

The radial wavefunction $\chi_l(kr) = rR_l(kr)$, which must satisfy equation (10.20), takes the forms

$$\chi_l(k_1 r) = rD_l j_l(k_1 r) \qquad r < a \tag{10.52}$$

$$\chi_l(k_2 r) = r[A_l j_l(k_2 r) + B_l n_l(k_2 r)] \qquad r \geqslant a \tag{10.53}$$

where

$$k_1 = [2\mu(E + V)]^{1/2}/\hbar \qquad k_2 = (2\mu E)^{1/2}/\hbar. \tag{10.54}$$

First of all we will consider the case of s-wave scattering. Then equations (10.52) and (10.53) become

$$k_1 \chi_0(k_1 r) = D_0 \sin k_1 r \tag{10.55}$$

and

$$\begin{aligned} k_2 \chi_0(k_2 r) &= A_0 \sin k_2 r - B_0 \cos k_2 r \\ &= C_0 \sin(k_2 r + \delta_0) \end{aligned} \tag{10.56}$$

where, as in equation (10.27),

$$C_0^2 = A_0^2 + B_0^2 \qquad \tan \delta_0 = -B_0/A_0. \tag{10.57}$$

Continuity of the logarithmic derivative of $\chi_0(kr)$ at $r = a$ requires

$$k_1 \cot k_1 a = k_2 \cot(k_2 a + \delta_0). \tag{10.58}$$

Inverting both sides, expanding the tangent on the right and rearranging we find

$$\tan \delta_0 = \frac{k_2 \tan k_1 a - k_1 \tan k_2 a}{k_1 + k_2 \tan k_1 a \tan k_2 a} \tag{10.59}$$

and the total s-wave scattering cross section is, from equation (10.40),

$$\sigma_0 = \frac{4\pi}{k^2} \sin^2 \delta_0. \tag{10.60}$$

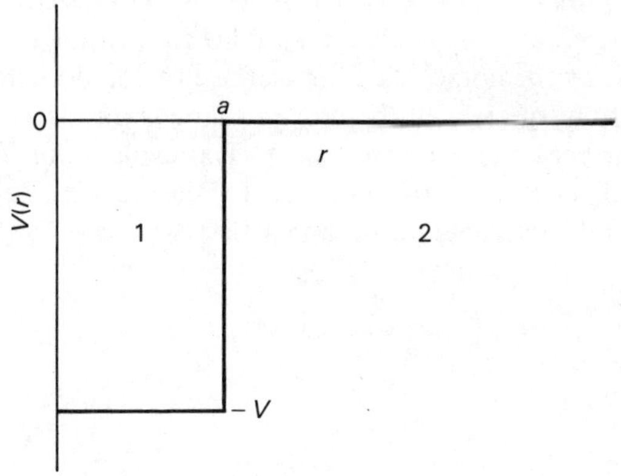

Figure 10.5 A spherical square well. As usual for scattering events, the potential is set to zero at infinity.

An interesting result which emerges immediately from equation (10.59) is that $\tan \delta_0$ and hence $\sin \delta_0$ and the s-wave scattering cross section will be zero if

$$k_2 \tan k_1 a = k_1 \tan k_2 a. \tag{10.61}$$

When this occurs $\delta_0 = n\pi$. If this condition is satisfied at low energy, when partial waves with $l > 0$ do not contribute, we have the remarkable result that the scatterer becomes 'invisible'!

For low-energy scattering $k_2 \to 0$ and hence $\tan k_2 a \to k_2 a$. In this limit equation (10.59) reduces to

$$\tan \delta_0 = k_2 a\left(\frac{\tan k_1 a}{k_1 a} - 1\right) \approx \sin \delta_0 \tag{10.62}$$

and hence, from equation (10.40),

$$\sigma_0 = 4\pi a^2\left(\frac{\tan k_1 a}{k_1 a} - 1\right)^2. \tag{10.63}$$

This shows that the cross section is zero when $\tan k_1 a = k_1 a$. This condition is a special case of equation (10.61), when $k_2 \to 0$.

Transmission peaks were discovered experimentally by C W Ramsauer in 1921, and independently by J S Townsend in 1922, in the scattering of low-energy electrons by noble-gas atoms. For certain energies the atoms appear nearly transparent. For example, a beam of 0.7 eV electrons is almost unattenuated on passing through argon, krypton or xenon gas. Figure 10.6 shows the variation of cross section with electron energy. The minima in the cross sections are very marked. Although the potential well presented by the noble-gas atom to the electron does not bear much resemblance to a square well the precise shape is immaterial provided it introduces the required $n\pi$ phase shift. The explanation for the effect was suggested by Bohr and confirmed by the detailed calculations of Holtsmark, and represented a significant early success for the quantum theory.

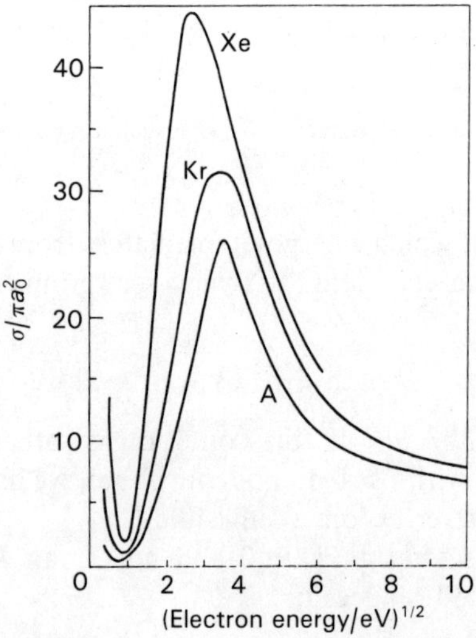

Figure 10.6 The Ramsauer effect. The electron scattering cross sections of the noble gases fall almost to zero in the neighbourhood of 1 eV.

Equation (10.63) reveals that, in general, the s-wave cross section tends to a finite value at low energy, but there are exceptional cases for which it is infinite! This phenomenon is referred to as resonance scattering and occurs whenever

$$k_1 a = \tfrac{1}{2} n\pi \qquad n \text{ odd.} \tag{10.64}$$

In this approximation E is fixed at approximately zero, so $k_1 \simeq (2\mu V)^{1/2}/\hbar$, and equation (10.64) is effectively a condition to be satisfied by the parameters of the well. The condition has an interesting interpretation. Comparison with equation (7.126) reveals that it is the condition for the square well to have a bound level at zero energy. $n = 1$ corresponds to the well being just strong enough to support one bound state, $n = 3$ corresponds to two bound levels, and so on. For very-low-energy scattering we can imagine that the particle is nearly trapped in the bound state at the top of the well, rebounding many times in the potential for which its wavefunction is a nearly perfect fit.

For the general case of l-wave scattering we must return to equations (10.52) and (10.53), and require the continuity of the logarithmic derivative at $r = a$. This condition is

$$k_1 a \frac{j_l'(k_1 a)}{j_l(k_1 a)} = \beta_l = k_2 a \frac{j_l'(k_2 a)\cos\delta_l - n_l'(k_2 a)\sin\delta_l}{j_l(k_2 a)\cos\delta_l - n_l(k_2 a)\sin\delta_l}. \tag{10.65}$$

The parameter β_l is the value of the logarithmic derivative of $R_l(r)$ at the boundary $r = a$, multiplied by a to make it dimensionless. Primes indicate differentiation with respect to argument before evaluation at $r = a$. Equation (10.65) can be rearranged to give

$$\tan\delta_l = \frac{k_2 a j_l'(k_2 a) - \beta_l j_l(k_2 a)}{k_2 a n_l'(k_2 a) - \beta_l n_l(k_2 a)}. \tag{10.66}$$

Detailed analysis of this expression will not be undertaken here. However, it appears that either the numerator or denominator may go to zero, leading to zero or resonant partial scattering cross sections, respectively. Zeros for $l > 0$ are not particularly interesting, since partial waves of lower order will usually contribute to the total cross section. The resonance behaviour is similar to the case $l = 0$ in that a resonance occurs at zero energy in the lth partial wave whenever conditions in the well are such that a level of angular momentum l is just bound at zero energy. A new feature is that resonance may also occur for small positive energies. The physical reason for this can be understood with reference to figure 10.7 which shows the effective potential for a square well, when

$l \neq 0$. The centrifugal barrier now provides the opportunity for a particle to be temporarily trapped. We can think in terms of quasi-bound states of positive energy. Such states have a finite lifetime because the particle can tunnel out of the trap. Scattering will be especially strong for particles whose energies match one of these levels. The large maxima in the range 8–15 eV in figure 10.6 arise from resonances in the $l = 2$ scattering. Calculations show that the δ_2 phase shifts are respectively $\frac{1}{2}\pi$, $\frac{3}{2}\pi$ and $\frac{5}{2}\pi$ for Ar, Kr and Xe.

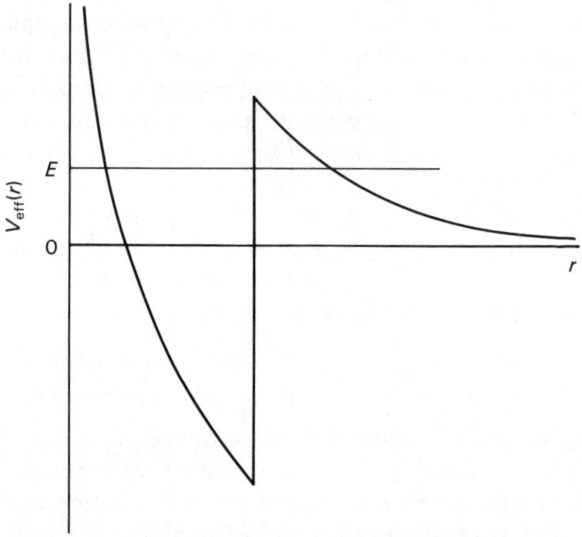

Figure 10.7 The effective potential-energy function for a spherical square well includes the effect of angular momentum. The infinite 'centrifugal barrier' prevents access to the origin. The deep notch in the potential provides an opportunity for a particle of positive energy to be trapped temporarily behind the outer barrier, through which it will eventually escape.

Although equation (10.66) was derived with reference to the spherical square well, it is of much wider applicability. Whatever the form of the potential, provided a lies in the asymptotic region, where the potential has become negligible, then the second equality in equation (10.65) must hold and hence equation (10.66) is true. In other words, the form of the potential determines the value of β_l at the radius a, and in turn β_l determines the particular combination of $j_l(k_2 r)$ and $n_l(k_2 r)$ in the asymptotic region and hence the phase shift δ_l. The first equality in (10.65) tells us what β_l is specifically for the square-well potential.

Computer exercise 10.1: scattering in three dimensions
The program 'Scattering in 3-D' is similar to the other programs dealing with three-dimensional spherically symmetric potentials, and the potential is defined in the same way. The program plots the partial wavefunctions for any value of l together with the partial wave which would exist in the absence of a scattering potential. This permits direct observation of the phase shifts. Another option permits the accurate numerical determination of the phase shifts and a third provides tabulations of the wavefunctions and also calculates the value of β_l at each integration step.

(*a*) Running the program with the default parameters results in a potential well of radius 2 au and depth 10 au. The scattered and unscattered s waves are both plotted together. The latter is simply the $l = 0$ partial wave component of the incident plane wave in the absence of a scattering potential. (The relative scales will not be correct unless you adjust them to be so by making the amplitudes the same in the asymptotic region.) Notice how the effect of the potential is to draw in the scattered wave relative to the unscattered one. Because the particle has more kinetic energy within the confines of the well, k_1 is greater than k_2 and the function oscillates more rapidly. The result is a phase advance; δ_0 is positive.

(*b*) Run the program again, changing to $V = +10$ au, with $E = 15$ au. This illustrates scattering by a repulsive potential. Now the potential pushes out the scattered wave; the phase shift is negative.

(*c*) Revert again to a potential well, with $V = -10$ au, $a = 1$ au. Reduce the energy to 1 au and increase the range RF to 12 au. Note that the phase shift is almost exactly π. The scattered wave has one extra zero crossing near the origin, but all the other zero crossings coincide with those of the unscattered wave (but have the opposite sense). This is an example of the Ramsauer effect. If the phase shift is a multiple of π the existence of the scatterer is undetectable (with particles of energy 1 au) in the asymptotic region, which is where the experimental observations must be made. Now take the 'Zeros and Phase-Shift' option. This lists the zeros in the two partial waves in sequence. It is possible to see when the asymptotic region has been reached since there the zeros become equally spaced. In this case $l = 0$ and the asymptotic region begins abruptly at $r = a$, because the spherical Bessel and Neumann functions are simply $\sin k_2r/k_2r$ and $-\cos k_2r/k_2r$, respectively. It will be found that the phase shift is very close to 1 (in units of π). It may be made even closer to 1 on adjusting E. It should be verified that the value you find satisfies equation (10.61).

(*d*) The Ramsauer effect will not be observed unless the scattering amplitudes of all partial waves for $l > 0$ are also negligible. Using the parameters of (*c*) above increase l to 1 and then 2. Observe the form of the effective potential, which now has a prominent centrifugal contribution. You will need to adjust the scale factors of the wavefunctions to obtain suitable traces. Notice that for both these waves the phase shift remains close to π, so there will be little contribution to the scattering. Now set $l = 3$. The phase shift drops to almost exactly zero! The scattering potential produces only a minor perturbation upon the wavefunction. Clearly this will be the case for all higher order partial waves, and so this is a true example of the Ramsauer effect.

The discussion at the end of section 10.4 indicated that the scattering amplitude of the lth wave will be small if $k_2 a < [l(l+1)]^{1/2}$, or $E < l(l+1)/2a^2$ in atomic units. In this case $E = a = 1$ and so, according to this argument, we should expect the phase shift of the d wave to be small. In fact, we have just shown that we need to go to the f wave before the phase shift becomes negligible. Clearly the argument cannot be rigorous, since it takes no account of the depth of the well. Even a small penetration into a very deep well will result in a significant phase shift, which is the case here. Run the program again with $l = 2$ but with $V = -5$. Now the $l = 2$ phase shift has dropped from π to zero, although the range of the potential is the same.

(*e*) In order to observe resonance scattering we must adjust the parameters of the potential well so that it has a virtual level at a small positive energy, which is less than the peak of the centrifugal barrier. One approach is to choose values of a, l and E, and then to vary V until a phase shift of approximately $\frac{1}{2}\pi$ is observed on plotting the scattered and unscattered waves. Then small adjustments to E can be made until a precise $\frac{1}{2}\pi$ phase shift is obtained, as measured in the asymptotic region.

Since we know, from the example of (*d*) above, that $E = 1$, $a = 1$, $l = 2$ and $V = -10$ leads to a phase shift of $\approx \pi$, we are led to try the same parameters with a reduced well depth, $V = -8$ au say, in the expectation of a smaller phase shift. This reduces the phase shift to less than $\frac{1}{2}\pi$ and after some trials with various values of V we find that $V = -8.2$ au gives a phase shift of nearly $\frac{1}{2}\pi$ as judged by eye, in the near field at $r \approx 10$ au. (It is necessary to adjust the various scale factors in order to get satisfactory traces and this may be done most efficiently by using the tabulation option in order to calculate the required scales.) The phase shift becomes constant only in the asymptotic region, and can be measured there by using the 'Zeros and Phase-Shifts' option. First of all it is necessary to locate the asymptotic region itself. Increase RF to 40 au and observe the values of the zeros. The program tabulates the zeros ('rz') of both the scattered and unscattered waves, and also the successive intervals between zeros ('drz'). The asymptotic region has been reached when these intervals become sufficiently constant. In the last column the phase shift between the two waves is plotted in units of π. The value of E should now be adjusted until this number tends to 0.5 in this asymptotic region. The phase shift is now exactly $\frac{1}{2}\pi$, and the functions should be replotted. It is particularly useful to plot them with scale factors such that their amplitudes are asymptotically equal. The resonance nature of the scattering is then spectacularly revealed by the huge increase in amplitude of the scattered wave relative to the unscattered one in the interior of the well. Figure 10.8 illustrates the effect for the example discusssed here.

If the phase shift is determined for a range of energies in the region of the resonance value, and plotted as a function of k_2 as in figure 10.9, the

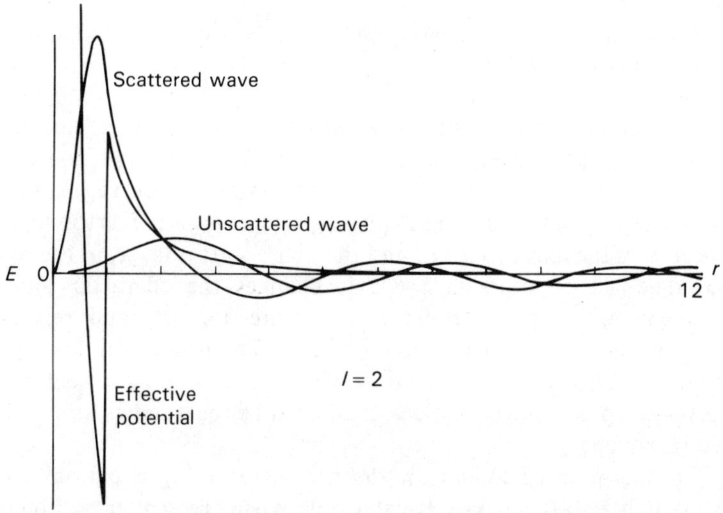

Figure 10.8 An example of resonance scattering for $l = 2$. The partial wave which would exist in the absence of any scattering centre is shown together with the wave as modified by the scattering potential. Note the greatly increased probability density in the region of the notch in the potential and the $\frac{1}{2}\pi$ phase shift between the two waves in the asymptotic region.

s-shaped curve typical of all resonance phenomena is obtained. The sharpness of resonance is a function of the effectiveness of the trap. Clearly this will depend upon the value of E compared with the height of the barrier. The above exercise should be repeated, trying to adjust E and V so that the barrier is effectively wider and/or higher, when a narrower resonance should be observed.

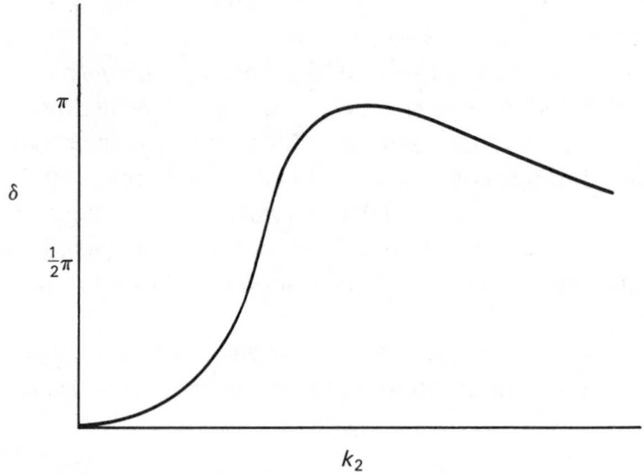

Figure 10.9 The scattering phase shift δ as a function of k_2, the momentum of the incoming particle.

By examining the form of the wavefunction for energies above and below the resonant energy it becomes apparent why the phase shift changes so rapidly at resonance. If E is too small, the curve just fails to cross the axis, resulting in a small phase shift, whereas if E is too high the curve crosses the axis too soon, and the phase shift is large, in the region of π. The act of crossing the axis changes the character and the subsequent behaviour of the function quite abruptly and resonance occurs precisely in this transitional region. The process of finding the resonance energy is almost identical with that of finding an eigenvalue for a bound state. Indeed, an eigenstate can be regarded as an infinitely sharp resonance.

(*f*) It was pointed out in the discussion following equation (10.28) that the phase-shift analysis is valid only when the scattering potential falls to zero more rapidly than r^{-1}, because otherwise the phase shift continues to increase with r. This can be observed by entering a Coulomb potential $(-1/R)$, a large value of RF and selecting the 'Zeros and Phase-Shifts' option. It will be found that the phase continues to increase, and that this arises from the behaviour of the scattered wave, since of course the unscattered wave shows the usual asymptotic behaviour with a constant shift between zeros. Exponential or Gaussian potential functions on the other hand give perfectly satisfactory results. (But remember that the program cannot deal with potentials more singular at the origin than r^{-1}.)

〉 10.7 Scattering of identical particles

When scattering occurs between identical particles new features arise in both classical and quantum mechanical scattering. Such events are most simply described in the centre of mass (CM) system. There the total linear momentum is zero and the incident and target particles approach from opposite directions (figure 10.10). If we consider the counting rate of the detector, we find that it is increased for identical particles, because incident particles scattered through the angle θ and target particles recoiling through the angle $\theta - \pi$ both find their way into the detector. In the classical case it is possible to distinguish the particles by following their trajectories, but in the quantum case identical particles are indistinguishable when their wavefunctions overlap, and this leads to further effects which we will now consider.

We first recall that the asymptotic form of the wavefunction for particle scattering is

$$u = \mathrm{e}^{\mathrm{i}kz} + f(\theta)\frac{\mathrm{e}^{\mathrm{i}kr}}{r} \qquad r \to \infty \tag{10.67}$$

where $f(\theta)$ is the amplitude of the scattered wave. It is related to the differential scattering cross section and hence the counting rate at the detector by

$$\frac{\mathrm{d}\sigma}{\mathrm{d}\Omega} = |f(\theta)|^2. \tag{10.68}$$

Allowing for the effect described above, for identical particles we might expect that the differential scattering cross section would become

$$\frac{\mathrm{d}\sigma}{\mathrm{d}\Omega} = |f(\theta)|^2 + |f(\pi - \theta)|^2 \tag{10.69}$$

but we will find that this is incomplete. The reason is that the wavefunction (10.67) is not correctly symmetrized for identical particles. Although this function apparently relates to a fictitious single particle of reduced mass, in reality it is a two-particle wavefunction; the spatial variables are components of $\boldsymbol{r} = \boldsymbol{r}_1 - \boldsymbol{r}_2$, the relative coordinate, and k is related to the reduced momentum, $\mu\dot{r}$ (equation (A3.17)).

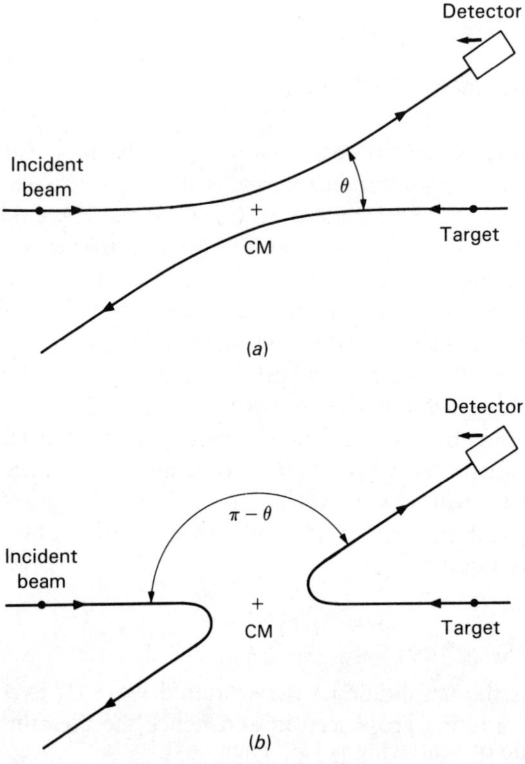

Figure 10.10 Scattering of identical particles, as seen in the CM system. The counting rate is enhanced because incident particles scattered through an angle θ (a) and target particles recoiling at an angle $\pi - \theta$ (b) are both captured by the detector.

Let us examine the behaviour of the wavefunction under an exchange of particle coordinates. Clearly if $\boldsymbol{r}_1 \Leftrightarrow \boldsymbol{r}_2$ then $\boldsymbol{r} \to -\boldsymbol{r}$, so in this case the particle exchange and parity operations become equivalent. In spherical polars the parity operation of reflection through the origin results in $r \to r$, $\theta \to \pi - \theta$ and $\phi \to \phi + \pi$. Also $z \to -z$, and so equation (10.67) becomes

$$u = \mathrm{e}^{-\mathrm{i}kz} + f(\pi - \theta)\frac{\mathrm{e}^{\mathrm{i}kr}}{r} \qquad r \to \infty. \tag{10.70}$$

Evidently this wavefunction has neither definite parity nor exchange symmetry. However, we know that the overall wavefunction must have a definite symmetry, depending on the type of particle, and we can construct suitable ones by superposition, in this case of wavefunctions (10.67) and (10.70). If the particles are spinless bosons the wavefunction must be symmetric, and we will take the sum

$$u = \mathrm{e}^{\mathrm{i}kz} + \mathrm{e}^{-\mathrm{i}kz} + (f(\theta) + f(\pi - \theta))\frac{\mathrm{e}^{\mathrm{i}kr}}{r}. \tag{10.71}$$

This equation now describes the interaction in the CM frame very clearly, since we can recognize the incoming particle fluxes from both directions. We also note that there is now complete symmetry about the $\theta = \frac{1}{2}\pi$ plane. A factor of $2^{-1/2}$ has *not* been included on making the superposition, because we still wish to relate the scattered flux to the flux in *one* of the incoming beams.

The differential scattering cross section is thus found to be

$$\begin{aligned}\frac{\mathrm{d}\sigma}{\mathrm{d}\Omega} &= |f(\theta) + f(\pi - \theta)|^2 \\ &= |f(\theta)|^2 + |f(\pi - \theta)|^2 + [f^*(\theta)f(\pi - \theta) + f^*(\pi - \theta)f(\theta)] \\ &= |f(\theta)|^2 + |f(\pi - \theta)|^2 + 2\,\mathrm{Re}[f^*(\theta)f(\pi - \theta)].\end{aligned} \tag{10.72}$$

The third term on the right represents an interference between the scattering amplitudes $f(\theta)$ and $f(\pi - \theta)$. As an example of the effect of this term, consider the case of scattering between α particles (spin zero) when the angle of scattering is $\frac{1}{2}\pi$. Then

$$\frac{\mathrm{d}\sigma}{\mathrm{d}\Omega} = |f(\tfrac{1}{2}\pi)|^2 + |f(\tfrac{1}{2}\pi)|^2 + 2|f(\tfrac{1}{2}\pi)|^2 = 4|f(\tfrac{1}{2}\pi)|^2. \tag{10.73}$$

The cross section is quadrupled. One factor of two arises even in the classical case as we have seen, and the other has no classical analogue.

These results, first obtained by Mott in 1930, have been verified in low-energy Coulomb scattering of α particles and other spinless bosons including ^{12}C (Bromley *et al* 1960, Chadwick 1930). The interference term is a function of θ and leads to large oscillations of the differential cross section with angle, which occur at all energies.

The scattering of particles with spin $\frac{1}{2}$ leads to further complexity. The overall wavefunction must be antisymmetric. Hence if the particle spins are parallel (triplet scattering) then the spin function is symmetric (see section 8.2) and so the spatial function must be antisymmetric:

$$u = \mathrm{e}^{\mathrm{i}kz} - \mathrm{e}^{-\mathrm{i}kz} + (f(\theta) - f(\pi - \theta))\frac{\mathrm{e}^{\mathrm{i}kr}}{r}. \tag{10.74}$$

The differential scattering cross section is then

$$\frac{\mathrm{d}\sigma}{\mathrm{d}\Omega} = |f(\theta)|^2 + |f(\pi - \theta)|^2 - 2\,\mathrm{Re}[f^*(\theta)f(\pi - \theta)] \tag{10.75}$$

leading to a different form of interference. For example, the scattering cross section at $\theta = \frac{1}{2}\pi$ is reduced to zero! If, on the other hand, the spins are antiparallel (singlet scattering) the cross section will again have the form of equation (10.72). In practice the usual experimental arrangement employs unpolarized incident and target particles, so what is actually observed is a weighted average over the relative spin orientations. (The situation is further complicated if the forces between the particles are spin dependent.) These predictions have been verified in proton–proton scattering experiments (Tuve *et al* 1936).

Another interesting consequence of the symmetry requirements stems from the relationship between the symmetry and the parity of the wavefunction, for a system of two identical particles. If the spatial wavefunction is symmetric, and hence of even parity, only even-order partial waves can contribute to the scattering and, if it is antisymmetric, only odd-order partial waves contribute. This follows from the parity of the spherical harmonics as specified in equation (6.56). This leads to the interesting observation that, for low-energy scattering between spin-$\frac{1}{2}$ particles, where only s-wave scattering can occur, the cross section vanishes for triplet scattering, since the spatial wavefunction in that case is antisymmetric.

〉Problems

10.1 Calculate the probability flux density as a function of r for the functions $Ah_0^{(1)}(kr)$, $Ah_0^{(2)}(kr)$ and $Aj_0(kr)$. (See tables 10.1 and 7.5.)

10.2 A wave $\exp[\mathrm{i}(kz - \omega t)]$ is incident upon a scattering centre, and the scattered wave is found to have the form

$$\psi = A \cos \tfrac{1}{2}\theta \, \frac{\exp[\mathrm{i}(kr - \omega t)]}{r}.$$

Show that the probability flux density in the scattered wave is

$$j = \frac{\hbar k}{mr^2} \,|\, A \,|^2 \cos^2 \tfrac{1}{2}\theta \; \hat{r}$$

and hence that the scattering cross section is

$$\sigma = 2\pi \,|\, A \,|^2.$$

10.3 A particle of mass m is scattered at a spherical potential barrier defined by

$$V(r) = V \qquad r \leqslant a$$
$$V(r) = 0 \qquad r > a.$$

The particle energy E is such that $k_2 a = a(2mE)^{1/2} < 1$. Find the s-wave phase shifts δ_0 for the cases $E > V$ and $E < V$. Check that when $E > V$, $\delta_0 \to 0$ as $V \to 0$ and that when $E < V$, $\delta_0 \to -k_2 a$ as $V \to \infty$, corresponding to the hard-sphere case treated in section 10.5.

Evaluate the phase shift and cross section for electron scattering when $a = 0.1$ nm, $V = 10$ eV and $E = 1$ eV. You can check your answer for the phase shift using the program '3-D Scattering'. (Hint: adapt the analysis of section 10.6 for scattering from a potential well, making the appropriate approximations and substitutions.)

10.4 The interaction between a neutron and a proton (triplet state) can be represented by a spherical square-potential-well of range 2×10^{-15} m and depth 36.25 MeV (see problem 7.7). Calculate the cross section for very-low-energy scattering of neutrons by protons.

〉 Appendix 1

〉 The Postulates of Quantum Mechanics

There is not a unique set of postulates in quantum mechanics, but the following six form a commonly chosen self-consistent set. All of them are invoked in this text apart from number six, which requires a deeper knowledge of classical mechanics than is presumed of the reader.

1. The wavefunction

For a particle moving under the influence of an external potential there is an associated wavefunction. That wavefunction determines everything that can be known about the particle and is a single-valued complex function of the coordinates of the particle and the time. The squared modulus of the function, evaluated at a particular point and time, is proportional to the probability density of the particle, at that point and time.

2. The Schrödinger equation

The wavefunction satisfies the time-dependent Schrödinger equation at all times.

3. Operators

Every dynamical variable may be represented by a Hermitian operator. The eigenvalues of these operators represent the possible results of measurements of the values of these variables. Immediately after such a measurement the wavefunction of a system will be that eigenfunction of the relevant operator which corresponds to the eigenvalue obtained as a result of the measurement.

4. Position and momentum operators
The operators representing the position $\boldsymbol{r}$ and momentum $\boldsymbol{P}$ of a particle are

$$\hat{\boldsymbol{r}} = \boldsymbol{r} \qquad \hat{\boldsymbol{P}} = -\mathrm{i}\hbar\nabla \tag{A1.1}$$

respectively. Operators representing other dynamical quantities are related to these operators in the same way that the classical quantities are related to the classical position and momentum variables.

5. Expansion postulate
The eigenfunctions of an operator form a complete (generally infinite) set of linearly independent functions such that any acceptable wavefunction can be expressed as a linear superposition of them. If a measurement of the variable represented by the operator $\hat{Q}$ is performed on a system with normalized wavefunction

$$\psi = \sum_n a_n\phi_n \tag{A1.2}$$

where the ϕ_n are normalized eigenfunctions of $\hat{Q}$ with eigenvalues q_n, then the probability of obtaining the result q_n is $|a_n|^2$.

6. Poisson brackets
The commutators of the operators representing dynamical variables in quantum mechanics are proportional to the corresponding classical Poisson brackets. The constant of proportionality is $\mathrm{i}\hbar$. For example,

$$[\hat{Q}, \hat{R}] \Leftrightarrow \mathrm{i}\hbar\{q, r\}. \tag{A1.3}$$

Appendix 2

Airy Functions

These functions were first studied by Airy in the analysis of diffraction near caustic surfaces, in which context they appeared as integrals. Thus the regular Airy function takes the form

$$\mathrm{Ai}[\pm(3a)^{-1/3}y] = (3a)^{1/3}\pi^{-1}\int_0^\infty \cos(at^3 + yt)\,\mathrm{d}t. \qquad \text{(A2.1)}$$

This integral may be evaluated by the technique of contour integration which is beyond the scope of this book (see Jeffreys and Jeffreys (1972)). However, it may also be shown that the integral satisfies the Airy differential equation:

$$\mathrm{d}^2u/\mathrm{d}y^2 - yu = 0 \qquad \text{(A2.2)}$$

which may be solved by the same technique as is used for all the other differential equations studied in this book, namely by finding a power series which satisfies the equation. On substituting the series

$$u = \sum_{s=0}^{\infty} a_s y^s \qquad \text{(A2.3)}$$

we find the following condition which must be satisfied:

$$\sum_{s=0}^{\infty} a_s s(s+1)y^{s-2} + \sum_{s=0}^{\infty} a_s y^{s+2} = 0. \qquad \text{(A2.4)}$$

This must be satisfied for each power of y separately, consequently

$$a_2 = 0 \qquad \text{(A2.5)}$$

$$a_3 = -\tfrac{1}{6}a_0 \qquad \text{(A2.6)}$$

and

$$a_s = -a_{s-3}/[s(s+1)]. \quad \text{(A2.7)}$$

The recurrence relation (A2.7) relates every third term, but since $a_2 = 0$ we have two rather than three independent series. Using our freedom to choose two arbitrary constants in the solution of a second-order equation, and initially letting these be $a_0 = 1$ and $a_1 = 1$, these series take the forms

$$f(y) = 1 + \frac{1}{3!}y^3 + \frac{1.4}{6!}y^6 + \frac{1.4.7}{9!}y^9 + \ldots \quad \text{(A2.8)}$$

$$g(y) = y + \frac{2}{4!}y^4 + \frac{2.5}{7!}y^7 + \frac{2.5.8}{10!}y^{10} + \ldots. \quad \text{(A2.9)}$$

Although these are convergent series both $f(y)$ and $g(y) \to \infty$ as $y \to \infty$. Neither is therefore immediately suitable as a wavefunction. However, a linear combination of them can be found which decays exponentially with large positive values of y. To show that this is possible consider the superposition

$$\mathrm{Ai}(y) = c_1 f(y) - c_2 g(y). \quad \text{(A2.10)}$$

To be a satisfactory wavefunction we require both $\mathrm{Ai}(y)$ and $\mathrm{Ai}'(y)$ to tend to zero as $y \to \infty$. Hence

$$\frac{c_1}{c_2} = \lim_{y \to \infty} \frac{g(y)}{f(y)} \qquad \frac{c_1}{c_2} = \lim_{y \to \infty} \frac{g'(y)}{f'(y)} \quad \text{(A2.11)}$$

and therefore

$$\lim_{y \to \infty}\left[\frac{g(y)}{f(y)} - \frac{g'(y)}{f'(y)}\right] = 0. \quad \text{(A2.12)}$$

Now it may be verified by direct substitution of the series for $f(y)$ and $g(y)$, that the following relation is true for *all* values of y:

$$g(y)f'(y) - f(y)g'(y) = -1$$

and so

$$\frac{g(y)}{f(y)} - \frac{g'(y)}{f'(y)} = -\frac{1}{f(y)f'(y)}. \quad \text{(A2.13)}$$

Since both $f(y)$ and $f'(y) \to \infty$ as $y \to \infty$, the right-hand side of (A2.13) $\to 0$ as $y \to \infty$, and so equation (A2.12) is true.

Although this proves that a suitable linear combination is possible, there is unfortunately no elementary way of determining the constants c_1 and c_2. They emerge on performing the contour integration of (A2.1), and turn out to be

$$c_1 = 3^{-2/3}/\Gamma(\tfrac{2}{3}) = 0.355\,028 \tag{A2.14}$$

and

$$c_2 = 3^{-1/3}/\Gamma(\tfrac{1}{3}) = 0.258\,819. \tag{A2.15}$$

The irregular Airy function is another, linearly independent, combination of the two series:

$$\mathrm{Bi}(y) = 3^{1/2}[c_1 f(y) + c_2 g(y)]. \tag{A2.16}$$

This function still diverges as $y \to \infty$. Both series oscillate for negative y (see figure 3.10). The effect of the factor $3^{1/2}$ in the expression for $\mathrm{Bi}(y)$ is to make the oscillations of $\mathrm{Ai}(y)$ and $\mathrm{Bi}(y)$ tend to the same amplitude as $y \to -\infty$.

The values of y for which $\mathrm{Ai}(y)$ and $\mathrm{Ai}'(y)$ equal zero determine the eigenvalues of the quantum bouncer and some other potential wells with linearly varying potentials, and the first few are listed in table A2.1 for convenience.

Table A2.1

n	$-y(\mathrm{Ai}=0)$	$-y(\mathrm{Ai}'=0)$
1	2.338 107 41	1.018 792 97
2	4.087 949 44	3.248 197 58
3	5.520 559 83	4.820 099 21
4	6.786 708 09	6.163 307 36
5	7.944 133 59	7.372 177 26
6	9.022 650 85	8.488 486 73
7	10.040 174 34	9.535 449 05
8	11.008 524 30	10.527 660 40
9	11.936 015 56	11.475 056 63
10	12.828 776 75	12.384 788 37

The mathematical properties of the Airy functions are tabulated in Abramowitz and Stegun (1965). Various quantum mechanical applications of the functions are discussed in Aguilera-Navarro *et al* (1981), Delbourgo (1977) and Gibbs (1975).

〉 Appendix 3

〉 Two-Body Systems: Reference Frames

〉 A3.1 Two-body systems

Many of the systems studied in this book consist of two particles. Examples are the electron and nucleus of a one-electron atom, the two atoms of a diatomic molecule and the probe and target particles in a scattering event. Provided the interaction between the two particles is independent of their *absolute* positions, which is the case for any isolated system, a considerable simplification to the analysis is possible, which essentially reduces it to a one-particle problem.

We consider the problem first from a classical viewpoint. Figure A3.1 shows two particles of masses m_1 and m_2. Their position vectors with respect to an arbitrary origin are $\boldsymbol{r}_1$ and $\boldsymbol{r}_2$. The only forces acting are $\boldsymbol{F}_{12}$ and $\boldsymbol{F}_{21}$ resulting from their mutual interaction. However, from Newton's third law we know that

$$\boldsymbol{F}_{12} = -\boldsymbol{F}_{21} = \boldsymbol{F}. \tag{A3.1}$$

The equations of motion of the two particles can therefore be written

$$m_1\ddot{\boldsymbol{r}}_1 = \boldsymbol{F}_{12} = \boldsymbol{F} \tag{A3.2}$$

and

$$m_2\ddot{\boldsymbol{r}}_2 = \boldsymbol{F}_{21} = -\boldsymbol{F}. \tag{A3.3}$$

Hence

$$m_1\ddot{\boldsymbol{r}}_1 + m_2\ddot{\boldsymbol{r}}_2 = 0. \tag{A3.4}$$

It proves very convenient to change the variables to $\boldsymbol{r}$ and $\boldsymbol{R}$, where

$$\boldsymbol{r} = \boldsymbol{r}_1 - \boldsymbol{r}_2 \tag{A3.5}$$

and

$$\boldsymbol{R} = \frac{m_1\boldsymbol{r}_1 + m_2\boldsymbol{r}_2}{m_1 + m_2}. \tag{A3.6}$$

$\boldsymbol{r}$ is the position vector of m_1 relative to m_2 and $\boldsymbol{R}$ is the centre of mass of two particles. It follows immediately from equations (A3.6) and (A3.4) that

$$(m_1 + m_2)\ddot{\boldsymbol{R}} = M\ddot{\boldsymbol{R}} = 0 \tag{A3.7}$$

where $M = m_1 + m_2$ is the total mass of the system. This result tells us that in the absence of external forces the centre of mass of the system exhibits uniform motion.

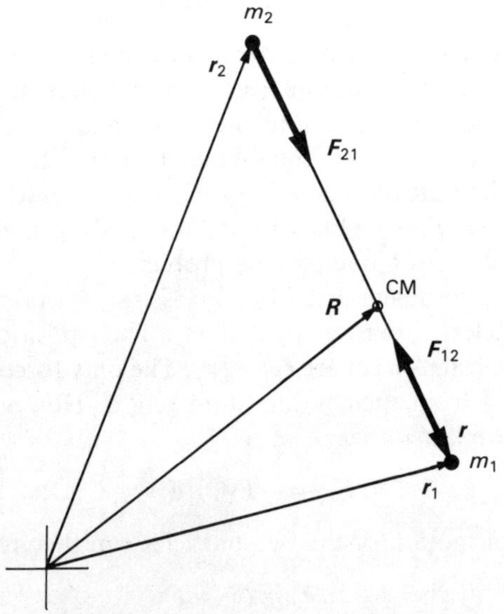

Figure A3.1 A system of two particles.

From equations (A3.2), (A3.3) and (A3.5) we find

$$\ddot{\boldsymbol{r}} = \ddot{\boldsymbol{r}}_1 - \ddot{\boldsymbol{r}}_2 = \frac{\boldsymbol{F}}{m_1} + \frac{\boldsymbol{F}}{m_2} \tag{A3.8}$$

which we will write as

$$\mu\ddot{\boldsymbol{r}} = \boldsymbol{F}. \tag{A3.9}$$

μ, the reduced mass, is defined as

$$\mu = \frac{m_1 m_2}{m_1 + m_2}. \tag{A3.10}$$

Solving equations (A3.5) and (A3.6) for $\boldsymbol{r}_1$ and $\boldsymbol{r}_2$ we find

$$\boldsymbol{r}_1 = \boldsymbol{R} + \frac{m_2}{M}\boldsymbol{r} \qquad \boldsymbol{r}_2 = \boldsymbol{R} - \frac{m_1}{M}\boldsymbol{r}. \tag{A3.11}$$

Equation (A3.9) shows that the analysis is equivalent to that of a particle of mass μ and position vector $\boldsymbol{r}$, acted upon by the mutual force $\boldsymbol{F}$. Once this equation has been solved the motions of the individual particles can be found from (A3.11).

The total kinetic energy of the system is

$$T = \tfrac{1}{2} m_1 \dot{\boldsymbol{r}}_1^2 + \tfrac{1}{2} m_2 \dot{\boldsymbol{r}}_2^2 \tag{A3.12}$$

and using equations (A3.10) and (A3.11) this can be written

$$\begin{aligned} T &= \tfrac{1}{2} m_1 \left(\dot{\boldsymbol{R}} + \frac{m_2}{M}\dot{\boldsymbol{r}}_1\right)^2 + \tfrac{1}{2} m_2 \left(\dot{\boldsymbol{R}} - \frac{m_1}{M}\dot{\boldsymbol{r}}_2\right)^2 \\ &= \tfrac{1}{2} M\dot{\boldsymbol{R}}^2 + \tfrac{1}{2}\mu\dot{\boldsymbol{r}}^2. \end{aligned} \tag{A3.13}$$

Similarly the angular momentum of the system is

$$\begin{aligned} \boldsymbol{J} &= m_1 \boldsymbol{r}_1 \times \dot{\boldsymbol{r}}_1 + m_2 \boldsymbol{r}_2 \times \dot{\boldsymbol{r}}_2 \\ &= m_1\left(\boldsymbol{R} + \frac{m_2}{M}\boldsymbol{r}_1\right) \times \left(\dot{\boldsymbol{R}} + \frac{m_2}{M}\dot{\boldsymbol{r}}_1\right) + m_2\left(\dot{\boldsymbol{R}} - \frac{m_1}{M}\dot{\boldsymbol{r}}_2\right) \times \left(\dot{\boldsymbol{R}} - \frac{m_1}{M}\dot{\boldsymbol{r}}_2\right) \\ &= M\boldsymbol{R} \times \dot{\boldsymbol{R}} + \mu\boldsymbol{r} \times \dot{\boldsymbol{r}}. \end{aligned} \tag{A3.14}$$

Equations (A3.13) and (A3.14) show that the conserved mechanical quantities T and $\boldsymbol{J}$ can each be partitioned into two parts, one of which relates to the motion of the centre of mass, the other to the relative motion.

The quantum mechanical treatment of the two-body problem starts by examining the corresponding Hamiltonian operator, which is

$$\hat{H} = \frac{\hat{p}_1^2}{2m_1} + \frac{\hat{p}_2^2}{2m_2} + V(r). \tag{A3.15}$$

$\boldsymbol{p}_1$ and $\boldsymbol{p}_2$ are the individual particle momenta. However, we have seen that the kinetic energy can be partitioned as in equation (A3.13), and expressing this in terms of momenta we have

$$T = \frac{P^2}{2M} + \frac{p^2}{2\mu} \tag{A3.16}$$

where

$$\boldsymbol{p} = \mu \dot{\boldsymbol{r}} \tag{A3.17}$$

and

$$\boldsymbol{P} = M\dot{\boldsymbol{R}}. \tag{A3.18}$$

Equation (A3.15) then becomes

$$\begin{aligned} \hat{H} &= \frac{\hat{P}^2}{2M} + \left(\frac{\hat{p}^2}{2\mu} + V(r)\right) \\ &= \hat{H}_{\mathrm{cm}} + \hat{H}_{\mathrm{rel}}. \end{aligned} \tag{A3.19}$$

Since the Hamiltonian can be partitioned into parts relating to centre of mass motion and relative motion we expect solutions of the TISE which take the form $u(\boldsymbol{r}, \boldsymbol{R}) = u_{\mathrm{cm}}(\boldsymbol{R})u_{\mathrm{rel}}(\boldsymbol{r})$, and write the equation as

$$(\hat{H}_{\mathrm{cm}} + \hat{H}_{\mathrm{rel}})u_{\mathrm{cm}}u_{\mathrm{rel}} = Eu_{\mathrm{cm}}u_{\mathrm{rel}}. \tag{A3.20}$$

Dividing by $u_{\mathrm{cm}}u_{\mathrm{rel}}$ we find

$$\frac{1}{u_{\mathrm{cm}}}\hat{H}_{\mathrm{cm}}u_{\mathrm{cm}} + \frac{1}{u_{\mathrm{rel}}}\hat{H}_{\mathrm{rel}}u_{\mathrm{rel}} = E. \tag{A3.21}$$

Since the two terms on the left are independent of each other they must each be constant, and so we find

$$\hat{H}_{\mathrm{cm}}u_{\mathrm{cm}} = E_{\mathrm{cm}}u_{\mathrm{cm}} \qquad \hat{H}_{\mathrm{rel}}u_{\mathrm{rel}} = E_{\mathrm{rel}}u_{\mathrm{rel}} \tag{A3.22}$$

where

$$E = E_{\mathrm{cm}} + E_{\mathrm{rel}}. \tag{A3.23}$$

We therefore find that the TISE can be separated into equations for the centre of mass motion and the relative motion. The former equation takes the form

$$\frac{\hat{P}^2}{2M}u_{\mathrm{cm}}(\boldsymbol{R}) = E_{\mathrm{cm}}u_{\mathrm{cm}}(\boldsymbol{R}) \tag{A3.24}$$

which is the TISE for a free particle. Solutions to this equation are the momentum eigenstates: they relate to the external force-free motion of the entire system. The TISE for the relative motion is

$$\left(\frac{\hat{p}^2}{2\mu} + V(r)\right) u_{\text{rel}}(\boldsymbol{r}) = E_{\text{rel}} u_{\text{rel}}(\boldsymbol{r}) \qquad \text{(A3.25)}$$

and is the equation we have used throughout the text.

〉A3.2 Centre of mass reference frames

When considering a bound system it is natural to adopt a reference frame in which its centre of mass is at rest (the CM frame), and in many cases this will be identical with the laboratory (lab) frame. For a scattering event it is more usual for the target to be stationary in the lab frame initially, in which case the centre of mass will move in the direction of the beam. Nevertheless, it is usual to treat even a scattering event in the CM frame in the first instance since the analysis is simpler. It is therefore necessary to be able to relate the event as described in the CM frame to the observations which are made in the lab frame. To illustrate the process we will consider a simple elastic collision between particles of mass m_1 and m_2. In the lab frame m_2 is initially at rest. In the CM frame the momenta of the two particles are at all times equal and opposite and, since for an elastic collision the overall kinetic energy is unchanged, it follows that in this frame the individual momentum magnitudes are unchanged after the collision. Figure A3.2 shows the collision as viewed in the two frames. In the following discussion capitals denote the centre of mass itself, the subscript L denotes a quantity measured in the lab frame and the subscript C denotes a quantity measured in the CM frame. Primes denote values *after* the collision.

In the laboratory frame the centre of mass is moving to the right with constant velocity $\boldsymbol{V}$. In the CM frame before the collision m_2 is moving to the left with velocity $-\boldsymbol{V}$ and, since in that frame the total momentum is zero, it follows that

$$m_1 v_{1\text{C}} - m_2 V = 0. \qquad \text{(A3.26)}$$

The velocity $\boldsymbol{V}$ must be added vectorially to the velocity $\boldsymbol{v}'_{1\text{C}}$ of the observed particle in the CM frame to obtain its velocity $\boldsymbol{v}'_{1\text{L}}$ in the lab frame, as shown in figure A3.3(a, b). Hence

$$\boldsymbol{v}'_{1\text{L}} = \boldsymbol{v}'_{1\text{C}} + \boldsymbol{V}. \qquad \text{(A3.27)}$$

Taking components of this equation we find

$$v'_{1L} \cos \theta_L = v'_{1C} \cos \theta_C + V \tag{A3.28}$$

and

$$v'_{1L} \sin \theta_L = v'_{1C} \sin \theta_C. \tag{A3.29}$$

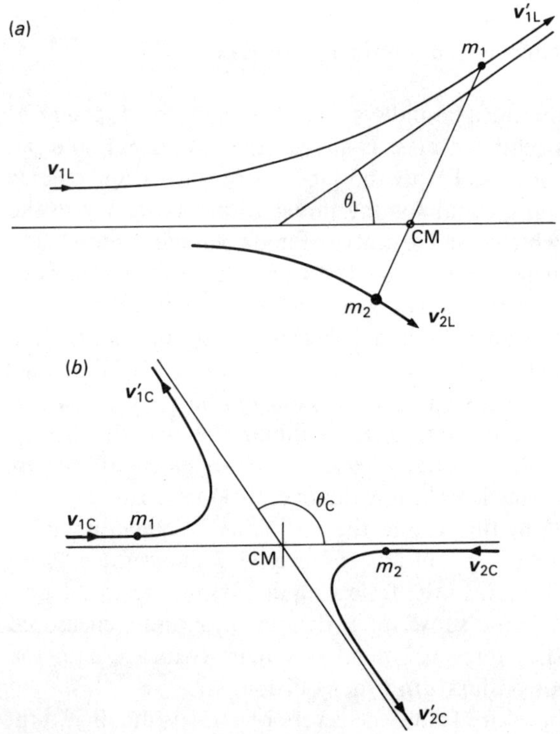

Figure A3.2 (*a*) A collision seen in the laboratory frame of reference. Particle 1 is initially at rest. (*b*) The same collision viewed in a frame in which the centre of mass is at rest.

Dividing these two equations and making use of equation (A3.26) and the fact that $v_{1C} = v'_{1C}$ we find

$$\tan \theta_L = \frac{\sin \theta_C}{\cos \theta_C + m_1/m_2}. \tag{A3.30}$$

This relation shows that if m_2 is very large, the scattering angles in the two frames are the same, as we would expect since in that case the two frames become identical.

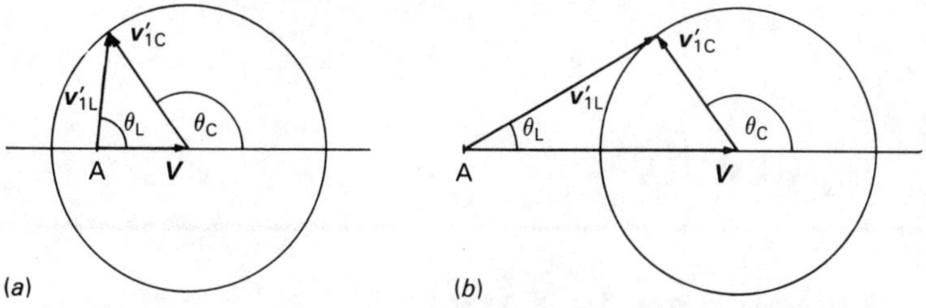

Figure A3.3 Relation between θ_L and θ_C. Two cases arise: (*a*) $V < v'_{1C}$ and (*b*) $V > v'_{1C}$. In (*a*) the range of θ_L is $0 \rightarrow \pi$. In (*b*) θ_L has a maximum equal to $\sin^{-1}(v'_{1C}/V)$.

We can use equation (A3.30) to relate the differential cross sections in the two frames. The essential point to grasp here is that the rate at which particles are scattered into a detector cannot depend upon the frame in which we choose to describe the events. Also the incoming flux of particles at the target is unchanged by a uniform translation of the entire system, and hence from equation (9.1) we must have

$$d\Omega_L(\theta_L, \phi)\left(\frac{d\sigma}{d\Omega}\right)_L = d\Omega_C(\theta_C, \phi)\left(\frac{d\sigma}{d\Omega}\right)_C$$

or

$$\left(\frac{d\sigma}{d\Omega}\right)_L \sin\theta_L \, d\theta_L \, d\phi = \left(\frac{d\sigma}{d\Omega}\right)_C \sin\theta_C \, d\theta_C \, d\phi. \tag{A3.31}$$

Note that the azimuthal angle ϕ is invariant between the two frames.

The ratio

$$\frac{\sin\theta_L \, d\theta_L}{\sin\theta_C \, d\theta_C} = \frac{d(\cos\theta_L)}{d(\cos\theta_C)} \tag{A3.32}$$

can be obtained by recasting equation (A3.30) as

$$\cos\theta_L = \frac{\gamma + \cos\theta_C}{(1 + 2\gamma\cos\theta_C + \gamma^2)^{1/2}} \tag{A3.33}$$

where

$$\gamma = m_1/m_2. \tag{A3.34}$$

The result is that

$$\left(\frac{d\sigma}{d\Omega}\right)_L = \left(\frac{d\sigma}{d\Omega}\right)_C \frac{[(1 - \gamma^2 \sin^2\theta_L)^{1/2} + \gamma\cos\theta_L]^2}{(1 - \gamma^2 \sin^2\theta_L)^{1/2}}. \tag{A3.35}$$

〉Appendix 4

〉Using the Programs

The notes accompanying the program disc and/or a text file on the disc itself explain how to run the programs on your particular computer. Here we discuss matters common to all implementations of the programs.

Editing parameters: default values

After one or more introductory screens the main Options screen appears. One of the choices is 'Edit Parameters'. This permits all the parameters that define the model to be viewed and edited. The programs provide default values of all parameters, and these will be adopted if you simply press RETURN. Alternatively, they can be edited by overtyping them. Press RETURN when you are satisfied. All input can be in the form of valid BASIC expressions if desired. Input should be in capital letters.

Integration step size

Nearly all the programs involve the numerical integration of a differential equation and require you to set a 'step size'. The optimum step size is a compromise between the desirability of vanishingly small steps and the limited precision with which the computer calculates. Thus if the step is too small there will be huge errors in the small increments which the computer calculates. In addition, the time taken to produce a plot may become unacceptably long. On the other hand, the step size should take account of the scale over which the potential changes. However, this becomes impossible when discontinuous potentials such as square wells and barriers are studied. Accurate results can still be obtained provided the step size is such that step boundaries fall at the discontinuities. Some of the programs ensure that this is the case, effectively resetting the boundaries if incompatible values are input. This method is ineffective in the case of programs which use the Numerov method of integration, since this is a two-step process. The Numerov method is preferable for smooth potentials and is used for the hydrogen ion program. Both Numerov and Runge–Kutta variants are provided for the general 3D potential. The latter should be used for discontinuous potentials.

Units

Where appropriate, programs provide a choice of 'Atomic' or 'Practical' units. With practical units, lengths must be entered in 0.1 nm (ångström) units and energies in eV, and it is assumed that the particle in question has the charge and mass of an electron. With atomic units (and more generally 'natural' or 'dimensionless' units) the values of the fundamental constants $\hbar$, e, m, c and $4\pi\varepsilon_0$ are all set to unity. The effect of this is that a_0, the radius of the first Bohr orbit in hydrogen, also becomes unity, since

$$\begin{aligned} a_0 &= 4\pi\varepsilon_0\hbar^2/me^2 \\ &= 1\ \text{au} \qquad (\equiv 0.052\,917\,72\ \text{nm}). \end{aligned}$$

It is therefore natural to adopt this as a unit of length, when we are dealing with an electron in a system on an atomic scale. If, however, we were dealing with some other particle, a muon say, we set *its* mass equal to unity. The conversion factor from natural units to metres would change, but the physical equations describing the two particles would be identical. Similarly a quark, with a charge of $\frac{1}{3}e$, would lead to another natural unit of length.

Combinations of the fundamental constants can be found with the dimensions of other physical quantities, and natural units chosen for those also. Thus the natural unit of energy becomes

$$E = me^4/\hbar(4\pi\varepsilon)^2$$

which equals 27.211 65 eV for an electron.

By making use of dimensionless units the computer programs can be adapted to deal with systems of very different scale, whilst keeping the numbers involved within comfortable limits. Ultimately these need to be translated into practical values for comparison with the real world, by applying the appropriate conversion factors. However, this process will be performed for you automatically, provided the particle modelled is an electron, if you select the practical unit option.

Wavefunction plotting

Most of the programs provide a plotting option. Often the starting conditions for the integration are provided automatically by the computer. However, the computer plots as it calculates, rather than preparing the entire plot in advance, so it may well be necessary to adjust the scale to get a sensible display, and then to replot. The tabulation option (see below) can be of assistance here, especially as it is appreciably faster in execution than the plotting routines. The range of tabulated values which is visible on the screen lies between ± 1.

Several of the programs permit the superimposition of several plots with the same axes, so that wavefunctions can be compared, or wavefunctions and probability densities plotted together. The necessary input for this takes place at the bottom of the graphics screens. Usually the display includes a plot of the effective potential function. If this is changed, as for example if the angular momentum is changed, it is then replotted.

Wavefunction tabulation and normalization

Many of the programs provide an option to tabulate the wavefunction. This is useful for a variety of purposes. It permits an accurate determination of the positions of nodes and maxima, and the calculation of an appropriate scale factor for plotting. In addition the computer performs a simple integration of the probability density as it tabulates, thereby allowing the wavefunction to be normalized, if required. Note that the integration extends over whatever range the computer tabulates, which is determined by the setting of XF or RF. Normalization to equivalent values is essential if one wavefunction is to be compared in any quantitative way with another.

Finding eigenvalues

Several of the programs contain routines for finding energy eigenvalues. In the case of the 1D square well the routine uses the Newton–Raphson algorithm to solve the analytic equations (3.54) and (3.55), whose roots are the eigenvalues. The eigenvalues of the multiple well with periodic boundary conditions (Kronig–Penney model) are also obtained by numerical solution of an analytic equation (9.123). More generally analytic equations are not available to be solved, and a method is used in which the wave equation is repeatedly integrated with trial values of E. If E is not an eigenvalue the wavefunction will eventually tend to either $\pm\infty$. To start the routine two values of energy must be found for which the wavefunctions diverge in opposite directions. Referring to figure A4.1, one wavefunction, with energy ENX, must approach the x axis and then turn away without crossing. The turning point is at $x =$ xNX. The other, with energy EX, must cross the x axis (at xX). The routine works by successive interpolations of E between these two values until a sufficiently close approximation to the eigenvalue is found. Note that EX will always be greater than ENX; the program is aware of this so you do not have to specify which energy is EX and which ENX. Of course, wavefunctions frequently exhibit nodes and turning points which are inherent and have nothing to do with divergence. We are talking here of the *asymptotic* behaviour. The program needs to know when the asymptotic region has been reached, so you must enter a value of $x =$ xA, smaller than the turning/crossing points xNX or xX, but larger than any previous turning point which is intrinsic to the eigenfunction, such as xT. It is a point beyond which no further oscillatory behaviour is expected of the eigenfunction and is referred to in the programs as the 'x value for start of asymptote'. (Note that in the case of radial wavefunctions it is the χ function whose turning points are relevant here.) The necessary input information is best determined by trial and error using the plotting routines. The only other information required is the precision required of the eigenvalue, expressed as parts in 10^n, where n must not be greater than 9.

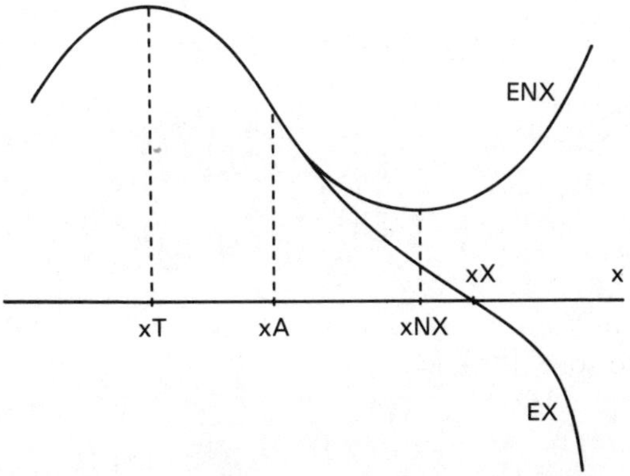

Figure A4.1 Terminology used in eigenvalue-finding routine.

Once the program has this information the process is automatic. First of all it validates the two values of E that you have provided, and then proceeds with the interpolation, keeping you informed of its progress. Finding eigenvalues in this way is a very time-consuming process; it may take many minutes, depending on the precision required. Note that the eigenvalues found will depend on the step size used, and generally speaking will be more accurate the smaller the step. Precision can be as high as 1 ppm as you can verify using known eigenvalues such as those of hydrogen or the harmonic oscillator.

Printing

Printing facilities vary with the computer implementation. All versions allow text to be printed, e.g. wavefunction tabulations and eigenvalue lists. The BBC computer versions contain Epson compatible screen dumps for the graphics screens. For other computers see the information provided with the disc.

In order to print anything you must first set the 'output to printer?' option in the Edit Parameters screen. The computer will then print whatever is appropriate from the next task you set it to do, and then it will cancel the option automatically.

The 'Print Parameters' option in the main Options screen is provided principally in order that hard copy of the parameters used in a particular investigation can be obtained.

〉 Appendix 5

〉 Table of Constants

Reproduced with permission from a table published by the Royal Society jointly with the Institute of Physics and the Royal Society of Chemistry.

Quantity	Symbol	Value (with s.d. uncertainty)
Speed of light in a vacuum	c	$2.997\,924\,58 \times 10^{8}$ m s^{-1} (exact)
Permeability of a vacuum	μ_0	$4\pi \times 10^{-7}$ H m^{-1}
Permittivity of a vacuum, $1/\mu_0 c^2$	ε_0	$8.854\,187\,817... \times 10^{-12}$ F m^{-1}
Elementary charge (of proton)	e	$1.602\,177\,33(49) \times 10^{-19}$ C
Gravitational constant	G	$6.672\,59(85) \times 10^{-11}$ N m^2 kg^{-2}
Unified atomic mass constant	m_u	$1.660\,5402(10) \times 10^{-27}$ kg
Rest mass:		
of electron	m_e	$9.109\,3897(54) \times 10^{-31}$ kg
	m_e/m_u	$5.485\,799\,03(13) \times 10^{-4}$
of proton	m_p	$1.672\,6231(10) \times 10^{-27}$ kg
	m_p/m_u	1.007 276 470(12)
	m_p/m_e	1836.152 701(37)
of neutron	m_n	$1.674\,9286(10) \times 10^{-27}$ kg
	m_n/m_u	1.008 664 904(14)
Energy equivalence of rest mass:		
of electron, $m_e c^2/(e/\mathrm{C})$		0.510 999 06(15) MeV
of proton, $m_p c^2/(e/\mathrm{C})$		938.272 31(28) MeV
of neutron, $m_n c^2/(e/\mathrm{C})$		939.565 63(28) MeV
Planck constant	h	$6.626\,0755(40) \times 10^{-34}$ J s
$h/2\pi$	$\hbar$	$1.054\,572\,66(63) \times 10^{-34}$ J s
Rydberg constant, $\mu_0^2 m_e e^4 c^3/8h^3$	R_∞	$1.097\,373\,1534(13) \times 10^{7}$ m^{-1}
Fine structure constant, $\mu_0 e^2 c/2h$	α	$7.297\,353\,08(33) \times 10^{-3}$
$(\mu_0 e^2 c/2h)^{-1}$	α^{-1}	137.035 9895(61)

Continued

Quantity	Symbol	Value (with s.d. uncertainty)
Hartree energy, $2R_\infty hc$	E_h	$4.359\,7482(26) \times 10^{-18}$ J
Bohr radius, $h^2/\pi\mu_0 c^2 m_e e^2$	a_0	$5.291\,772\,49(24) \times 10^{-11}$ m
Electron radius, $\mu_0 e^2/(4\pi m_e)$	r_e	$2.817\,940\,92(38) \times 10^{-15}$ m
Compton wavelength:		
of electron, $h/m_e c$	λ_C	$2.426\,310\,58(22) \times 10^{-12}$ m
of proton, $h/m_p c$	$\lambda_{C,p}$	$1.321\,410\,02(12) \times 10^{-15}$ m
of neutron, $h/m_n c$	$\lambda_{C,n}$	$1.319\,591\,10(12) \times 10^{-15}$ m
Avogadro constant	L, N_A	$6.022\,1367(36) \times 10^{23}\ \mathrm{mol}^{-1}$
Faraday constant	F	$9.648\,5309(29) \times 10^{4}\ \mathrm{C\,mol}^{-1}$
Molar gas constant	R	$8.314\,510(70)\ \mathrm{J\,K^{-1}\,mol^{-1}}$
Boltzmann constant, R/L	k	$1.380\,658(12) \times 10^{-23}\ \mathrm{J\,K^{-1}}$
Stefan–Boltzmann constant	σ	$5.670\,51(19) \times 10^{-8}\ \mathrm{W\,m^{-2}\,K^{-4}}$
First radiation constant, $2\pi hc^2$	c_1	$3.741\,7749(22) \times 10^{-16}\ \mathrm{J\,m^2\,s^{-1}}$
Second radiation constant, hc/k	c_2	$1.438\,769(12) \times 10^{-2}\ \mathrm{m\,K}$
Bohr magneton, $e\hbar/2m_e$	μ_B	$9.274\,0154(31) \times 10^{-24}\ \mathrm{J\,T^{-1}}$
Nuclear magneton, $e\hbar/2m_p$	μ_N	$5.050\,7866(17) \times 10^{-27}\ \mathrm{J\,T^{-1}}$
Gyromagnetic ratio of proton	γ_p	$2.675\,221\,28(81) \times 10^{8}\ \mathrm{s^{-1}\,T^{-1}}$
	$\gamma_p/2\pi$	$4.257\,7469(13) \times 10^{7}\ \mathrm{H_2\,T^{-1}}$
(proton in pure H_2O)*	γ_p'	$2.675\,152\,55(81) \times 10^{8}\ \mathrm{s^{-1}\,T^{-1}}$
	$\gamma_p'/2\pi$	$4.257\,6375(13) \times 10^{7}\ \mathrm{H_2\,T^{-1}}$
Magnetic flux quantum, $h/2e$	Φ_0	$2.067\,834\,61(61) \times 10^{-15}\ \mathrm{V\,s}$
Magnetic moment of electron	μ_e	$9.284\,7701(31) \times 10^{-24}\ \mathrm{J\,T^{-1}}$
in terms of Bohr magneton	μ_e/μ_B	$1.001\,159\,652\,193(10)$
Magnetic moment of proton	μ_p	$1.410\,607\,61(47) \times 10^{-26}\ \mathrm{J\,T^{-1}}$
in terms of Bohr magneton	μ_p/μ_B	$1.521\,032\,202(15) \times 10^{-3}$
(proton in pure H_2O)*	μ_p'/μ_B	$1.520\,993\,129(17) \times 10^{-3}$
in terms of nuclear magneton	μ_p/μ_N	$2.792\,847\,386(63)$

* As measured in a spherical sample of pure water at 25 °C with the diamagnetic shielding effect of $(1 - \mu_p'/\mu_p)$ of $25.689(15) \times 10^{-6}$.

〉 References

Abramowitz M and Stegun I A (eds) 1965 *Handbook of Mathematical Functions* (New York: Dover)

Aguilera-Navarro V C, Iwamoto H, Ley-Koo E and Zimmerman A H 1981 Quantum bouncer in a closed court *Am. J. Phys.* **49** 648–51

Bromley D A, Kuehner J A and Almqvist E 1960 Resonant elastic scattering of C^{12} by carbon *Phys. Rev. Lett.* **4** 365–7

Chadwick J 1930 α–α scattering *Proc. R. Soc.* A **128** 114

Delbourgo R 1977 On the linear potential hill *Am. J. Phys.* **45** 1110–12

Gibbs R L 1975 The quantum bouncer *Am. J. Phys.* **43** 25–8

Greenberger D M 1983 The neutron interferometer as a device for illustrating the strange behaviour of quantum systems *Rev. Mod. Phys.* **55** 857–905

Jeffreys H and Jeffreys B S 1972 *Methods of Mathematical Physics* 3rd edition (Cambridge: Cambridge University Press) p 508

Merzbacher E 1962 Single valuedness of wave functions *Am. J. Phys.* **30** 237–47

Tuve M A, Heydenberg N P and Hafstad L R 1936 The scattering of protons by protons *Phys. Rev.* **50** 806–25

〉 Answers to the Problems

〉 Chapter 1

1 (a) 8.3×10^{-10} eV (b) 2.3×10^{-4} eV (c) 0.5 eV 7.5×10^{29}
2 (a) 2.48 μm 1.21×10^{14} Hz (b) 1.24 nm 2.42×10^{17} Hz
(c) 1.24 pm 2.42×10^{20} Hz 6.24×10^{14} 5.93×10^{-3} W
3 2.11×10^{-13} eV 2.48×10^{-31} kg m s^{-1}
4 4.34 μm
5 $\hbar^2/8L^2m$ (a) 0.95 eV (b) 5.2×10^4 eV (c) 95 MeV
6 $mc^2/32\pi^2$
7 $me^4/2\pi\hbar^3(4\pi\varepsilon_0)^2n^3$ $r = 0.53$ μm $\lambda = 5.98 \times 10^{-2}$ m
$\nu = 5.01$ GHz $f = 7.2$ GHz
8 $\Delta n \simeq 5$

〉 Chapter 2

3 $(\alpha^2 + \beta^2)^{1/2}/\alpha$
4 (a) 1.5×10^{-16} s (b) 88 m (c) 1.6×10^{26} m
6 $j(x) = (\hbar k/m)\,|\,f(x)\,|^2$

〉 Chapter 3

1 2 4.44 eV 15.87 eV 15.87 eV
3 $(n + \frac{1}{2})\hbar\omega$, $n = 1, 3, 5$ ∞
5 0.157
6 $(2/L)^{1/2} \sin n\pi x/L$, $n = 1, 2, 3, 4\ldots$ No parity

7 $n = 1 \quad a = \hbar^2/mK$
8 $n \simeq 10^{12}$

〉Chapter 4

1 $(n + \frac{3}{2})\hbar\omega$ six-fold degenerate
3 22.03 eV 116.1 eV 285.3 eV
4 6.69 eV 25.06 eV
5 (a) $(\pi a_0^3)^{-1/2}$ (b) a_0 (c) $\frac{3}{2}a_0$ (d) $1 - (2\alpha^2 + 2\alpha + 1)\mathrm{e}^{-2\alpha}$
(e) $\frac{4}{3}\alpha^3$ 3.6×10^{-14}

〉Chapter 5

7 $P(1) = 0.874$ $P(2) = 0.0528$ $\langle \hat{H} \rangle = 7\hbar^2/mL^2$
9 $P(1) = 0.855$ $\langle \hat{H} \rangle = -e^2/3a_0(4\pi\varepsilon_0)$
10 (a) $(2\alpha/\alpha')/[1 + (\alpha/\alpha')^2]$ 0.985 (b) 0 (c) $(\frac{1}{2}\hbar\omega)(\alpha'/\alpha)^4$
$\frac{1}{4}\hbar\omega$
11 0

〉Chapter 6

3 $m_1 h \cos\alpha$
5 0.112 rad

〉Chapter 7

4 (a) 0.0343 (b) 0.00366
5 $V(r) = 12.446\{1 - \exp[-0.200\,43(r - 11.881)]\}^2$
$n_{\max} = 24$ 0.9598 0.3558 eV 0.007 08 0.002 626 eV
6 $u_{111}(r, \theta, \phi) = -[u_{100}(x, y, z) + \mathrm{i}u_{010}(x, y, z)]2^{-1/2}$
$u_{110}(r, \theta, \phi) = u_{001}(x, y, z)$
$u_{11-1}(r, \theta, \phi) = [u_{100}(x, y, z) - \mathrm{i}u_{010}(x, y, z)]2^{-1/2}$
7 $\simeq 36$ MeV $\simeq 13$ MeV

〉Chapter 8

2 (a) $\lambda V/2a$ (b) 0 (c) $\lambda V/a$

〉Chapter 9

2 0.0102, 0.176, 0.72 $E \ll 150$ V
3 0.847, 0.0401, 2.26×10^{-85}, 2.3 pm

〉Chapter 10

1 $|A|^2\hbar/mkr^2 \quad -|A|^2\hbar/mkr^2 \quad 0$
3 $\cot\delta_0 = (1 + k_2^2a^2)/k_2a^3(k_1^2 - k_2^2)$
$\cot\delta_0 = (\varkappa \coth \varkappa a)/k_2(1 - \varkappa a \coth \varkappa a)$
4 $3.77 \times 10^{-28}\ \mathrm{m}^2 \quad 68 \times 10^{-28}\ \mathrm{m}^2$

〉Table of integrals

$$\int_0^\infty e^{-\alpha^2x^2}\,dx = \frac{\pi^{1/2}}{2\alpha} \qquad \alpha > 0$$

$$\int_0^\infty x^{2a+1}e^{-\alpha^2x^2}\,dx = \frac{a!}{2\alpha^{2a+2}} \qquad \alpha > 0,\ a = 1, 2, 3, \ldots$$

$$\int_0^\infty x^{2a}e^{-\alpha^2x^2}\,dx = \frac{1.3.5\ldots(2a-1)\pi^{1/2}}{2^{a+1}\alpha^{2a+1}} \qquad \alpha > 0,\ a = 1, 2, 3, \ldots$$

$$\int x^n e^{\alpha x}\,dx = e^{\alpha x}\left(\frac{x^n}{\alpha} - \frac{nx^{n-1}}{\alpha^2} + \frac{n(n-1)x^{n-1}}{\alpha^3} \ldots (-1)^n\frac{n!}{\alpha^{n+1}}\right) \qquad n \geqslant 0$$

$$\int_0^\infty x^n e^{-\alpha x}\,dx = \frac{n!}{\alpha^{n+1}} \qquad \alpha > 0,\ n = 1, 2, 3, \ldots.$$

〉 Index